AMERICAN EDEN

AMERICAN EDEN

DAVID HOSACK,
BOTANY, and MEDICINE
in the GARDEN of the
EARLY REPUBLIC

VICTORIA JOHNSON

LIVERIGHT PUBLISHING
CORPORATION
A Division of W. W. NORTON & COMPANY
Independent Publishers Since 1923
NEW YORK   LONDON

FIRST EDITION

For information about permission to reproduce selections
from this book, write to Permissions,
Liveright Publishing Corporation, a division
of W. W. Norton & Company, Inc., 500 Fifth Avenue,
New York, NY 10110

For information about special discounts for bulk purchases,
please contact W. W. Norton Special Sales at
specialsales@wwnorton.com or 800-233-4830

Manufacturing by Quad Graphics, Fairfield
Book design by Barbara M. Bachman
Production manager: Anna Oler

Library of Congress Cataloging-in-Publication Data

Names: Johnson, Victoria, 1969– author.
Title: American Eden : David Hosack, botany, and medicine in the garden of the
early republic / Victoria Johnson.
Description: First edition. | New York : Liveright Publishing Corporation, [2018] |
Includes bibliographical references and index.
Identifiers: LCCN 2018002489 | ISBN 9781631494192 (hardcover)
Subjects: LCSH: Hosack, David, 1769–1835. | Medical botanists—United States. |
Botany, Medical—United States.
Classification: LCC QK99.U6 J64 2018 | DDC 580.973—dc23
LC record available at https://lccn.loc.gov/2018002489

Liveright Publishing Corporation
500 Fifth Avenue, New York, NY 10110
www.wwnorton.com

W. W. Norton & Company Ltd.
15 Carlisle Street, London W1D 3BS

1 2 3 4 5 6 7 8 9 0

For my parents, who love cities and gardens,

and for Rebecca, who tended this garden

Picture a sea dotted with sails,

a lovely sweep of notched shoreline,

blossoming trees on greensward sloping down

to the water, a multitude of small,

artfully embellished candy-box houses

in the background. . . .

—ALEXIS DE TOCQUEVILLE,

NEW YORK CITY,

MAY 1831

CONTENTS

TO THE READER

DAVID HOSACK'S LAST NAME, WHICH IS OF SCOTTISH ORIGIN, was pronounced "Hozzick" by his contemporaries. The Elgin Botanic Garden was named for the town of Elgin, Scotland, and was pronounced "EL-ghin." I have preserved the original spelling and grammar of every person whose words I quote in this book, unless otherwise noted. Any emotions or opinions I attribute to them come from their own writings.

During Hosack's lifetime, many of the plants of North America had not yet been identified or named by naturalists, so he sometimes drew on the European names of related species as he tried to identify his specimens. The scientific term for nonnative plants used by many botanists, both past and present, is *exotic*. I have generally preserved the names used by Hosack and his botanical associates for each plant mentioned in this book. Current names for a selection of these plants can be found at americaneden.org.

I discuss many remedies Hosack and his colleagues used in their medical practices in the eighteenth and nineteenth centuries, but this book is in no way intended as a treatment guide.

AMERICAN EDEN

PROLOGUE

SEPTEMBER 1797. THE BOY WOULD BE DEAD BEFORE DAWN. He was fifteen, more handsome already than his famously handsome father, who had turned back toward New York City when he received the news. At best, the boy's father would arrive in time to hold his son's hand as the end came. At worst, he would arrive only in time to embrace his grieving wife. The young doctor in attendance had already sent her out of the death chamber.

The doctor was solidly built, neither tall nor short, with thick black hair that crowned a massive head. People often noticed his sonorous voice and his piercing eyes, which were so dark as to appear completely black. He carried himself with a trace of arrogance—at least his rivals thought so. Friends detected in his ramrod bearing only boundless energy and sound principles. He was a man built to take command of a sickroom, or a funeral procession.

The doctor's name was David Hosack, and he was just twenty-eight years old. He knew that his more experienced medical colleagues would try to bring down the fever—was it typhus? scarlatina?—with cold cloths pressed to the skin. But that measure had already proved useless. As the boy's father raced home, Hosack took one last, risky gamble. He chose heat instead of cold, drawing a steaming bath and mixing a botanical remedy into the water—a bitter powder called Peruvian bark, made from the cinchona tree, native to the Andes. The bark, which was later discovered to contain quinine, had been used for centuries by the Quechua people to cure malaria before the Jesuits imported it to Europe in the 1630s. It had become a staple medicine first for European doctors and then for American ones, and although it was used far more often for malaria, Hosack hoped against hope that it would bring down the fever. Next he poured several bottles of alcohol into the bathwater to stimulate the circulation. After the boy had been lowered in, he sprinkled in smelling salts. At first, the thin body lay still in the steaming water, but within minutes, the boy began to regain his senses and his pulse quickened. Hosack swaddled him in warm blankets and carried him back to the bed, where he slept deeply for several hours before awaking—delirious once again. The doctor prepared another bath, and then another. The fever slowly receded. The boy would survive.

Hosack refused to leave the house that night, but he permitted himself to doze in a nearby bedroom after the hours of anxious effort. As he later recalled the scene, he bolted awake to find the boy's father, Alexander Hamilton, at his bedside. Taking Hosack's hand, Hamilton said with tears in his eyes that he could not remain one moment longer in his own house without expressing his deepest gratitude. In that moment, Hosack became a trusted friend to one of the nation's most famous and powerful men. But his medical intuition that night did more than forge a bond between a Founding Father and a young physician. It moved him one step closer to an idea he had quietly been nursing for three years. A few weeks after saving Hamilton's son Philip, Hosack picked up a quill and composed a letter to the president and trustees of Columbia College, where he was a professor of medicine and botany.

He meant with his letter to move the earth. It would take time, years of patience etched into twenty acres as his fellow New Yorkers showered him with both accolades and scorn. Finally orchards would arise, of apple, pear, and apricot. Carnations and daffodils would dot the lawns. Medicinal plants—poppies, chamomile, feverfew, ginseng, and dozens more—would grow in tidy plots and along shaded walkways. A glass edifice nearly two hundred feet long would stretch across the land, a magnificent conservatory to shelter the plants of the world's deserts and jungles from icy New York winters. Hosack would gather into his island Eden more than two thousand species of plants, collected from correspondents around the globe and from the farms next door.

It was an American triumph: the first botanical garden founded for the new nation.

Because of his garden, Hosack became one of the most famous Americans of his time. His medical research there cemented his reputation as the most innovative physician in New York. When Alexander Hamilton and Aaron Burr needed an attending physician for their 1804 duel, they both chose David Hosack. Thomas Jefferson, Alexander von Humboldt, and Sir Joseph Banks sent Hosack plants and seeds for his garden and lavished praise on him. When the sixty-six-year-old Hosack suffered a stroke in 1835, newspapers from South Carolina to New Hampshire ran bulletins about his illness and offered prayers for his recovery. Even before this, he had been immortalized in paintings, in marble busts, on commemorative coins, and in the names of plant species. Some Europeans called him the Sir Joseph Banks of America. It was the highest honor imaginable for an American scientist.

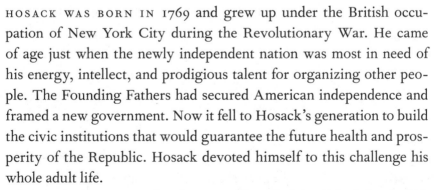

HOSACK WAS BORN IN 1769 and grew up under the British occupation of New York City during the Revolutionary War. He came of age just when the newly independent nation was most in need of his energy, intellect, and prodigious talent for organizing other people. The Founding Fathers had secured American independence and framed a new government. Now it fell to Hosack's generation to build the civic institutions that would guarantee the future health and prosperity of the Republic. Hosack devoted himself to this challenge his whole adult life.

Hosack was a complex figure. Although he relished family life and was affectionate with his many children, he pursued endless civic causes that tore him away from home. He loved remarking that the more a man took on, the more he accomplished every day. He slept as little as possible, avoided alcohol, and flirted with vegetarianism. He could be fiercely competitive one moment and generous the next. He cared deeply about his fellow Americans and spent his entire life trying to help them, but he could also be vain and snobbish. As a young doctor, he enjoyed warm relationships with several eminent mentors who loved him like a son, yet he didn't hesitate to break publicly with their antiquated medical theories—for which they forgave him.

Hosack devoted his life to bettering his nation, but he insisted on staying out of party politics, even when his friends tried to persuade him to run for office. He believed being a good doctor depended on neutrality. He enjoyed lifelong friendships with famous men who admired him while hating one another. He revered Alexander Hamilton and was shattered by his death in the 1804 duel. But Hosack was so close to Aaron Burr that after the duel he remained in touch with Burr throughout the years of his disgrace and exile. Hosack even sent one of his younger brothers, William Hosack, to Europe as Burr's traveling companion. While they were abroad, Hosack cared for Burr's adored daughter, Theodosia, during a terrible illness, keeping her father apprised of her condition in frequent letters.

Hosack's twin passions were medicine and nature. He was fascinated by the human body and saw it as an unknown country whose mysterious terrain called for intrepid new explorers. As a young medical student he risked his life to defend the controversial practice of corpse dissection because he knew it was the best chance doctors had to understand the diseases that killed Americans in droves every year. He became a mesmerizing public speaker and one of the most beloved medical professors in the United States, drawing crowds of students who hung on his every word and wrote down even his jokes in their notebooks. He performed surgeries never before documented on American soil and advocated smallpox vaccination at a time when many people were terrified of the idea. He pioneered the use of the stethoscope in the United States shortly after its invention in France in 1816. He published one innovative medical study after another—on breast cancer, anthrax, tetanus, obstetrics, the care of surgical wounds, and dozens of other subjects. In the early twentieth century, a medical journal paid tribute to Hosack's many contributions by noting that "there is perhaps no one person in the nineteenth century to whom New York medicine is more deeply or widely indebted than to this learned, faithful, generous, liberal man."

Yet although Hosack found surgery vital and exciting, he was convinced that saving lives also depended on knowing the natural world *outside* the human body. As a young man, he studied medicine and botany in Great Britain, and he returned to the United States convinced that it was at their intersection that Americans would find the most promising new treatments for the diseases that regularly swept the country. Hosack talked and wrote constantly about the natural riches that covered the unexplored North American continent. The health of the young nation, he argued, would depend on the health of its citizens, and thus on the skill of its doctors in using plants to prevent and treat illness. The political strength of the young nation would depend on its self-sufficiency, and thus on the ability of its farmers to select and raise the hardiest and most nourishing crops. The dignity of the young nation would depend on showing haughty Europe that the New World was abundantly blessed with natural wealth and scientific talent.

After the night he saved Philip Hamilton's life with the help of a medicinal plant—Peruvian bark—Hosack spent every spare second launching the new nation's first botanical garden. He lobbied friends, colleagues, and politicians for the financial support he needed to buy land, build greenhouses, hire gardeners, and collect plants from all over the continent and the world. Many people around him could not grasp what he was trying to do. Some mocked him publicly. He founded the garden anyway, at great expense to himself, eventually spending on it more than a million in today's dollars. He named his garden the Elgin Botanic Garden, after his father's hometown of Elgin, Scotland. Over the next decade, he assembled a collection of plants so diverse that his species lists make some of the most accomplished American botanists today shake their heads in amazement.

Many people now think of a botanical garden above all as a beautiful place to spend a day, but Hosack was driven by a much more urgent mission. All around him adults and children were dying of diseases

*David Hosack, with
his botanical garden
in the distance*

whose causes no one knew—yellow fever, scarlet fever, typhus, consumption, and others. The role of microbes in these diseases would not be discovered until the late nineteenth century and into the twentieth. When Hosack was a young man at the end of the eighteenth century, doctors relied on explanations for health and illness that made sense in their current frameworks. They argued about spasms of the nervous system, excessive blood in the circulatory system, jammed digestive tracts, and poisonous gases emanating from swamps near towns. They shared anecdotal case notes about their patients, but these couldn't possibly amount to a clear picture of the patterns and causes of illness and mortality. Close inspections of damaged organs and tissues through autopsies were one way doctors tried to discern how diseases executed their deadly work—the word *autopsy* means "see for oneself"—yet corpse dissections horrified many of Hosack's contemporaries. Sometimes available medicines and surgeries helped sick patients, but they often failed to stop the progress of a disease—or they even accelerated death. The discovery of new medicines was a scattershot process of trial and error.

The Elgin Botanic Garden was Hosack's brilliant answer to this desperate situation. Most of the medicines familiar to late-eighteenth-century doctors came from the plant world. There were hundreds of plant-based medicines known to European doctors and many more in use among native peoples around the world. Peruvian bark was the most effective such drug so far in use among eighteenth-century doctors, but it was also scarce and expensive to import. Hosack grasped with more clarity and energy than anyone else that if the United States didn't begin to grow and test plants from around the world, American medicine was doomed to chronic chaos. He announced the creation of his garden and began to collect thousands of specimens. He used his Elgin collections to conduct and supervise some of the earliest systematic research in the United States on the chemical properties of medicinal plants. He grew poppies to study opiates. He grew willow and spiraea, two species with painkilling and anti-inflammatory elements that would be combined into a new drug called aspirin at the dawn of the twentieth century. He experimented with medicinal plants from the tropics in his Manhattan greenhouse—for example, Madagascar periwinkle, an extract of which is now used to make anticancer drugs.

Surrounded by piles of plants and boiling pots filled with leaves, bark, and flowers, Hosack and his students isolated and compared the chemical properties of exotic and native plants as they sought to create a cheap local supply of plant-based drugs. As new waves of disease broke out around them, they used what they were learning to modify traditional treatment protocols and to understand more about the effects of specific medicinal plants on the human body. Seen from the vantage point of the twenty-first century, the Elgin Botanic Garden has less in common with a beautiful city park than with the National Institutes of Health, the Food and Drug Administration, the Centers for Disease Control, and CRISPR gene-editing laboratories.

Hosack was shaped by Enlightenment ideals. His elders had fought for self-government, founded hospitals, created a national banking system, experimented with electricity, and planned cities. He grew up surrounded by men and women who had deep faith in the ability of humans to better the world through their rational powers of thought and action. Hosack took this exhilarating idea and applied it to medicine, botany, and agriculture. He did as much as any man of his generation to keep the channels between American and European naturalists open, ensuring a constant influx of novel plant species to the United States along with all the latest information on their properties. Hosack's reputation grew so great that European scientists—including two of Napoleon's botanists—came to study with him at Elgin. He sent his own students to study in London and Paris so they would return home prepared to advance American science, just as he had done.

Working without the aid of germ theory, Hosack made frustrating mistakes and sometimes charged down dead ends as he strove to match drugs to specific diseases. But he was the leading pioneer in this American quest, and the Elgin Botanic Garden was his staging ground. He trained young Americans to ask life-and-death questions about nature and medicine and equipped them with the scientific methods to pursue the answers. He founded one of the nation's first medical journals to circulate his and others' research, and he helped build some of the first institutions ever to formulate professional and ethical regulations for American medicine. After Hosack's death his students—and their students in turn—went on to pioneer in the fields of surgery, etiology, and drug discovery.

Hosack's garden helped change how Americans saw the natural world. In 1785, when Hosack was still just an average New York teenager, a French botanist arrived in Manhattan on a mission from King Louis XVI and immediately complained in a letter home, "There are no informed people here, not even amateurs." At the time few Americans beyond well-heeled gentlemen with country estates and a handful of nurserymen and college professors thought plants were worthy of scientific study. But during the first two decades of the nineteenth century, Hosack taught at his garden and wrote with such excitement about the marvelous properties of plants, from the plainest to the showiest, that he helped ignite a craze for botany. Men and women flocked to lectures by Hosack and his students, snapping up the first American botany textbooks and guides to plant collecting in the wild. In 1817, when another French botanist visited New York and met Hosack and his circle, he came to a different conclusion about Americans' relation to nature: "Truly, this is a land of botanists."

Hosack died in 1835, less than a year before Ralph Waldo Emerson published his essay *Nature* and cofounded the Transcendental Club in Boston, among whose members was Emerson's young friend Henry David Thoreau. Emerson and Thoreau both wrote so compellingly about the spiritual and material meanings of the natural world to humans, and with such resonance for modern American environmentalism, that they have obscured our view of the generation of American naturalists who preceded them. When Emerson was born in 1803, Hosack was already tending the Elgin Botanic Garden and teaching young Americans there how to decipher the healing powers of nature hidden in the plainest plant. To Hosack and his students, botany was a heroic pursuit that took them into meadows and forests but also into the minute internal structures of each specimen. As they labored to name and classify their discoveries, they felt themselves peering into the globe-spanning, dizzying complexity of the natural world.

What Hosack loved about botany, Emerson discovered in Paris in 1833. It was a botanical garden—not a ramble in the woods—that propelled Emerson to his great celebration of the natural world as the site and the source of our highest intellectual and spiritual development. In the Jardin des Plantes, he had an epiphany about nature that helped give birth to American Transcendentalism. He saw the "grammar of botany" laid out all around him, "this natural alphabet, this green and yellow and crimson dictionary on which the sun shines." In that moment Emerson glimpsed a new way of understanding the relation of humans to nature. Nature did more than provide the farmer with crops and the poet with beauty. Nature, Emerson realized as he surveyed the organized beds of plant specimens, also reveals our quintessentially human drive to discover unifying principles and purpose in the world around us. Emerson's ideas, and Thoreau's, became an inheritance that many nature-lovers embrace today. Yet if earlier generations of Americans had had enough vision to support Hosack's garden, Emerson might not have had to cross the Atlantic for his epiphany. By the time Emerson visited Paris in 1833, the Elgin Botanic Garden had collapsed for lack of public and private funding—but only after Hosack had trained a generation of natural-ists and doctors there.

After Hosack's death, his botany students continued his work against all odds, relying on dried specimens and botanizing trips to train still more young botanists—Hosack's intellectual grandchildren—some of whom went on to found botanical gardens in cities across the country. Gradually these gardens transformed the physical and cultural land-scapes of places far beyond New York, fulfilling Hosack's century-old dream of a national network of botanical gardens that shared their research with one another. Today, there are hundreds of these insti-tutions in the United States. They are beautiful places to stroll, yes, but they also conduct botanical and environmental research and educa-tion. They can be found in all fifty states, the District of Columbia, and every borough of New York City—except one. The farmland Hosack tilled on Manhattan Island in the first decade of the nineteenth century now lies dormant beneath the limestone and steel of Rockefeller Center.

HOSACK'S LIFE SPANNED the precise moment when New York hov-ered between a bucolic past that unfurled over rolling fields and a cos-mopolitan future spiked with tightly packed skyscrapers. In Hosack's New York, cows wandered the Bowery, Greenwich Village was out of town, and the meadows of Harlem were filled with medicinal plants. At the same time, the tensions that governed city life in his era are familiar to many of us today, as old ways of interacting with the natu-ral world are transformed by human activity. Hosack was an urban pioneer who sought to learn from and celebrate nature in the face of skepticism and sometimes ridicule. He invited citizen-scientists to help map our native plant species before they were overrun by invasives. He called for a national system of agricultural stations eighty years before one was actually established. He advocated citywide tree planting for beauty and public health even as his fellow citizens were razing forests to the ground. He experimented with crops and medicinal plants just beyond the city's border. Today, people across the country and around the world are similarly reinventing what it means to be a city dweller.

People are keeping bees in Brooklyn and farming in Detroit. Hosack was the original urban gardener, experimenting with fruits and vegetables by day and mingling with his cosmopolitan friends at night.

His life tells the story of how Americans learned to think about both the natural world and their own bodies, but it is also the story of how one of the world's greatest cities became just that. During Hosack's childhood, Philadelphia—not New York—was the undisputed center of American culture and learning. New York was most famous for its shameless commercialism. By the 1820s, when Hosack was in his fifties, even the men in charge of Philadelphia's great institutions were conceding that New York had become the first city in the nation for the arts, sciences, and philanthropy. Hosack was one of the main forces in this change. In addition to creating the Elgin Botanic Garden, which drew international attention and praise, he founded or helped found many other civic institutions: the New-York Historical Society, the city's first museum of natural history, its first art museum, its first literary society, its first school for the deaf and mute, its first mental hospital, its first public schools, its first subsidized pharmacy for the poor, its first obstetrics hospital, and a homegrown version of the Royal Society of London.

The doctor who went down in history as a shadowy figure at the edge of the Hamilton–Burr duel was in fact one of the most influential Americans of his day. When he stood chatting with an acquaintance on a street corner in lower Manhattan, a surprising number of passersby would greet him by name. He was the kind of man who could save a dying child in the morning, deliver a riveting anatomy lecture in the afternoon, and throw a lavish party that night. His talent and ambition brought him face-to-face with some of the most interesting figures of his time. He socialized not only with Hamilton and Burr but also with Washington Irving, Thomas Cole, Clement Clarke Moore, Sir Joseph Banks, and the Marquis de Lafayette. He shared his research on the American prison system with Alexis de Tocqueville during the latter's visit to the United States, and he mentored a young medical student from Hyde Park named Isaac Roosevelt—Franklin D. Roosevelt's grandfather—at the Elgin Botanic Garden. By the time Hosack retired to his stunning Hudson River estate at Hyde Park (adjacent to the Roosevelt estate), he had been elected to the Royal Society and was being sought out in letters and visits by princes, presidents, and famous artists and scientists.

Even in retirement Hosack continued to shape the way Americans thought about nature. He was a champion of the Hudson River School of painters, who did so much to romanticize the American landscape and inspire generations of conservationists right down to this day. The landscape design Hosack implemented at his Hyde Park estate inspired the first professional landscape architect in American history, Andrew Jackson Downing, who helped launch the careers of Frederick Law Olmsted and Calvert Vaux, the designers of Central Park. Hosack is a lost link between eighteenth-century naturalists' formal, classificatory approach to nature and the artfully lush landscapes bequeathed to us by Olmsted and Vaux and their heirs.

Hosack's work influenced the way generations of younger Americans thought about nature, but his life also alters our inherited picture of the two elder statesmen who were his friends and his most famous patients—Alexander Hamilton and Aaron Burr. Thanks to the painstaking work of historians, we now know a great deal about how passionate Washington, Adams, Jefferson, and Madison were about botany (the study of plants) and horticulture (the art and science of cultivating plants). Hosack's relationships with Hamilton and Burr reveal that these two rivals—men whom we regard above all as political animals—were also fascinated by the natural world in general and plants in particular.

Burr shared Hosack's interest in medical botany, considering it so critical to the education of his young daughter, Theodosia, that he mailed her study questions on the topic while he was serving in the Senate in the 1790s. Like Hosack, Burr sensed that American progress in identifying and conquering specific diseases would come through sustained inquiry into the medical properties of plants. Hamilton, too, was interested in botany and horticulture. When he was laying out his country estate near Harlem in the first few years of the nineteenth century, he stopped at the Elgin Botanic Garden to talk with Hosack, who gave him plant cuttings and gardening advice. Burr's and Hamilton's shared interest in the study of the natural world allied them in support of an effort to found a museum of natural history in New York City, a campaign that unfolded from May to early July 1804—thus during the weeks of heated dispute that culminated in the infamous July 11 duel. Hamilton's life story ended with that duel, but Hosack's relationship with Burr continued. As Burr retreated from the national political stage, his interest in botany and natural history blossomed further. After he fled to Europe in 1808, he made stops at many of the continent's great gardens and arranged for specimens to be sent to Hosack for the Elgin Botanic Garden.

Seeing the world of the early Republic through Hosack's eyes changes the way we see the infancy of our nation. It shows us that nature was a constant and vivid presence not only for wealthy Virginia planters and homespun New England farmers. In the very heart of New York City, even gentlemen unmoved by the thought of striding in high summer through a sea of goldenrod could lapse happily into the minute examination of a single specimen on a library table. Through Hosack's eyes, we see Americans turning to nature for food, beauty, and the raw materials of commerce—but now we also see them placing in the natural world their best hopes for the discovery of lifesaving medicines. For Hosack and his circle, nature meant not only the continent's forests and prairies but also the humblest weed plucked from a ditch along Broadway and shoved in the pocket of an errant child.

Chapter 1

"TEAR IN PIECES
THE DOCTORS"

DAVID HOSACK WAS BORN IN NEW YORK CITY IN 1769. HE entered the world of British colonial America via his maternal grandparents' house at 44 Frankfort Street, very near today's City Hall. His father, Alexander, was a Scotsman who had left his country in the mid-eighteenth century to fight in the French and Indian War, then set up shop as a cloth merchant on William Street in New York. As a child Hosack learned the importance of supporting civic institutions from his father, who served as a volunteer fireman and was a booklover who borrowed piles of volumes from the local subscription library, the New York Society Library (founded in 1754 and still in existence today), including British novels, imperial histories, and adventure travelogues about Captain James Cook's voyages around the globe.

Hosack was a handsome boy with an upturned nose and large, dark eyes. In the fashion of the day, he tied his long black curls in a ribbon at the nape of his neck and wore a fluffy white cravat at his throat. Just after his seventh birthday, the American rebellion of 1776 brought redcoated soldiers marching into New York. Hosack spent the rest of his childhood in an occupied city. British ships anchored in the East River served as hellish prisons for captured patriots. A few blocks from Hosack's house soldiers pounded along the cobblestones bearing their wounded to the impromptu field hospital in the main building of King's College, on Church Street near the Hudson River. The orchards and forests of Manhattan were razed for firewood during the frigid winters.

When Hosack was twelve, General George Washington and his forces defeated the British at Yorktown with the help of the French; a year and a half later, the war was officially declared over. As New Yorkers celebrated in the spring of 1783, a son was born to a family called the Irvings, who lived on the same street as the Hosacks, and the baby's parents named him in honor of the victorious general. On November 25 of that same year, the British evacuated New York City. General Washington, accompanied by a contingent of officers, soldiers, and political leaders, rode his gray horse down the Boston Post Road and onto the Bowery with happy crowds swarming around him. As the British sailed out of the harbor, New Yorkers who had been away fighting in the war or had fled for calmer climes began to return and take stock of their homes. Two devastating fires had raged through New York during the war, leaving a quarter of the city in ruins. All across the tip of Manhattan Island, new houses and shops clattered together under pounding hammers. At the East River piers, dockhands heaved crates onto sailing ships as captains huddled over maps with their merchant masters. The raw materials of a continent— cotton, rice, flax, flour, furs—sailed out over the Atlantic again. King's College was renamed Columbia College, and its grand College Hall, now emptied of wounded British soldiers, was readied to receive the sons of the Sons of Liberty.

Hosack enrolled at Columbia in 1786 at the age of seventeen. He focused on classics, learning Greek and Latin, and also studied French. He would later tell his children that he had felt "naturally very dull" as a student and had to drive himself ferociously to complete even mundane tasks. He struggled with bouts of depression but discovered he could conquer his "gloomy reflections" as long as he kept his mind absorbed in learning new ideas. His self-discipline paid off, and he unearthed new strengths and talents. He won three awards in his college years, including one for public speaking.

At Columbia he also made new friends. His favorite was DeWitt Clinton, a young man of six feet three who loomed over almost everyone he met, including Hosack. DeWitt's uncle, George Clinton, was governor of New York and had been for almost as long as Hosack could remember. DeWitt's natural aura of dignity and his pride in his family connections made him seem stiff and haughty to some people, but Hosack detected a rare kindred spirit behind the off-putting façade— here was another teenage boy who took his education more seriously than anything else in his life. In fact, Clinton studied so singlemindedly that his mother wrote in exasperation, "Is my son Dewitt Dead or is he alive or has he forgot he has a mother." Thirty years later Clinton would write a letter to one of Hosack's own young sons telling him to "study—study—study—night and day."

Hosack and Clinton did differ in two important respects. While Hosack was acquiring a reputation for oratorical brilliance at Columbia, Clinton was perfecting the thudding monotone for which his public addresses would later become famous. And while Clinton was developing a passion for the law, Hosack was falling in love with medicine. He was electrified by the notion that a man could devote himself to learning how to save another person's life—maybe hundreds or even thousands of lives. The more he thought about it, the stronger his conviction grew that he could find no higher calling. It was obvious that New York needed more doctors. Although the war

Columbia's College Hall

was over, the eighteenth century's regular armies of disease, accident, and death continued their assaults. Mothers died in childbirth, or survived only to lose their infants soon thereafter. Men rowed over to Weehawken or Paulus Hook (in today's Jersey City) to defend their honor on the dueling ground. If they didn't perish on the spot, they were carried back ashen-faced and faint from gunshot wounds. Sailors suffered by the hundreds from syphilis. Paupers watched their fingers turn black from frostbite as they wintered in unheated quarters or under the open sky. Bodies of the drowned bobbed up against the wharves and came to rest on river beaches after drunken missteps or deliberate farewells. "How just is yr observation that in the midst of life we are in death," John Jay's wife, Sarah, mused to her mother in the spring of 1788.

Hosack longed to ease the illness and grief he saw everywhere he turned. Now his hard-won knowledge of ancient languages proved useful, for it gave him access to the most important treatises—not just those by American and British doctors but also the many medical works of the past and present in Greek and Latin. He learned that most doctors in Europe and America still relied on the idea that the human body was governed by substances and fluids known collectively as humors. The humoral framework had originated in the ancient world, had been refined by the Greek physicians Hippocrates and Galen, and then had persisted, despite periodic amendments, for two millennia. In the late eighteenth century, human health was widely believed to be affected by the ebb and flow of bodily substances such as blood, bile, phlegm, sweat, and saliva. Most doctors, even those who eschewed the old term *humors*, remained convinced that a person's health depended on keeping bodily substances in a careful balance—a balance that was in constant danger. Each breath of air, for example, sustained life but could also chill or overheat the body. Each mouthful of food brought nourishment but could increase or decrease the acidity of the stomach. The symptoms of illness—headaches, sores, tumors, and so on—were so many signals that the total bodily system was out of balance. It fell to doctors to try to restore equilibrium and with it health. One typical example from Hosack's studies found a doctor in New York attributing a young woman's inability to menstruate to the fact that she had once been forced to wade through a deep puddle during a rainstorm.

Doctors chose surgical and medicinal treatments that would regulate the flow of bodily substances. Bloodletting, for example, helped a patient thought to be suffering from an excess of blood, a common diagnosis when fever was present. Doctors sliced the skin with a lancet, catching the red gush in a waiting bowl—or instead of a lancet some doctors used a scarificator, a device with a spring-loaded row of blades that raked through the skin at lightning speed. There were many other surgical interventions doctors could choose from as they tried to restore equilibrium, and Hosack's medical books also discussed hundreds of medicinal remedies directed toward the same end. For example, some medicines voided a patient's bowels and were classed as cathartics or purgatives. A group of drugs known as emetics induced vomiting. If a doctor wanted his patient to sweat, he prescribed a diaphoretic or a sudorific, while a sialogogue would trigger saliva, an emmenagogue would promote menstruation, and so on. As the teenage Hosack learned about these medicines, he saw with mounting excitement how much remained unknown about their properties, both individually and in combination. He also learned that some doctors were trying to identify medicines they referred to as "specifics"—medicines that would target specific illnesses rather than overall equilibrium.

Hosack had grown up walking past druggists' and apothecaries' shops, the main precursors of modern drugstores. The druggists in New York catered mainly to doctors and ship captains and tended to specialize in imported medicines. Traders in South America, Africa, India, China, and elsewhere collected or bought medicinal plants and minerals and sold them to importers in Europe and the United States. At John Beekman's shop on Queen Street, for example, Hosack could step into a strange world of glass vials, ointment pots, and wooden boxes containing white crystals, greasy salves, gnarled roots, and crusty dried flowers. Beekman sold Epsom salts to purge the bowels, aloe from Barbados for dressing wounds, and Peruvian bark for fevers. Peruvian bark came in the form of a powder that was often dissolved in wine or liquor to mask the bitter taste. Although physicians had been arguing among themselves since the seventeenth century about how it worked and for which kinds of fevers, it had repeatedly proven its worth in hospitals and private practices, above all against malaria.*

When he walked into an apothecary shop Hosack could see some of the same remedies as at the druggists'. Apothecaries, however, catered not only to doctors and ship captains but also to the many other New Yorkers who kept medicines on hand at home. Here there was even more bustle as the head apothecary and his apprentices pounded, ground, boiled, and decanted their raw materials into pills, powders, salves, infusions, and other finished medicines. Customers could ask for a particular medicine, or they could buy a portable medicine chest filled with bottles of common remedies. Sometimes the farmers who brought their produce to the city stopped in before crossing the river back to the isolation of the Brooklyn countryside.

Most of the medicines Hosack saw came from plants and minerals, but some came from animals. Castor oil, prized for its antispasmodic properties, was derived from beavers' anal glands, which were harvested whole and smoke-dried. (Today's castor oil comes from a plant called castor bean, or *Ricinus communis*.) Spermaceti, a white substance harvested from the head of the sperm whale, was often used to make candles, but it was also a key ingredient of skin salves, including cold cream—a term already in use during Hosack's childhood. The shops also sold *sal ammoniac volatilis*, a smelling salt that came in little cakes and was made from animal bones or manure. With its horrible urine-like odor it could be waved under the nose or applied to the skin to revive a patient who was delirious or fainting.

Of all the medicines—animal, vegetable, or mineral—that Hosack saw in the New York shops, the most potent was mercury, usually administered in the form of calomel pills (a chloride of mercury). Although mercury is poisonous, it was popular then among doctors for its powerful purgative and saliva-inducing effects and was prized

---

* The active ingredient of Peruvian bark, quinine, would not be isolated until around 1820. The name *malaria* comes from *mala aria* or bad air, a reference to the association of fetid marshes and swamps with the disease. Researchers would not confirm mosquitoes as the vector for malaria until the late nineteenth century.

for treating fevers. Fever, in its many confusing varieties, was the most frightening sign that a human body was in peril. Some fevers seemed to strike just one person at a time, while others swept through whole districts in the same season each year. Yellow fever had killed hundreds of New Yorkers in repeated outbreaks in the eighteenth century, and no one knew when it might return or how it was transmitted. Smallpox also periodically besieged the city. During the Revolutionary War, a nationwide smallpox epidemic had killed more than four times as many Americans as had the British forces. Its victims ran a high fever and grew painful, pus-filled pockets under the skin that erupted and reeked like a rotting animal carcass. Those who survived bore the scars of their ordeal for life.

Some New Yorkers were beginning to embrace the practice of inoculation against smallpox. Inoculation, like many other medical practices in eighteenth-century America, had been imported in part from Europe, where professors and physicians—inspired by the Enlightenment's spirit of inquiry and recent improvements to the microscope—were conducting increasingly systematic experiments on the dead and the living alike. New York's doctors, some of whom had studied in Europe, weighed each new medical theory against their patients' actual symptoms and responses to treatments. They conducted risky surgeries at the New-York Hospital and visited the ailing poor at the almshouse. They wrote careful case notes about racing pulses, fizzy blood, and black vomit. They worked under a constant cloud of public suspicion that they prescribed their pills and bleedings for financial rather than medical reasons, and they faced competition from the dozens of druggists, apothecaries, and barbers who practiced medicine on the side. An essayist in 1786 captured the skeptical mood, writing that doctors "have good harvests and good times, when epidemical diseases and the small-pox prevail. Barbers wait impatiently for the agreeable spring, when people get themselves cupped and blooded, to prevent sicknesses."

Hosack was undeterred by the swirling confusion and criticism. He knew, however, that even if he read every medical treatise in the city he would not gain the clinical skills he craved—yet Columbia College offered no anatomy classes. He longed to study actual human bodies instead of written descriptions and pictures, so he apprenticed himself during his junior year to Richard Bayley, a former military surgeon. Bayley was in private practice with his father-in-law, John Charlton. Together they ministered to some of New York's most powerful families: Astors, DePeysters, Schuylers, Livingstons, and their social peers. Bayley also ran an anatomy laboratory at the New-York Hospital where he gave clinical demonstrations to aspiring young doctors. In this room smelling of dried blood and rotting flesh, Hosack discovered that he adored dissection—not because he relished the macabre scene but because he was finally seeing the miraculous world hidden beneath the human skin. The more he learned, the more "ardently attached" he felt to medicine.

Bayley had trained in London under John and William Hunter, two brothers who were pioneers in anatomy and surgery. The Hunters had taught that the only way to save the living was to study the dead. At the New-York Hospital, Hosack now watched in fascination as Bayley lectured with the aid of actual body parts. Though the humoral framework would not lose its hold for many decades, dissection was helping doctors to deepen their knowledge of the circulatory and nervous systems and to see how disease affected human tissue and organs. American medicine in the late eighteenth century was in the early days of a struggle, still underway today, between efforts to target specific diseases through the analysis of empirical evidence and efforts to improve overall well-being. Bayley's autopsies offered a hands-on way to gather empirical evidence, but Hosack also saw that Bayley faced a serious obstacle. Although the practice of using cadavers for medical research had gained some acceptance in Britain, it was still barely tolerated in the United States.

New York's medical students were responsible for supplying their teachers with bodies, and they felt safest plundering the fresh graves of criminals and enslaved people. The African Burial Ground was only a block or so from the hospital; after this cemetery was unearthed during the construction of a government building in 1991, one of the skeletons was found holding the sawed-off top of his skull in his arms.* When these devalued bodies were hard to come by, the students haunted the graveyards where wealthier New Yorkers were buried. Early in 1788, rumors of outrageous acts of grave robbing swirled through the city. In late February city officials offered $100 for information about the pillaging of a fresh grave at Trinity Church. Unsurprisingly, Hosack left no record about whether he participated in these excursions, but he was growing ever more committed to the scientific principles that moved Bayley to lower his knife toward a waiting corpse.

———— ⊗∞⊗ ————

IT WAS A THRILLING TIME to be young in New York. As Hosack traced the circuit between his home and his classes, he might at any moment cross paths with the most famous men in the country, men who were tussling over the fate of the national experiment. New York was the state capital (the move to Albany was a decade away), and since 1785 it had also been the seat of the Continental Congress. All around the city, pro-Constitution Federalists were arguing bitterly with the Antifederalists headed by Governor George Clinton about whether New York should ratify the Constitution at the convention slated to be held in Poughkeepsie in the coming summer. While Hosack studied with Bayley at the hospital, Alexander Hamilton, John Jay, and James Madison were drafting their series of newspaper essays advocating for ratification, which would come to be known as the *Federalist Papers*.

In April 1788, Bayley's laboratory was suddenly attacked. Some people later said it happened because one of the students had waved a severed arm out of the window at a group of children playing in the hospital yard, telling a boy the arm belonged to his recently deceased mother. This crude remark apparently reached the children's parents, because a group of armed men stormed into the hospital and up

---

* The remains of more than four hundred people were reburied, and their resting places now form part of the African Burial Ground National Monument, dedicated in 2007 and administered by the National Park Service.

the stairs to Bayley's laboratory. They were horrified at what they found there—"different parts of bodies, eyes, &c. coffins filled with excrements" and "a long table, bloody knives, and a horrid nauseous stench." As word of the carnage spread, people spilled into the streets. Hosack was not at the hospital at the time of the attack, but he set off running in a desperate bid to save some of the anatomical preparations housed at Columbia. On Park Place, a few blocks from College Hall, a man struck him on the forehead with a rock, knocking him to the ground. He feared he was about to be killed but a family friend who saw the attack hustled him to safety. Mayor James Duane and other officials ushered a group of doctors and medical students into the jail and locked them inside for their own protection, even as people beat on the doors with rocks and threatened to burn down the jail.

Meanwhile, John Jay was at home with his wife when General Matthew Clarkson appeared, asking to borrow a sword. Clarkson said he was on his way to the jail to make sure the insurgents would not "tear in pieces the Doctors who are confined there," as Sarah Jay reported in a letter to her mother. Jay armed Clarkson and himself and they went out into the rain. At the jail, Jay was bashed on the head with a rock and knocked unconscious. Dr. Charlton brought him home in a carriage with "two large holes in his forehead"—but to Sarah's relief no fracture of the skull. The next day her husband's eyes were swollen and discolored, and he soon came down with an illness she attributed to his exposure that day. He would not be drafting his pro-Constitution essays for some time to come. At sunset on the second day of the riot, city officials ordered the firing of cannon, alerting the militia to assemble. Dozens of guardsmen—one estimate put their number at a hundred and fifty—gathered at City Hall and marched in formation toward the jail, where they clashed with the rioters. At least three people were killed before an uneasy quiet settled over New York.

Hosack's resolve to become a doctor was unshaken by the violence, but he was among the medical students and professors who left town until tempers had calmed. He moved to Princeton to complete his undergraduate work at the College of New Jersey—the future Princeton University. However, he found it hard to stay away from New York for long. In the fall of 1788, with the Constitution now secure, his hometown was named the capital of the new nation. On April 30, 1789, George Washington placed his hand on a Bible on the balcony of the former City Hall, which was now known as Federal Hall. The United States had its first president.

Hosack next pursued advanced studies in anatomy, surgery, chemistry, and midwifery, working with three men who were to loom large in his adult life. Samuel Bard would come to love Hosack like a son, Nicholas Romayne to persecute him, and Wright Post to bear witness to his most bitter disappointment. For now, though, Hosack saw them simply as three of the most accomplished doctors in New York—Bard especially. On June 17, 1789, at the presidential mansion on Cherry Street, Bard dug a lancet into George Washington's unanesthetized thigh and cut out an abscess. The president remained calm and mute.

THE CITY HOSACK LOVED was now the proud capital of a new republic, but in some ways the place was as European as ever. He could still hear Dutch spoken here and there. He could walk along streets with names such as Roosevelt, Cortlandt, and the Bowery (from the old Dutch word for farm, *bouwerij*). When he glanced toward the harbor, he could catch sight of a jagged, step-gabled Dutch façade upending the straight British rooflines. At the northern end of the island, beyond a village long ago named for a distant city called Haarlem, members of the stalwart old Dyckman family were about to commence another century of farming.* In the smoky bars of lower Manhattan, men drew on their cigars, a Dutch habit that one Frenchman who visited New York in 1788 found disgusting.

Hosack could see even more of Britain than of Holland around him. New Yorkers had fought hard to throw off the British yoke, but the wealthiest among them still preened each morning before gilt-edged mirrors imported from London and donned suits sewn for them by the

---

* The farmhouse that the Dyckmans built in the 1780s can still be visited today at Broadway and 204th Street, in the Inwood neighborhood of Manhattan.

English tailor on Water Street. Rich women rolled imported silk stockings up their calves and stepped into slippers finished with gold buckles. Delicate little hands reached for porcelain teacups across damask tablecloths, and they grazed the keys of harpsichords that had been hoisted from damp cargo holds down at the docks. The wealthy young ladies of the city cultivated such expensive tastes that potential suitors had begun to drag their feet a little. "Luxury forms already, in this town, a class of men very dangerous in society—I mean bachelors," observed the visiting Frenchman Jacques-Pierre Brissot de Warville. At one New York dinner party, Brissot noted the gaudy costumes of the female guests, two of whom "had their bosoms very naked." He was "scandalized at this indecency among republicans."

Republicans? Perhaps not. At least not in the eyes of those fellow citizens who couldn't afford to cultivate upper-crust British tastes or who simply found such habits undemocratic. To be sure, many New Yorkers rich and poor, opulent and austere, welcomed the day the Constitution became the law of the land. In late July 1788, a month after couriers had ridden into the city to report that New Hampshire had tipped the balance toward ratification, thousands of New Yorkers had poured into the streets to celebrate. But united as they stood in their distaste for foreign tyranny, they had also seen during the ratification process that deep fissures divided them. Many of the city's wealthier citizens counted themselves Federalists—a loose political grouping whose members were attracted to aspects of the British model of government, especially its strong central authority. And many of these Federalists wished the United States to maintain cordial ties with Britain, a sentiment particularly strong in New York City, where more than a few merchants and bankers owed their well-appointed townhouses and fine clothing to the lucrative seaport trade with Britain and its West Indian markets.

The Federalists' standard-bearer was Alexander Hamilton, a human whirlwind who argued with eloquence that only a powerful central government would be capable of unifying the states and preserving the fragile young nation. Hamilton had acquired star status among New York City's Federalists for his tireless defense of the Constitution and his decisive role in securing its ratification by the state of

New York. A war hero who had turned in a daring performance at the decisive Battle of Yorktown, Hamilton still carried himself with the panache of an officer as he strode the streets of postwar Manhattan. Hamilton's young wife, Elizabeth Schuyler, a dark-eyed beauty from a prominent Albany family, was lively but modest. Hamilton thought that Eliza was "unmercifully handsome" and full of "good nature, affability, and vivacity." Eliza, in turn, was devoted to the affectionate Alexander, whom she had met and married in 1780, when he was still a colonel in the Revolutionary Army. Now, as the 1780s drew to a close, she and Hamilton occupied the very center of New York high society. During the heady days following ratification, some people had suggested the whole city simply be rechristened in her husband's honor: Hamiltoniana.

This apotheosis would have unleashed howls of fury from the growing coalition of politicians known as Antifederalists, who had fiercely opposed ratification and regarded Hamilton as a traitor to the ideals of the Revolution. In their view, the new Constitution endowed the federal government with sweeping powers inimical to both states' and individuals' rights. When New York's diehard Antifederalist governor, George Clinton, had tried to mobilize his supporters statewide in an effort to block ratification, he had managed to enrage a good swath of New York City, where it was feared Congress would pack up and leave town should the state delegates junk the Constitution. At the state Constitutional Convention in Poughkeepsie in the summer of 1788, Governor Clinton bristled publicly at Hamilton, while around the same time in the city one New Yorker wrote to a friend, "I do not believe the life of the Governor & his party would be safe in this place."

Hamilton and his supporters eventually won this battle, but for all the shared exuberance at Washington's inauguration and the anointing of New York as the nation's first capital, more trouble lay ahead. James Madison, one of the staunchest defenders of the Constitution, which he had so decisively shaped, had begun to worry along with Secretary of State Thomas Jefferson about the prospect of an eternal unholy alliance between politicians and speculators. "The Coffee House," Madison wrote Jefferson, "is in an eternal buzz with the gamblers."

Madison and Jefferson soon found a way to rescue the nation's political leaders from the seething den of financial iniquity on the Hudson. On June 20, 1790, Jefferson staged a dinner party at his house on Maiden Lane at which he helped broker a course-changing deal between Madison and Hamilton. According to Jefferson's account of the evening, the two men came to an agreement that in a matter of months the government would leave New York for Philadelphia, to meet there while the permanent capital city was built closer to the South, on the Potomac. (In the years to come, New York would remain the financial capital of the United States, thereby creating at least the appearance of distance between financial speculation and the nation's political business.) For his part, Hamilton—now the first secretary of the treasury—would notch up a victory in his campaign to see the federal government assume the Revolutionary War debt of the states. And so it was that in the autumn of 1790 Hamilton and Jefferson, along with the rest of the government, moved to Philadelphia, where they continued their quarrels over finance, centralized authority, and foreign policy.

Hosack went, too.

HE WENT NOT FOR POLITICS but for medicine. Much as he appreciated his New York mentors, Philadelphia was home to the country's best medical professors, who were teaching at the University of Pennsylvania's medical school and on the College of Philadelphia's medical faculty (the two institutions would merge in 1791). Among them was Benjamin Rush, the most famous physician in the nation. Rush's approach to medicine was iconoclastic, empirical, and restlessly curious, making him the perfect teacher for the inquisitive Hosack.

Rush had been teaching and practicing medicine in Philadelphia since 1769. He had earned a reputation as the kind of man who was unafraid to shout his opinions from the rooftops. In 1773, he had published a blistering pamphlet on the evils of slavery. "Hear their cries, and see their looks of tenderness at each other upon being separated. . . . Let us fly to them to step in for their relief." In 1776, he had signed his name to the Declaration of Independence, just above

Benjamin Franklin's. To Hosack's joy, this eminent American scientist agreed to become his mentor, and he spent hours at Rush's side—not only in class and in clinical settings but also as a frequent guest at the family dinner table. Hosack listened awestruck as Rush spoke in a "perpetual stream of eloquence." Rush made the study of natural history sound romantic—it was, he said, "the first study of the father of mankind, in the garden of Eden." Hosack memorized his mentor's personal and professional habits and tried to emulate them. To maximize time and energy for teaching, medicine, and charity work, Rush arose early, ate sparingly, never touched alcohol, and read late into the night. He was no longer formally active in politics—insisting to Hosack that he was now a "mere spectator of all public events"—but in fact his home was always filled with government officials, naturalists, painters, and physicians. John Adams and Jefferson were Rush's close friends, and he also knew President Washington well.

One towering figure among Rush's friends was gone: Rush had been at Benjamin Franklin's bedside as the latter lay dying in the spring of 1790. Hosack had arrived in Philadelphia six months too late to meet this legendary American. Still, nearly every subject he would study that year bore the imprint of Franklin's curiosity and his passion for civic improvement. The Pennsylvania Hospital, where Hosack would receive clinical instruction, had been launched by Franklin in 1751. The medical school of the University of Pennsylvania—a university that owed its origins in large part to Franklin—held its lectures in a building that also housed Franklin's celebrated American Philosophical Society, which brought together men keen on promoting "all philosophical Experiments that let Light into the Nature of Things" and thus "tend to increase the Power of Man over Matter, and multiply the Conveniencies or Pleasures of Life." Franklin, spurred by the botanist John Bartram, had first floated the idea in 1743. The United States, Franklin argued, needed a forum where learned men could share their inquiries in botany, physics, chemistry, medicine, math, agriculture, and many other fields. By the time Hosack arrived in Philadelphia, the members included some of the nation's most eminent men, among them Washington, Jefferson, and—very soon—Hamilton. Hosack began to dream of one day being elected to its ranks.

Philadelphia in 1790 was a noisy place fired by a new breed of citizens: the free leaders of a country they had forged themselves. Washington had gathered around him political figures matched in wits and clashing in vision. As these men hammered out the new national institutions, Philadelphia's meeting halls, taverns, and printing houses rang with the shouts and laughter of a lively political culture. Hamilton, as secretary of the treasury, sought to strengthen the federal government through the creation of a powerful national bank, but Secretary of State Jefferson fought Hamilton hard, fearing the rise of a tyrannical central government that would trample the rights of the states. During the winter of 1790 to 1791, as Hosack hunched over his books and cadavers, Hamilton and Jefferson battled for George Washington's approval—and the soul of the nation—just a block or two away, with Jefferson observing that he and Hamilton were "daily pitted in the Cabinet like two cocks."

Hamilton prevailed over Jefferson, at least for the time being—in February 1791, the president signed Hamilton's bank bill. Jefferson was despondent. By mid-May, he had arranged with Hamilton's former ally Madison to make a temporary escape from Philadelphia on a botanical tour of New England. Jefferson loved plants. "What a field we have at our doors to signalize ourselves in!" he had recently exulted to the president of Harvard. "The botany of America is far from being exhausted." He sketched out an itinerary for his trip with Madison that would take them up the Hudson River, then north into Vermont, and south again via Massachusetts. They met in New York and left town toward the end of May. When they reached the spot north of the city where the Hudson widens into a body of water that Dutch settlers had described as a sea (a *zee*), Jefferson began a travel journal: "May 21. Toppan sea." The Tappan Zee.

In Vermont Jefferson was delighted to find sugar maple trees growing "in vast abundance." He and Rush had been exploring the idea of replacing cane sugar, imported from the slave-owning West Indies, with what Rush liked to call "innocent maple sugar." Since at least 1788 Rush had been arguing that maple sugar was preferable in every way to cane sugar, and Jefferson had become an enthusiastic supporter of his efforts. Before Jefferson left Philadelphia for his New England trip, he and Rush had sat over cups of coffee sweetened with maple sugar and discussed the possibilities. Jefferson, a slave owner, was interested in maple sugar's promise for increasing the agricultural self-reliance of the United States. Now, in early June 1791, as Jefferson and Madison surveyed a Vermont hillside covered with the trees, the scheme seemed within reach, even as Rush in Philadelphia worked on revisions to his sugar-maple pamphlet, which he would present to the American Philosophical Society in August.

When Jefferson and Madison turned south again, they traveled down the Connecticut River, coming out into Long Island Sound between the villages of Old Lyme and Old Saybrook before traveling the length of Long Island to New York City. They paused at the village of Flushing (in today's Queens) long enough for Jefferson to purchase the entire stock of sugar maples from the Prince family's nursery there.

IN MAY 1791, just as Jefferson and Madison were about to leave on their New England trip, Hosack submitted a dissertation on cholera to the University of Pennsylvania medical faculty. He opened the work with a martial tribute to the stomach. "This organ, in a natural and healthy state, like a faithful well-armed Sentinel, is always on the watch, who when attacked, either repels the enemy, or deprives him of his arms, without which he is now incapable of defence." Hosack dedicated his dissertation to his former teacher Bayley, a man who knew something about being attacked. It would be another sixty years before the London doctor John Snow would discover that cholera was a waterborne illness, and ninety years before the German scientist Robert Koch would isolate the cholera bacillus. Hosack's analysis of cholera was firmly rooted in the humoral understanding of the body, yet his bold criticism of theories advanced by his famous new mentor, Rush, suggests how comfortable their relationship was—and how confident Hosack was growing in his own ideas.

His degree complete, he married a young woman he had met in Princeton, Catherine Warner, whom he called Kitty. They moved to Alexandria, Virginia, close to where the nation's permanent capital city was being built. While Hosack worked at establishing a medical practice, Kitty bore their first child, a boy they named Alexander, after Hosack's father. Soon, however, it became clear that the new capital was emerging at a glacial pace, and although Hosack was attracting some patients, he worried about supporting Kitty and Alexander. After agonizing deliberation, he concluded that both his professional prospects and his clinical skills would benefit from the most advanced and prestigious medical education in the Western world. The American professors whom he admired had all studied at the University of Edinburgh, a city that in the second half of the eighteenth century was home to some of Europe's most brilliant philosophers, naturalists, writers, and medical men. It was the heart of the Scottish Enlightenment.

Hosack found it "painful to think of leaving" Kitty and Alexander, but in the late summer of 1792 he settled them with his parents on William Street in New York and boarded a ship bound for Britain. It must have been a disorienting way to mark his twenty-third birthday that August 31—surrounded by strangers on the open Atlantic. When he disembarked at Liverpool, his first stop on the way to Edinburgh, a family friend treated him to an unforgettable evening. Hosack found himself sipping hot toddy by a cozy fire as a young Scotsman named Robert Burns sang odes to the beautiful land where Hosack's own father had been born. He couldn't have dreamed up a more romantic prelude to his new adventure. When he finally arrived in Edinburgh, he was dazzled by the turreted city—"that dream in masonry and living rock," Robert Louis Stevenson would later call it. At one end of the Old Town sat Holyrood Palace, while at the other end Edinburgh Castle perched like a galleon atop the billowing cliffs. To the north was the neoclassical New Town, laid out by city elders in the middle of the eighteenth century, where fine masonry-work façades lined green squares.

At the university Hosack signed up for a punishing regimen. He attended classes twelve hours a day, racing from anatomy to midwifery, from pharmacology to chemistry, from clinical observation to dissection. Arsenic sizzled on red-glowing copper. Blade scraped on bone. In the evenings he shadowed the city's surgeons as they walked the wards of the Royal Infirmary. Through it all, he soaked up the accumulated knowledge of centuries of medical instruction and practice. The animating spirit of the faculty was William Cullen, a professor who had died two years earlier but still practically ran the place from his grave, so great was his intellectual influence. Nearly everything Hosack learned about medicine that year—and a good deal of what he had already learned in the United States—traced its roots to Cullen in one way or another. After joining the Edinburgh faculty in 1755, Cullen had taught thousands of future European and American doctors his novel understanding of the origins of disease, which he saw as lying in what he called "spasms" of the nervous system. Cullen—whose nickname among the students was reportedly "Old Spasm"—had also upended the way *materia medica* (soon to be called pharmacology) was taught. The handbook that became Hosack's constant companion, the *Edinburgh New Dispensatory*, owed its novel structure to one of Cullen's insights.

For centuries before Cullen, reference books on the *materia medica* had listed all the known medicinal substances in alphabetical order within the huge categories of animal, vegetable, and mineral. It was a confusing system for a doctor trying to find a remedy for a particular ailment, and Cullen had pointed out that it made much more sense to organize medicines by their effects on the body—emetics, sudorifics, purgatives, and so on. By the time Hosack arrived in Edinburgh, the current edition of the *Edinburgh New Dispensatory* partly followed this revolutionary scheme, and for generations to come most new pharmacology textbooks would follow it, too. For this insight and so many others, Benjamin Rush, who had studied with Cullen at Edinburgh in the 1760s, wrote after the latter's death in 1790 that "while Astronomy claims a Newton, and Electricity a Franklin, Medicine has been equally honoured by having employed the genius of a Cullen."

In addition to teaching Rush, Cullen had trained many of the Edinburgh professors in whose classrooms Hosack now sat trying to absorb every new bit of knowledge. He studied the way they classified fevers, their most recent breakthroughs in chemistry, and how they prepared pills, infusions, tinctures, syrups, and every other form of medicine. As he listened to his professors, Hosack took notes not only about the content of their lectures but also about their oratorical styles. He did the same when he sat in church on Sundays listening to the Presbyterian minister. Should he ever find himself in front of a classroom filled with captive young men, he planned to avoid boring them. Hosack's close relationship with Rush brought him to the attention of his Edinburgh professors, and he wrote to ask how Rush would feel about having some of his recent writings published in a British edition. There was strong interest, he noted, in Rush's pamphlet on maple sugar.

Hosack missed Kitty and Alexander, yet for a young man like him—prodigiously energetic and hungry to learn—these were exhilarating months. One of his favorite classes was midwifery, a specialty practiced by male doctors as well as by female midwives and "man-midwives." By an odd coincidence, Hosack's midwifery professor was named Alexander Hamilton, although he looked nothing like his elegant American counterpart. Professor Hamilton had a stocky body, a jutting chin, and a squashed nose. He turned out to be a generous, sociable man, and he invited his new American student home to family dinners. Professor Hamilton also enjoyed hosting students and colleagues at a garden he kept on the outskirts of Edinburgh, including Hosack on at least one occasion. As Hosack strolled through the garden listening to the casually erudite conversation around him, he began to panic. Botanical expertise, it seemed, flowed in British veins. He normally felt at ease in the most exacting company, but now he was the odd man out, the uncultivated American. Standing there tongue-tied before Professor Hamilton and his guests, Hosack felt, as he later put it, "very much mortified by my ignorance of botany."

Eighteenth-century Britain was a land obsessed with plants—growing them, collecting them, studying them, admiring them. The whole kingdom was covered in gardens. There were grand palace gardens, tidy public gardens in the cities, beautifully landscaped gardens around country homes, and countless little flower and vegetable gardens. For Professor Hamilton and his colleagues, though, it was the curative properties of plants that were most fascinating, because the vast majority of medicines known in the eighteenth century, as in every century before, had been ferreted out of bark, roots, stems, leaves, seeds, petals, and fruits. The leaves of mugwort (*Artemisia vul-*

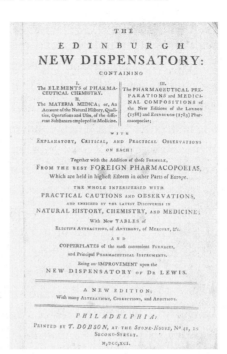

Edinburgh New Dispensatory, 1791 edition

garis), for example, a member of the daisy family that dotted the lanes and fields of England, could be boiled into a medicinal tea believed to reduce menstrual bleeding. An infusion from leaves of menyanthes (*Menyanthes trifoliata*), a flowering plant known more commonly as bogbean, soothed herpes sores. The resin of arborvitae trees (*Thuja occidentalis*) produced an excellent cough syrup that was thought to have prevented scurvy in the British army at Boston during the American Revolution.

The study of how to use plants to make medicine was called *medical botany*. It was an ancient field, yet almost two millennia after the Greek physicians Theophrastus, Dioscorides, and Galen had written influential treatises, Western doctors were still struggling to determine which plants alleviated which illnesses—and how they did so. Today, scientists know that plants produce four main kinds of antimicrobial substances, but in the late eighteenth century, these compounds were as little known as the role of microbes themselves. Doctors and botanists did their best with the tools they had, reading through thousands of plant descriptions in ancient, medieval, Renaissance, and contemporary texts in search of the keys that would reveal what made a given species medically potent. In both Europe and the United States, physicians fretted over the scarcity and expense of imported drugs like Peruvian bark, and they worked hard to find comparable native species. They traded case notes with one another and argued over medicines, dosages, and side effects. They conducted autopsies on patients whom their remedies could not save—or killed outright—and published the results, in the hope of nudging medical knowledge along the path of progress. Sometimes they left their desks, libraries, and surgical theaters to study under the open sky in botanical gardens.

Europe's botanical gardens had originated in medieval medicinal gardens attached to monasteries. During the Renaissance the study of plants had gradually shifted from monasteries to universities, and botanical gardens were established in the sixteenth and seventeenth centuries in university towns such as Pisa, Padua, Leiden, and Oxford. In the age of European colonial expansion, botanical gardens were becoming increasingly important to imperial powers eager to hoard and cultivate lucrative plant species such as cacao, coffee, and spices. For doctors and professors, botanical gardens served as encyclopedias, laboratories, and classrooms. They lugged their *materia medica* books to the gardens and compared written descriptions of plants to living specimens. They dug up roots to slice them into sheer slivers and examine them under microscopes. They boiled leaves in distilled water and mixed the resulting liquid with ammonia, sulphuric acid, and other chemicals to try to determine their composition. They pinched off flower petals and ground them to a pulpy mess with their mortars and pestles, making medicinal teas and salves that they tested on themselves and their patients.

Hosack had learned about the importance of plant-based medicines during his medical studies in New York and Philadelphia, but he had thought of them as supplies to be purchased from druggists and apothecaries. It wasn't until he embarrassed himself in Professor Hamilton's garden that he glimpsed a world in which garden rakes could unlock the saving power of nature. Then and there, he resolved to learn everything he could about plants. The best place in Edinburgh to do so was the botanical garden across the bridge from the university. It had started life as a small medicinal garden run by two local physicians near Holyrood Palace, but by the time Hosack arrived it had moved to a larger site on Leith Walk, where a professor of botany from the university lived in a beautiful "Botanic Cottage" surrounded by thickly planted herbs, flowers, shrubs, and trees. Medical students from the university paced the garden paths trying to memorize species names as their professors lectured about the ailments they thought (not always correctly) treatable by medicines made from these plants. Although Hosack left no record of visiting it, his newfound interest in botany must have led him to Edinburgh's famous botanical garden from time to time.

Suddenly his studies were interrupted by tragic news from the United States. On December 29, 1792, when Hosack had been in Edinburgh for about three months, his son, Alexander, died in New York. American infants perished regularly from diphtheria, smallpox, cholera, pneumonia, and a host of other diseases; perhaps one of these had killed Alexander. Hosack's father paid for a coffin in January. Hosack remained in happy ignorance for many weeks as the news made its way to Scotland. When he learned of the death, he may well have felt a measure of guilt about his decision to leave Kitty and Alexander. From that moment forward, his devotion to medicine would be tinged with the melancholy wisdom of personal loss. His son had survived only six months.

Chapter 2

"AN ENDLESS SOURCE OF
INNOCENT DELIGHT"

Most of Scotland lies to the north of Edinburgh, drap-
ing in green-velvet layers over soft mountains and pausing briefly on
the shores of glassy lochs. Travelers on the moss-walled country lanes
could find themselves in a rainstorm one moment and in a shimmering,
backlit valley the next. In the spring of 1793, Hosack set out by stagecoach on a journey through this landscape. He was bound for Elgin,
the town his father had left behind at midcentury to sail for America
and the French and Indian War. Hosack reached the northern coast
after more than a week of travel. As the sea appeared, the landscape
flattened out and gave way to an enormous expanse of sky. Here swallows skimmed the earth at dusk and curly-horned sheep grazed in sil-
houette against the blue-green curtain of the sea.

Elgin was a handsome town lined with medieval arcades. A five-
minute walk from the central marketplace took Hosack to the vault-
ing ruins of Elgin's thirteenth-century cathedral. At the end of the
fourteenth century, an adulterous nobleman named Alexander Stewart (also known as the Wolf of Badenoch) led a marauding band of
Highlands men to Elgin to avenge Stewart's excommunication by
its bishop. They burned the town together with its cathedral. When
Hosack arrived four hundred years later, the empty windows of the
ruined towers still traced high arcs against the sky.

Hosack loved the north. At Elgin, he made the acquaintance of two of his uncles, and through them, he met the Brodie family, who lived just outside Elgin in a sand-colored castle. Visitors like Hosack who rode up the long, straight carriage drive to Brodie Castle saw a cluster of towers, chimneys, and battlements rising before them, the whole surrounded by well-tended lawns and forest trees. When Hosack arrived in the spring of 1793, the laird of the clan was James, 21st Brodie of Brodie. James welcomed his young American guest with moving warmth, insisting that he stay in the castle with the family. Hosack found himself dining not only with a laird but also with a duke and a duchess, for James's niece had married the Duke of Gordon, and the couple were in residence during Hosack's visit.

He lingered at Brodie Castle for a week. James was an ardent botanist and horticulturist who loved to ramble in search of new plants. He adored the Scottish countryside so much, in fact, that after being elected to Parliament in the year following Hosack's visit, he declared he was "really unhappy" in London and decided to return to Scotland. The grounds of James's castle contained formal French gardens, a pond, and a ninth-century slab carved with Pictish symbols. Lindens and beech trees spread their boughs over broad walkways. Nearby, Hosack could stroll along the banks of the Muckle Burn, a pretty stream crisscrossed by arched stone bridges, or he could follow the River Findhorn as it wound through the meadows toward the North Sea. These idyllic days reinforced Hosack's desire to learn all he possibly could about plants, as his resolve to study medical botany became pleasantly jumbled with the glamour of a botany-mad nobleman and the picturesque seaside setting. He tore himself away from the Brodies and Elgin to travel south again, and in May 1793 he left Scotland for the botanical capital of the Western world.

LONDON.

Hosack took lodgings in the city with another young man who had studied with Benjamin Rush in Philadelphia. Then, probably on the recommendation of his Edinburgh professors, he set out for a neigh-

borhood called Brompton, on the leafy western outskirts of town. He
was looking for a man named William Curtis, whose entire existence
had been devoted to acquiring exactly what Hosack now coveted:
intellectual dominion over the kingdom of flora.

Curtis was forty-seven years old, a short and thickset man with
a round face that turned red when he tramped through the fields at
his customary breakneck pace. In spite of his sunny disposition, Curtis could sometimes seem aloof, but this was only because he felt shy
conversing with men he thought knew more than he about science.
He needn't have worried—he was the greatest practical botanist in
the whole of Britain.

From earliest boyhood, Curtis had loved roaming through the
pretty Hampshire countryside where he had been born. To him the
natural world was "an endless source of innocent delight," and he had
been miserable when his father sent him away to study medicine. He
could still recall thirty years later how badly he had longed to bolt
back outdoors, away from "the putrid air of the dissecting room" and
"the noxious fumes of the laboratory." After this awful experience, he
decided to become a gardener and botanist so he could spend the rest of
his life surrounded by trees, birdsong, and the scent of flowers.

For six years, Curtis worked as the Demonstrator of Plants at the
Chelsea Physic Garden, a medicinal garden on the Thames run by the

William Curtis

Society of Apothecaries. His job was to use the garden's collections to teach aspiring apothecaries and doctors. Curtis's clinical experience mingled perfectly with his excellent horticultural instincts. Whereas many previous British botanists had thought of plants as so many opportunities to beautify the kingdom's gardens and bring delicacies to its tables, Curtis also saw an opportunity to improve British medicine. He was determined to learn about the possible *virtues*—the eighteenth-century term for the healing powers of a plant—that might lie hidden in the heaps of new species constantly arriving from around the globe. But he was equally interested in discovering the unknown medicines growing wild along England's country lanes. He dreamed of creating his own garden filled with native and foreign plants of all possible species in order to conduct systematic research of a kind unprecedented in Britain. Finally, Curtis waved farewell to his steady salary and in 1789 founded a botanical garden of his own in Brompton.

By the summer of 1793, when Hosack came looking for Curtis, the Brompton Botanic Garden was brimming with life and color. At the southern edge of Curtis's garden, Hosack first encountered a porter's lodge and a gate marked BOTANIC GARDEN, OPEN TO SUBSCRIBERS. He entered and found himself on a wide gravel path. All around him was a tidy patchwork of flowers, grasses, shrubs, and trees, but Hosack's feeble grasp of botany turned Curtis's orderly kingdom into an inscrutable mess. To his left was a cluster of rugged-looking plants and beyond them were dozens more unfamiliar species. Nearby lay a parcel of what appeared to be grass. At the far end of the garden was a small greenhouse filled with specimens from all over the world. But it was in the very first plot to Hosack's right that he found the plants he had come for: dozens of medicinal specimens, their labels inscribed with the Latin names familiar from his Edinburgh medical textbooks.

Hosack summoned up his considerable powers of persuasion and introduced himself to Curtis. The latter was no longer offering private instruction, burdened as he was by the work of tending his garden, publishing on botany, and exchanging plants with naturalists around the globe. He must have found this young American's enthusiasm irresistible, however, because he agreed to take him on as a private student. So began a summer Hosack would remember with pleasure for the rest of his life. He signed up as a Brompton subscriber and received a copy of the garden regulations, which included a ban on dogs and also on "inconsiderate or designing" visitors who made "bouquets of the flowers." His new teacher had rather formal manners, yet at the same time he was pleasant and helpful.

As the two men strolled along the gravel paths, Curtis explained how he had organized his three and a half acres of garden. It was only recently that most European botanists had adopted a unified system for naming and classifying plants. For hundreds of years, people had identified new plants using unwieldy and incommensurate systems. Then, in the middle of the eighteenth century, Carl Linnaeus, a Swedish botanist who had initially trained as a physician, swept away the old botanical order with a series of pathbreaking books: *Systema Naturae* in 1735, *Philosophia Botanica* in 1751, and *Species Plantarum* in 1753. Most plants, Linnaeus argued in *Systema Naturae*, could be divided into groups according to the varying numbers of their stamens and pistils—their sexual organs. This new sexual system was a simple, powerful scheme that botanists could henceforth use for each plant they newly identified. But Linnaeus also realized that the thousands of plants already known to botanists would now need reclassifying according to the sexual system. In the preface to *Philosophia Botanica*, he confessed to being "frightened" by the scale of his own project, but he overcame both his anxieties and a severe attack of gout to produce *Species Plantarum*. In this mammoth work, he advocated a simple two-word Latin combination for each plant (first its genus, then its species) and proposed new names for around six thousand known plants. He also laid out a system for naming future discoveries and urged it on his fellow botanists, whom he undiplomatically accused of favoring "pompous expressions and flowers of rhetoric."

Some fellow botanists were offended by Linnaeus's tyrannical tone. In France, especially, Linnaean botany did not take hold easily. French botanists preferred the "natural system," an approach articulated by Bernard de Jussieu (and later his nephew Antoine-Laurent de Jussieu) that grouped plants together on the basis not only of their sexual organs but also their flowers and fruits. In Britain, too, some botanists were reluctant to adopt the Linnaean system—but not William Curtis, who loved the clarity of Linnaean nomenclature and the ease with which new plants could be classified by the numbers of their stamens and pistils. Curtis was so enthusiastic that he published his own easy-to-follow illustrated guide to the sexual system in 1777, and he taught Linnaean botany to all his students, including Hosack.

Curtis also launched two ambitious publishing projects that brought Linnaean botany to laypeople across Britain and made him famous. The first was a lavishly illustrated compendium of all the plants in the London region entitled *Flora Londinensis*, which Curtis financed through subscriptions. Beginning in the late 1770s, his subscribers periodically found among their letters and parcels a bundle of hand-colored engravings about the size of a large tea tray. Curtis could not draw very well himself, but he had a reputation for being able to spot even the tiniest inaccuracy in other people's plant drawings. He found his finicky standards completely exceeded by a gifted young artist named James Sowerby, one of several he hired to execute illustrations for the *Flora*. Curtis accompanied each plate with the plant's Linnaean and common names and its usual habitat—near sandy beaches, deep in the woods, and so on. He also included chatty descriptions of each species. For example, writing of an herbaceous plant (*Arum maculatum*) whose roots were used to treat asthma, Curtis noted that it was sometimes called "Lords and Ladies" because its flowering parts resemble copulating genitalia and thus "attract the notice of children in the spring." In true Enlightenment fashion, Curtis exhorted his readers to put down their botany books and go outside to make their own observations. "To know the Fungi well, we must watch them daily and yearly; in short, *we must live with them.*"

The volumes of the *Flora Londinensis* were groundbreaking and beautiful, but they were also terribly expensive to publish. Curtis could never seem to sign up quite enough subscribers to underwrite the project. A friend described the project as the "poor dear *Fl. Londinensis*— *your glory*, your *best friend* your everything." But then Curtis hit upon a much more lucrative way to initiate laypeople into the delights of Linnaean botany. In 1787 he began publishing a one-shilling monthly called the *Botanical Magazine*, filled with pictures of the exotic ornamentals British gardeners were increasingly eager to cultivate at home.

The craze for Curtis's new magazine brought him two thousand sub-scribers a month.* By the summer of 1793, when Hosack arrived at the Brompton Botanic Garden, Curtis was the undisputed master of the Linnaean system in Britain.

Hosack was thrilled to be studying with the famous naturalist whose publications had escorted botany out of the botanical garden and onto the grounds of Britain's country estates. But to Hosack the most exciting aspect of Linnaean botany was not that troops of ladies and gentlemen could now rattle off Latin binomials as they picked their way through the woods. It was that Linnaeus had flung open a door to the discovery of new medicines.

Hosack knew from his Edinburgh teachers that physicians and apothecaries had for centuries relied on clumsy and inaccurate rules to try to divine the medicinal properties of specific plants. People kept holding out hope, for example, that plants of a particular color or smell or shape would have similar virtues and could thus treat similar ill-nesses. Linnaeus pointed out in *Philosophia Botanica* that these assump-tions had proved "false and delusive." He had therefore come up with a new framework, which he captured in a few simple new rules that mapped neatly onto his equally neat plant classification system. Plants of the same *order*, he predicted, would likely turn out to have at least some overlapping medical properties with one another. Plants of the same *class* would share most of the same virtues with one another. And plants of the same *genus* would tend to overlap entirely in their virtues. By way of example, Linnaeus noted that the various known species of *Convolvulus*, a genus in the bindweed family that included morning glories, all appeared to have purgative effects on the body.

This was a bold new approach to medical botany, but Linnaeus acknowledged that there was one hurdle. Anyone who was interested in discovering the virtues of an unfamiliar species of plant would need to know what properties were associated with other plants in the same class, order, or genus. Curtis, leafing through his volumes of Linnaeus, was inclined to agree. He also thought that asking medical students to memorize these properties from mere written descriptions of plants

---

* It is still in circulation today as *Curtis's Botanical Magazine*.

was at best inefficient and at worst fatal. It was only in a botanical garden like Brompton, Curtis thought, that medical students could learn enough about medicinal plants and their internal structures to go forth and find new ones. To help students like Hosack understand the exciting promise of Linnaean medical botany, Curtis labeled each of his medicinal plants not only with its Latin and common names but also with the page number from the *Edinburgh New Dispensatory* where that particular plant's virtues were discussed. This would help the students commit to memory the link between the medicinal uses of the plant and the way it looked when it was actually growing. Then they would be prepared to compare that knowledge to the new plants they would encounter—maybe even discover—in the future.

All that summer at Curtis's side, Hosack memorized plant names, studied their medical properties, and worked on his classification skills. For the medicinal beds, Curtis had chosen dozens of native plants commonly used by British physicians as well as many other plants from across Europe and around the world. Hosack already knew from his own studies that different medicinal plants had different effects on the body. Some plants, for example, functioned mainly to ease pain—such as Solomon's seal (*Convallaria multiflora*), with its ladders of bell-like white flowers dangling along the stem, shiny leaves, and roots that could be mashed into a poultice to reduce inflammation. Other plants could be used to help purge a sick body of offending fluids, a critical function in the humoral understanding of the body. Which of these plants a doctor decided to prescribe would depend on his patient's particular imbalance or ailment. When faced with a constipated patient, for example, he might choose blue succory (*Cichorium intybus*). Sometimes known as chicory, it had leaves, stalks, and roots that yielded a juice capable of inducing mild diarrhea. Pot marigold (*Calendula officinalis*) likewise acted to relieve constipation, but as the *Edinburgh New Dispensatory* noted, it was also "celebrated in uterine obstructions"—meaning that it flushed out miscarriages.

Curtis was also growing plants used by doctors to keep unruly excretions *inside* the body. For example, the leaves of common agrimony (*Agrimonia eupatoria*), a spindly plant with fragrant yellow flowers, possessed antidiarrheal properties helpful to patients afflicted with

wasting diseases such as scurvy, which involved a "lax state of the solids." Most of Curtis's medicinal plants could be used in treating more
than one illness, but some had long been linked with a single disease.
He was raising a pretty perennial called great blue lobelia (*Lobelia
siphilitica*), whose smelly, vomit-inducing roots Native Americans used
to treat venereal disease. He had also planted the widely useful vervain (*Verbena officinalis*) in his medicinal beds, but he laughed off the
idea that vervain could cure scrofula (an infection of the lymph nodes)
when tied with a white satin ribbon around the patient's neck. This was
nonsense, he said, "even making the greatest allowance for the powers of the imagination." Curtis could scoff at old medical myths, but
the painful truth was that no one truly knew yet just *what* made a particular plant effective against a given illness. And no one—not even
Linnaeus—had yet come up with a reliable way to discover and isolate
new plant-based medicines.

Hosack was so fascinated by what Curtis was teaching him that
he started going to the Brompton garden every day. Although medical botany had first led him there, he was now discovering what a
small county medicine occupied in the sprawling botanical kingdom. Curtis had dedicated another of his many beds, for example,
to agricultural crops such as oats, barley, and flax. He argued that
sound farming practices were every bit as vital to a great nation as
military prowess and cultural achievements. A farmer should be
on intimate terms with "his vegetable foes as well as his vegetable
friends." Curtis so deplored the way British landowners ruined their
soil with poor planting habits that he wrote an entire book on how
to use various kinds of grasses—"a much neglected tribe," he called
them—to restore exhausted soil. To illustrate his point Curtis sowed
a small meadow's worth of grasses, rushes, and sedges at Brompton. (He immediately found himself at war with "one or more hares"
that ate his *Juncus niveus* to the ground, despite his past experience
that British hares "preferred the *Poa procumbens*.") He also planted
floating foxtail grass (*Alopecurus geniculatus*) and sweet vernal grass
(*Anthoxanthum odoratum*), the last of which he adored for its "agreeable scent of new made hay." In summertime, any Brompton visitor
who dared to break the garden rules and stray from the central path

could wade nearly knee-deep through a tiny parcel of the famously picturesque English countryside.

Yet in the very loveliness of the English countryside, Curtis explained, lethal poisons lurked. A child reached for handfuls of shiny black berries and expired in convulsions hours later. A poor farmer collected wild greens to boil for food and then watched members of his family die after the meal. Curtis took care to include poisonous plants in his garden beds and also showed in the *Flora Londinensis* and the *Botanical Magazine* how to recognize them. At Brompton, for example, he was growing a poisonous herb called *Aethusa cynapium*, whose leaves looked so much like common parsley that it was known as fool's parsley. It caused vomiting, delirium, convulsions, and death. Nearby, Curtis had planted a pretty plant of lavender and pale green sometimes called sea bluebells (*Pulmonaria maritima*). One would hardly suspect, he wrote, that "poison lies under such an elegant form."

As Hosack circumnavigated the Brompton garden that summer, he learned that each bed had been as thoughtfully stocked as the cabinets in a natural history museum. Yet Curtis was too inspired a gardener not to fringe his Enlightenment rigor with decorative hues and scents. He had edged his long central path with a flower border and then bisected it with a bower stretching from one side of the garden to the other. At the end of a long winter, he always took heart at the "brilliant and exhilarating" colors of his spring crocuses (*Crocus vernus*), although he lamented that the sparrows loved the crocuses as much as he did. He recommended placing "the skin of a cat properly stuffed" nearby to scare the sparrows away.

Between the Brompton flower beds, Curtis had planted rows of Lombardy poplars, which he was pruning to create screens of foliage that would give visitors the sensation of moving from room to room in a house whose ceiling was the sky. For Hosack, the garden did become a new home. What he couldn't find in the soil, he found in books, spending hours in the little library in the back corner of the garden. Curtis had installed an aviary just next door, and with the birds singing and the leaves rustling on the shrubbery outside, a reader curled up in a library chair felt, as one visitor put it, "a thousand miles from London." Among the books Hosack could pull from the shelves were dozens of works by Linnaeus and other botanists and horticulturists, including Curtis himself—his *Flora Londinensis*, his *Botanical Magazine*, his book on agricultural grasses, and his handy guide to Linnaeus's sexual system. Curtis had also stockpiled such volumes as *Hints to Gentlemen of Landed Property*, *Directions for Bringing Seeds and Plants from Abroad*, and *The Theology of Insects, or Demonstration of the Perfections of God in All That Concerns Insects*. Early on, Curtis had allowed visitors to check out the library books and take them home, but after "much inconvenience" he had suspended the practice.

Hosack wrote to his father for additional funds and began to visit the London booksellers for himself. He purchased a five-volume edition of Linnaeus's masterpiece, *Systema Naturae*. Setting aside the first two volumes, which covered the animal kingdom, he opened the volume on plants to its title page and in his finest possible handwriting laid claim to the whole botanical world. "E. Libris David Hosack, Lond. 1793." He was ravenous for knowledge. He collected William Speechly's *Treatise on the Culture of the Vine*, James Bolton's *History of Funguses Growing about Halifax*, Lord Kames's *The Gentleman Farmer*, and a gilt-trimmed English translation of Jean-Jacques Rousseau's *Letters on the Elements of Botany Addressed to a Lady*. Rousseau offered comforting words for a botanical novice: "Do not however be terrified at the undertaking. . . . Nothing is required but to have patience to begin with the beginning." In truth, Hosack seemed to possess infinite patience for his new obsession. Bending over the Brompton beds, he labored for hours to connect the dense columns of Latin names printed in *Systema Naturae* with the way each plant looked as it grew.

Walking from one end of the garden to the other was like caravanning across whole empires, but instead of consulting a map Hosack tracked his progress around the globe with treatises on the plants of Japan, Ceylon, Peru, Chile, and other faraway lands. Curtis loved his homegrown British plants and chafed at criticism that he gave them too much space in the *Botanical Magazine*, yet at Brompton he lavished attention on hundreds of foreign specimens. It was in Brompton's greenhouse and hothouse that Hosack encountered the rarest of these treasures. In the greenhouse, for example, Curtis was raising a dark flower from Spain called melancholy toad-flax (*Antirrhinum triste*) and a woolly cineraria (*Cineraria lanata*), probably from the Canary Islands, with pinkish-purple blooms. His prize specimen in the stove-heated hothouse was a plant that "every person who can boast a hot-house will be anxious to possess": the scarlet fuchsia (*Fuchsia coccinea*), a Brazilian flower whose allure sent him into a fit of poesy. Writing in the *Botanical Magazine* in 1790, Curtis said its central petals looked like "a small roll of the richest purple-coloured ribband."

The Brompton Botanic Garden was an enormous horticultural and botanical achievement, and Curtis's fame was spreading across Europe. Nevertheless, he had failed to achieve his greatest ambition: compiling a complete natural history of the British Isles. To his friends he sometimes seemed depressed—"sunk in spirits" and "indolent." Once a week that summer, however, the clouds cleared from Curtis's brow and he became sunny again. He had decided to offer a weekly course of "herbarizing excursions" aimed at medical students, practicing physicians, and country squires—the last of whom he joked might "pursue the foxtail-grass with more advantage than the fox." Curtis was close to fifty now, but he still got as excited about collecting in the wild as he had when he was a boy.

Hosack signed up for the summer session of 1793, along with five other men. Their excursions began in the midmorning, often at the edge of Hampstead Heath at a sixteenth-century tavern called the Spaniards Inn.* Each student was instructed to arrive with a small notebook and be prepared for hours of botanizing; latecomers would be left behind as they marched off into the fields together like a tiny regiment. There was one other American in the class, a doctor from New York with whom Hosack would be able to trade gossip about life back home. He would also have plenty to discuss with a young Englishman named Robert Thornton, who had recently earned his medical degree from Cambridge with a thesis on oxygen, body heat, and blood. Thornton was rather handsome, but he tended to sweat profusely because his research had convinced him that the best way to stay cool in the scalding heat was to race about in a heavy woolen

---

* This tavern, which would later figure in Charles Dickens's *The Pickwick Papers* and Bram Stoker's *Dracula*, still welcomes guests today.

vest. Within a few years of the Brompton summer, Thornton would be launching the most ambitious botanical publication project Britain had ever seen, an expensive illustrated series that nearly ruined him. It was called the *New Illustration of the Sexual System of Linnaeus*, although he soon published a chunk of it under a snappier title: *The Temple of Flora, or Garden of Nature*.

There were several other British men in the class, but the most illustrious was the Right Honourable Charles Francis Greville. The second son of the Earl of Warwick, he had spent his childhood roaming the halls of Warwick Castle, his family's ancestral home. He was Member of Parliament for Warwick and an accomplished mineralogist who had been a Fellow of the Royal Society for nearly two decades.

As Hosack and his new classmates walked through the countryside, Curtis imparted to them his life's greatest pleasure—the power "to read in that book which to the generality of mankind is a mere blank." Curtis searched for plants in bloom so he could show the students the critical structures of growth and reproduction. His face lit up when he found an interesting species. He carefully plucked each specimen and

Gentlemen botanizing, from the frontispiece to
Curtis's Flora Londinensis

placed it between two pages of his notebook, directing the students to do the same with specimens of their own. Then he charged onward in search of the next prize.

Hosack discovered that meeting plants in the wild this way, where they never sprouted helpful Latin labels, was testing and deepening his store of botanical knowledge. Although Curtis fumed that a "rage for building" at the edges of London was wiping out many species he used to see there, Hosack now began to recognize the conditions of soil and sunshine that particular English plants sought out when left to their own devices. Sneezewort, the toothache remedy that Curtis was raising at Brompton, flourished alongside the ditches in the Battersea meadows, while the sweet-scented vernal grass that Curtis so loved grew in dense meadows in the bottomlands. The bright blooms of the scarlet pimpernel (*Anagallis arvensis*) sprinkled themselves across the cornfields, a contrasting effect Curtis admired.

After several hours of walking and collecting, Curtis would lead his tired band back to the tavern. He personally found long meals a waste of good plant-hunting time and ate only to fuel his hikes, but he allowed his students to refresh themselves before he rounded them up for the second part of the day's lesson at the tavern table. With the dishes cleared away, the men got out their notebooks and followed along as Curtis held up each plant in turn, pronouncing its Latin and common names and pointing out the sexual organs and other features that would help the students identify it in the wild. He explained the plant's uses, if any, for medicine or agriculture. He taught Hosack and the other men how to press and dry their own specimens at home, then mount and label each one on a large sheet of cream-colored paper. Each student would thus have the beginnings of his own herbarium— a collection of dried specimens that they could consult and expand in the years to come. As Curtis talked about each plant, his world-weary air fell away and he "glowed with youthful fire." His passion for the natural world ignited Hosack's imagination.

Chapter 3

"RIPPING OPEN
MY BELLY"

THAT AUTUMN, HOSACK CEASED HIS DAILY VISITS TO CURTIS. The cooler weather that was now thinning out the Brompton beds was excellent for another kind of harvest: the bodies of the dead collected from gibbets and poorhouses for the use of medical gentlemen. This was the start of the dissection season, when corpses retained their freshness longer than in the summer heat.

London, Hosack had discovered, was not only the capital of botany. It was also the capital of anatomy. He had studied the structures of the human body to the best of his ability in New York, Philadelphia, and Edinburgh, but it was in London that physicians were practicing anatomy with the most sophistication, thanks to William and John Hunter, the brothers who had taught Hosack's former New York teacher Richard Bayley. Hosack refused to leave for the United States before he had studied with London's great anatomists. He could put his growing knowledge of medical botany to better use for his future patients if he knew much more about blood, nerves, tissue, and organs. He now trained his attention on the slippery inner world of the human body.

In London, medical instruction mostly took place in hospitals and private medical schools—unlike in Edinburgh, with its great university. In 1746, before dissection bore even the thin patina of respectability it had acquired by the time Hosack arrived in London, William Hunter had opened a private medical school where each student was allowed to practice on a corpse of his own. Two years later, William's younger brother John, a carpenter by trade, joined him at the school as an assistant. John soon surpassed everyone, including William, in the painstaking art of anatomical preparation. For the next four and a half decades, John Hunter, sometimes working with his brother, sometimes alone, conducted hundreds of surgeries, dissections, and vivisections in an effort to understand the functioning of human and animal bodies. John Hunter worked and lived in a mess of pickled tumors and live animals. He sawed the heads off corpses and discovered the nerves responsible for the sense of smell. He injected milk into the intestines of a living dog to show that it did not enter the bloodstream, thus proving decisively that the lymphatic system is distinct from the circulatory system. He taught hundreds of medical students anatomy and surgery over the course of the eighteenth century. Among them, in the early 1770s, was a young man named Edward Jenner, who would perform the first documented cowpox inoculation twenty-five years later. Together, the Hunter brothers did more than anyone else in history to make dissection central to medical education.

Now, several times a week, Hosack wound his way through Holborn to a little dead-end lane called Thavies Inn, where he was taking classes at Dr. Andrew Marshal's medical school. Marshal had learned his craft from the Hunters and by 1785 had set up his own medical school. Although he was considered a poor public speaker, his talents as an anatomist came shining through during his in-class demonstrations. A fellow student of Hosack's later recalled that when Marshal lectured on a brandished body part, "you saw it with your own eyes, exactly as he described it, and noticed every peculiarity of its shape and appearance."

Marshal's specialty was the brain. He dissected the heads of the insane with particular relish and was always ready with a memorable example, like the case of Tommy Pearson, a fourteen-year-old butcher's apprentice who had gone mad after a stray dog bit him in the summer of 1787. Froth had sputtered out of the boy's mouth as he screamed at his nurse that Satan had entered his hospital room. "I see two girls, and a black boy; and the boy's belly is ripped up." The invisible fiend had then turned on him. "It cuts me; cuts Tommy's hand off! . . . Ripping open my belly!" When Marshal dissected the boy's body a few days later, he found excess moisture between the outer layers of the brain, an anomaly he had noticed in many other dissected brains of the insane.

Most riveting for his students were the dissections he conducted before their very eyes. In 1793, the year Hosack was studying with him, Marshal peeled apart the brain of a demented fifty-two-year-old woman who had died at the St. Clement's workhouse. He also sawed into the tall, muscular corpse of a former marine sergeant who had been permanently hospitalized with violent fits. The sergeant had been prone to disobedience unless an orderly barked at him, "All hands a-high." Then he sprang to attention and followed orders. During the autopsy, Marshal squeezed the sergeant's brain and found it "very hard" and "the ventricles distended with water." The autopsies Hosack watched and conducted at Marshal's school helped him better understand how different diseases affected the organs and tissue. They didn't provide a fail-safe guide to treatment, but now when presented with a sick patient he would have a clearer picture of what was happening inside the body and how to try to address it. For example, swollen

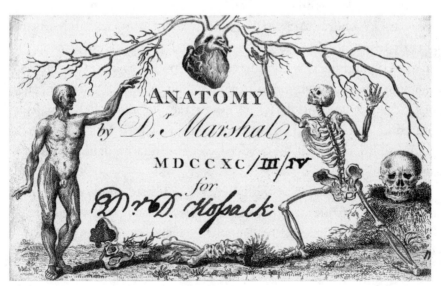

Hosack's entrance ticket to Dr. Andrew Marshal's anatomy course

ankles were filled with water and would call for a diuretic plant rem-edy, while a turgid abdomen needed a purgative. Fever accompanied with a racing pulse might be eased with a sudorific, to induce sweating.

Much as Hosack loved medical botany, he knew that plant-based remedies would not be enough in dire situations. Broken bones, pus-filled lungs, gunshot wounds, cancerous breasts—these and so many more disasters would require surgery, about which he knew little as yet. London's surgeons had the great advantage over American doc-tors of their thorough anatomical training. Hosack began shadow-ing Dr. James Earle, chief surgeon at the storied St. Bartholomew's Hospital in Holborn. He watched intently as Earle and his colleagues examined their nervous patients and then opened leather pouches to select the right tools for the job: knives, saws, bone nippers, perhaps a trephine for drilling a hole in the skull. There was a scoop especially designed to extract a bullet. And there was a "curved trocar for punc-turing the bladder per rectum"—as advertised by a leading London purveyor of surgical instruments.

Earle's specialty was a condition called a hydrocele, a swelling of the testicle sometimes caused by injury and often involving intense pain. Hosack would be the first to perform Earle's revolutionary treat-ment for it in the United States. Most hydroceles required surgical intervention, although Earle also saw hydroceles that had been burst open by falls from horseback or as they bounced against saddles. Of the available treatments, he shrank from the "exquisitely painful" pro-cess of slicing off the hydrocele, as well as from the "horror" of searing it with caustic. Instead, he punctured the testicle with a sharp, tubular trocar in order to drain the fluid. Then he injected the testicle with a mixture of alcohol and water. Earle performed this innovative proce-dure many times in the early 1790s. In March 1792, for example, he was summoned to treat a Frenchman who had just fled the Revolution for England. When the man disrobed, Earle saw he was suffering from a hydrocele "larger considerably than his head," which he had managed to keep hidden under his clothes by bandaging it backward between his thighs. Earle pierced the inflamed testicle with his trocar and siphoned out more than six pints of blood-tinged water. On another occasion Earle injected the hydrocele of an Indian servant boy who had been brought for treatment by his master. He also drained "clear straw coloured fluid" from an unlucky man with two hydroceles—one on each side. Hosack memorized Earle's methods for treating hydroceles, to take back with him to the United States.

⎯⎯ ⚬⚬⚬ ⎯⎯

HOSACK FOUND ANATOMY and surgery irresistible, almost godlike, but all that year he also continued to study botany. At the Brompton Botanic Garden, Hosack had so deeply impressed William Curtis with his drive and talent that in the fall of 1793, just after their summer together, Curtis nominated him as a Foreign Member of an exclusive society for natural history.

The Linnean Society of London operated from an elegant townhouse on Great Marlborough Street that belonged to its president, James Edward Smith. Smith and some friends had founded the Linnean Society in the 1780s to safeguard Carl Linnaeus's collections, which Smith had purchased with the help of his wealthy father. (The spelling *Linnean* was in honor of the title Linnaeus received upon being ennobled by the Swedish king for his services to science: Carl von Linné.) Hosack, on the strength of Curtis's nomination, was officially elected a Foreign Member of the Linnean Society in December 1793, about seven months after he had arrived in London. Finding himself welcomed into a luminous circle of scientific camaraderie, he was surprised and delighted by the hospitality Britons seemed willing to show their erstwhile foes. Hosack quickly became just as devoted to daily study at the Linnean Society as he had been at the Brompton garden.

Linnaeus, in his will—which he grandly called the "Voice from the grave to her who was my dear wife"—had with complete justice declared his trove of specimens to be "the greatest the world has ever seen." The crates Smith had received in October 1784 from Sweden included approximately fifteen thousand plants. British botanists were elated that they could consult specimens that had been labeled by Linnaeus himself, which would help them pinpoint the Linnaean names of unknown specimens in their own collections. More exciting still was the prospect of clapping eyes on many exotic species that British botanists had previously encountered only in books. In short order the Royal Society had unanimously elected Smith a Fellow, and over the next decade, Smith had climbed ever closer to the pinnacle of British botany. After he was engaged to teach botany to Queen Charlotte and her daughters Princess Augusta and Princess Elizabeth, one of his friends jokingly called him the "King of Botany."

Smith was just ten years older than Hosack, and the two men forged a warm friendship. They would correspond for the rest of their lives, with Hosack later sending many London-bound friends to meet Smith, among them a young Washington Irving. But Hosack felt humble and nervous in the presence of another member of the Linnean Society: Sir Joseph Banks. In 1768, at the age of twenty-five, following botanical studies at the Chelsea Physic Garden, the handsome and wealthy Banks had flung himself headlong into natural history. He had abandoned a hopeful young woman and a comfortable London life for three years' passage aboard Captain James Cook's *Endeavour.* Banks had almost single-handedly burnished botany with heroism and glamour while pursuing plant specimens across the globe. On the day of Hosack's birth, he had been sailing through the South Pacific. He had nearly been shipwrecked on the Great Barrier Reef before returning home laden with species of flora and fauna Britain had never seen. Even the grumpy Linnaeus had sung his praises, while King George III placed him at the head of the botanical collections at Kew Palace. In 1778, the Royal Society of London elected Banks president, a position of scientific glory he had held for fifteen years by the time Hosack arrived in London.

Despite their great divide in station and age, Banks was very kind to Hosack, who would later write Banks thanking him for his generosity "when I was a youth and totally unworthy." How thrilling for Hosack to be able to tell his father, who had once borrowed a book about Captain Cook's travels from the library in New York, that he was regularly socializing with Cook's most famous travel companion. Hosack listened with particular attentiveness to Banks's remarks on botany and one day heard him say something that made a lasting impression: "Even an imperfect dried specimen is preferable to the best painting."

In the spring of 1794, as the days grew longer and the first hints of green appeared in London's parks, Hosack began following Smith's public botany lectures at the Linnean Society, given on Tuesdays, Thursdays, and Saturdays. At other times, Hosack studied on his own in the collections, and so it was that he came to be standing one day before three strange-looking green cabinets in the library on Great Marlborough Street. Each was taller and thinner than a coffin, and each was fronted with double doors.

For centuries, botanists had bound their dried, mounted specimens in books. This system privileged physical convenience over scientific logic, because when new specimens were collected, they had to be filed separately from fellow members of their genus or simply shoved loose between the relevant pages. More vexing still, when revolutionary thinkers—above all Linnaeus himself—came along and upset the very principles underlying natural classification, collectors whose specimens were bound in volumes found themselves hopelessly stuck in the past. The three cabinets now facing Hosack were Linnaeus's own solution to the problem. Each shelf corresponded to a single class of plants in his sexual system, and on each shelf lay piles of unbound herbarium sheets. Each of these sheets, in turn, was mounted with just a single specimen. Many had been annotated in Linnaeus's own hand. Less than a year after Hosack had first looked into the mysteries of botany with Curtis as his guide, he was standing before the shrine of its high priest.

For four months he studied the Linnaean specimens. They looked brittle and faded compared to the plants he had beheld in full flower at Brompton, but with a little effort he could conjure up their living colors in his mind's eye. The breadth of the collection was astonishing—thousands of plants from Russia, Lapland, Sweden, Norway, Spain, Italy, France, Portugal, Grenada, Germany, Java, China, India, Egypt, Arabia, Siberia, Anatolia, Palestine, Persia, and many other countries and regions. Some of the sheets were decorated with intricate, hand-drawn vases from which the real stems of the attached specimens appeared to rise. On other sheets, round whorls of petals had subsided into the most delicate of fans, and flattened bulbs clung sheerly to the paper, their filaments trailing like watery tentacles. As he worked his way through the piles of specimens, he traveled down the valleys and over the mountains of a whole planet. Then, months before he boarded the ship that would take him back across the Atlantic, Hosack suddenly found himself home again. Linnaeus's green-painted cabinets contained many plants native to North America.

Linnaeus had never been to America, but his prize student, Pehr Kalm, had. From 1748 to 1751, by order of the Swedish Academy of Science, Kalm had ranged from Philadelphia up through New Jersey and New York to Canada in search of medicinal, agricultural, and commercial plants. He returned to his elated teacher in Uppsala with hundreds of specimens, many of which later made their way to the Linnean Society. As Hosack studied the Linnaean specimens, he could not have missed the careful letter "K"—for Kalm—written by Linnaeus at the bottoms of dozens of the sheets. Kalm's travels meant that Linnaeus's collection contained many specimens of plants that Hosack would have seen growing up on Manhattan, from the grandest oaks to the plainest grasses. Other collectors, especially John Bartram in Pennsylvania and Mark Catesby in the Carolinas, had also contributed North American plants to Linnaeus, either directly or via his British correspondents.

It had happened to Hosack at Brompton, too—this encounter with American plants. Among his medicinals, for example, Curtis was growing American native species such as great blue lobelia (*Lobelia siphilitica*) and white vervain (*Verbena urticifolia*). When Hosack had walked the country lanes of his Manhattan boyhood, they had been right there all along. Now, in London, his thoughts bounced constantly from botany to surgery and back to botany. Were plants or lancets, he wondered, the better approach to treating illness? While he studied at the Linnean Society that year, Hosack was continuing his anatomical research and writing a paper on the musculature of the eye. He dissected the eyeballs of fish, sheep, rabbits, oxen, and humans, pulling apart the viscous material in search of tiny muscles. He experimented on himself as well, fixing one eyeball open with a *speculum oculi* and pressing hard on it as he held objects up at varying distances. "By what power," he asked, "is the eye enabled to view objects distinctly at different distances?" He concluded—contrary to the theory put forth in

a Royal Society paper the previous year—that it was not a set of muscles embedded in the lenses that was responsible for the ability to see
clearly at different distances. Instead, he argued, the answer lay in the
external muscles of the eye, which he also praised for allowing humans
to convey their emotions. People so often remarked upon Hosack's
own large, expressive eyes, which sometimes signaled impatience but
at other times warmth or excitement.

Banks suggested that Hosack submit the paper to the Royal Society, and it was accepted. Because Hosack was not a Fellow, however,
one of his professors presented it for him on May 1, 1794. Banks, as
president of the Royal Society, sat on a raised dais at the front of the
society's sumptuous hall in Somerset House, with an enormous golden
mace resting on the table before him. The Fellows were ranged in rows
facing Banks as they listened to Hosack's paper. Soon it would be published in the *Philosophical Transactions of the Royal Society*. Hosack had
soared to the heights of British science in his one year in London, and
he wasn't yet twenty-five years old. Decades later, an eminent American physician would credit this paper with first making Hosack's name
"widely known on both sides of the Atlantic."

Girded with this triumph and all he had learned in Edinburgh
and London, Hosack finally felt ready to return home. After two
years away, he would sail back to the United States as a vastly more
skilled physician and one of the best-trained Americans in Linnaean
botany. He packed his books, papers, and clothing—and very likely
new medicines and surgical tools, too. His British friends sent him
off with a flourish. Smith gave him a set of specimens from Linnaeus's own collections, duplicates that he could spare without damaging the integrity of the Linnean Society's holdings. It was a touching
expression of his faith in Hosack's promise as a botanist. And not long
before Hosack left London, he was nominated for promotion in the
Linnean Society from Foreign Member to Fellow. This had never
been done before—making a foreigner a full-fledged Fellow—but
Hosack would have to leave London without knowing his fate. The
Fellows of the Linnean Society wouldn't be assembling for the vote
until later in the summer.

He would sail on an American ship named the *Mohawk*. Captain

Howard Allen had sailed her over to England that spring, arriving in early May 1794 at Dover. In London, Captain Allen had met with merchants, seen to it that their cargo was stowed aboard, and purchased supplies for his crew and passengers. By late June, all was ready. Hosack boarded, got settled into his berth, and looked over the ship that would be his home for the next six weeks. She had been fashioned of pine and iron the previous year in the shipyards at Hudson, New York. Like other ships of her class, the *Mohawk* probably had two decks, three masts, and square-rigged sails.

The *Mohawk* was one of the first American vessels to leave London since John Jay had arrived in mid-June to conduct a delicate diplomatic mission with Great Britain. Copies of Jay's initial dispatches regarding his negotiations with the British government were on board. "Kiss our little ones for me," Jay had written to his wife on May 12 before embarking at New York to the salute of cannonfire and the cheers of the immense crowds gathered on the Battery to wish him well. In the fall of 1793, the British had made an official policy of seizing American vessels on the West Indies trade route. When this news had reached the United States in March 1794, long-simmering tensions had threatened to boil over into war.

Meanwhile, the British were struggling to dominate Revolutionary France on the high seas—especially after the young artillery commander Napoleon Bonaparte helped expel the British fleet from the French port of Toulon in December 1793. The British, desperate for manpower, were hauling American sailors from their ships and conscripting them into the Royal Navy. With tempers flaring, Congress appropriated funds in late March 1794 for American seaports to gird themselves against potential British attacks. In May, President Washington sent Jay to London as a special envoy in an effort to defuse the tension. Governor George Clinton soon took the precaution of ordering "all Vessels of War belonging to Foreign Nations" to drop anchor at least a mile south of Governors Island, putting New York City beyond the reach of incoming cannonfire.

Toward the end of June, less than a week after Jay had arrived in London, he was in the middle of writing a letter to his wife when word arrived that an American vessel would be sailing shortly for New

York. Jay quickly drafted letters to both President Washington and Secretary of State Edmund Randolph—Jefferson had resigned on the last day of 1793—and entrusted copies to Captain Allen for transport aboard the *Mohawk*.

Hosack would be sharing the first-class accommodations with a group of men and several women, among them Americans who, like Hosack, were returning home after a polishing sojourn abroad. They had all led quiet lives, however, compared with the Englishman Thomas Law, who was thirty-seven years old when he boarded the *Mohawk*. Law had high cheekbones and a delicate mouth, all of which united to give him a dreamy appearance. Born to an important and wealthy family in Cambridge, England, he worshiped finance, poetry, adventure, and women, and he had spent his adult years in pursuit of them all. (Within a few years after arriving in the United States, Law would marry George Washington's step-granddaughter Eliza Custis.) He was traveling on the *Mohawk* that summer with a companion, a military man named William Mayne Duncanson.

Law had the look about him of a man who could not reach for a quill pen without a fluttering of soft white cuffs. Duncanson cut a sharp figure of martial virility, and his heavy-lidded eyes gave him a haughty air. Like Law, Duncanson had spent time in northeastern India, and the two men may have begun their friendship there. Law was also traveling with at least one of his several illegitimate young sons, whose mother he had left behind in India. With such worldly, colorful men as Law and Duncanson at the table, dinner in the captain's quarters would be a diverting affair, even if the food was the usual dessicated shipboard shoe leather. Captain Allen was by all accounts a warm and generous host.

As Hosack became better acquainted with his new companions, the captain guided the ship around the bulging belly of eastern England toward Newhaven, a town near Brighton on the southern coast. At that same moment, several dozen families from Sussex were floating on a barge down the River Ouse, also making for Newhaven, where they would board the *Mohawk*. The men on the barge were blacksmiths, carpenters, and other artisans who hoped to find work building the new capital city on the Potomac. They boarded the ship and settled into the steerage compartments—the sort of cabins notable only for their cupboard-size bunks, dirty mattresses, and fetid air. Now the *Mohawk* made for open sea.

Even aboard ship, Hosack could not tamp down his restless scientific curiosity. While they were still close to the coast of England, he attached a flaming wick to a cork, floated the cork in a basin full of water, and then lowered an empty wine decanter over it. He was trying to determine whether there was any basis for the conventional wisdom that ocean voyages were good for patients with consumption, and also for what he described as the "old German practice of placing consumptive patients in cowhouses where the air had lost a portion of its oxygene." As the fire consumed the oxygen in the air, he measured the change in the height of the water inside the decanter and took careful notes. He planned to repeat his experiment when the *Mohawk* was halfway across the Atlantic—if they made it that far. They were, after all, sailing into a war zone. All that summer, French frigates had been chasing down and raiding British and American ships. Revolutionary France was desperate for food, dry goods, and military supplies, and the hold of the *Mohawk* was filled with wares that would warm the heart of any French admiral.

Three weeks into the voyage, Captain Allen and the passengers spotted an American ship in the distance. When the *Mohawk* hove to, they learned that she was the *Industry*, and that she had sailed from the mouth of the Delaware River en route to Denmark. Her captain shared news of a rumored British attack on General Anthony Wayne's forces on the northern edge of the United States. For his part, Captain Allen could report on John Jay's safe arrival in London and the opening of his negotiations with Britain. After a brief exchange with the *Mohawk*, the *Industry* continued on her way to Denmark, but she never made it. A week after crossing paths with the *Mohawk*, she was captured by a French Revolutionary frigate. The *Industry* thus joined the lengthening list of ships taken by the French that summer, seizures that American and British newspapers were tallying for their readers with hair-raising frequency. At any hour of the day or night, Hosack and his fellow passengers, too, might suddenly find themselves prisoners of French Revolutionary forces. The *Mohawk*'s crew members were also contending with the unnerving specter of impressment by the British navy. All eyes were on Jay's negotiations in London, but it would be many months before Americans would learn whether they would succeed.

Hosack hoped for a peaceful outcome that would not divide Americans. While living abroad, he had noticed that many Britons still held out hope that the American experiment would fail. He would complain in a letter to Rush right after his *Mohawk* trip that the British "have the peculiar talent of embellishing every story" that gave the slightest hint of political conflict in the United States. Still, there was no ignoring the fact that Americans disagreed violently among themselves on whether the current tensions with Britain presaged disaster. Any Democratic-Republicans on board the *Mohawk* would have been angered by Jay's effort to establish cordial ties with Britain at the expense of France, even as the Federalists were loudly applauding Washington's wisdom in sending Jay to London. It was not until after they landed at New York that they would all learn of the sensational events unfolding in France in late July. On July 27, Robespierre was arrested while trying to seize the floor during a legislative session. He went to the guillotine the next day. Opponents of his rigid Revolutionary ideals breathed a collective sigh of relief.

As for Hosack, he thought physicians should steer clear of politics. Somewhere in the middle of the Atlantic, he set up his decanter and took a second oxygen reading.

IT WAS PROBABLY around the time of Robespierre's death that illness broke out on the *Mohawk*. Jail fever, war fever, ship fever—typhus. Hosack later described the terrible symptoms of typhus, among them the "frothy and offensive discharges from the bowels, discoloured lips ... [and] a cadaverous and offensive smell of the whole body." Transmitted by body lice, the bacteria killed the lice but lived on for days, infesting the raw, itchy sores of prisoners, soldiers, sailors, the poor. A mother gathered up dirty bedclothes in a sickroom, or a doctor lifted a sweat-drenched shirt away from clammy skin, and the disease silently moved to its next victim. How did it stow away on the *Mohawk*? Perhaps it had boarded with a sailor from London or a laborer from Sussex. Perhaps it had crossed the Atlantic on Captain Allen's last voyage, hidden in a moldering mattress. It was a disease that made its presence known subtly at first—in pale faces, light chills, a general weariness. Its name was descended from *typhos*, the Greek word for the delirious stupor that settled over the minds of the infected. In the close confines of the steerage compartments, the sickening scent of gangrenous flesh would have mingled with the stench of diarrhea.

Before long, some of the first-class passengers also fell ill. The whole ship was a stinking hospital. For the first time in his life, Hosack faced a medical catastrophe on his own, far from the surgical theaters and clinical wards where he had been able to rely on his professors' experience. Now, after climbing down to the steerage quarters to examine the sick, he had only his memory and his medical texts to consult. At the University of Edinburgh, the preference was for strong purgatives to clean out the bowels, especially during the late stages of the disease. But Hosack felt deeply mistrustful of drastic measures such as bleeding and purging, which he suspected of exhausting patients exactly when they needed to marshal their feeble energies. He would later condemn these conventional treatments for typhus as "dangerous" and "indiscriminate," advocating instead the use of gentle purgatives like rhubarb or magnesia as well as nervous-system stimulants like Peruvian bark, snakeroot, and bitters. It seems highly likely that Hosack had these staple medicines with him on the *Mohawk* after his time in London, and he could also ask Captain Allen to commandeer some of the ship's stores of liquor (another stimulant Hosack would later recommend for weakened typhus patients). The captain was genuinely concerned about his steerage passengers and worked tirelessly alongside Hosack to care for them. Although Hosack left no record of how he and Captain Allen contrived not to lose a single life, the New York papers later noted that all were saved.

WHILE HOSACK AND THE CAPTAIN toiled to contain the fever ripping through the *Mohawk*, the president of the United States was facing
a crisis of his own. For several years, Treasury Secretary Hamilton's
unpopular excise tax on the production of distilled liquor had incensed
frontiersmen in western Pennsylvania. Now, in early August 1794,
they gathered by the thousands near Pittsburgh, raising the specter
of a major rebellion against the federal government. As the news fil-
tered back east, coffeehouses and taverns buzzed with talk about the
disturbances. Washington, the valiant general who had helped defeat
an empire and shepherd a new nation into the world, now confronted a
grievous threat from within. One newspaper worried "that the western
disturbances have risen to so great a height, that the ordinary powers
of government are altogether inadequate to the suppression of them."
Hamilton agreed completely with these grim appraisals. Over the pro-
tests of Secretary of State Randolph, he was pressing Washington to
raise a special militia of twelve thousand men. Jefferson, Randolph's
predecessor, found such measures tantamount to suppressing the right
of citizens to assemble peacefully.

As the president neared a decision about sending Americans to fight
Americans, the *Mohawk* approached the coast of the United States.
Hosack took one last oxygen reading with his decanter. His experiments seemed to support the notion that there was less oxygen at sea,
although he later lamented to Benjamin Rush, "I could wish the experiments were repeated and were made under more various circumstances." On August 26 Captain Allen guided the *Mohawk* into New
York Harbor, passing between Governors Island and an unremarkable
little pile of sand and rock called Oyster Island. The surprising changes
recently wrought on Governors Island would have been obvious to
Hosack and any other New Yorkers waiting on the *Mohawk*'s deck as
she sailed past. All that summer, citizens had been boating over to volunteer alongside paid laborers as they erected a monumental structure
on the foundations of the old Continental Army fortifications. It was
a new fort meant to defend New York City from the much-anticipated
British attack. New Yorkers had banded together in work teams—

grocers, teachers, carpenters, cartmen, coopers, blacksmiths, bakers, deacons, Irishmen, Columbia students, and so on—and spent months moving wheelbarrows full of rocks. Many of the volunteers had trumpeted their Democratic-Republican opinions as they planned their outings to Governors Island, whose new fort they viewed as a thumb in the British eye. If the familiar old foe tried anew to penetrate the harbor, the plan went, her ships would crash into a wall of cannon fire launched from Governors Island to the east and from Oyster Island and nearby Bedloe's Island to the west.

As the *Mohawk* traversed the upper harbor and made for the mouth of the East River, Manhattan came into view: the tree-lined Battery, the spires of Trinity and St. Paul's, and beyond them the fields galloping northward as far as the eye could see. To starboard lay the green expanse of Brooklyn, stitched to Manhattan over and over by the ferryboats that carried the harvests and farmers of Long Island to the city. The *Mohawk* dropped anchor near Crane's Wharf at the foot of Beekman Street around seven o'clock in the evening, with all of her passengers safe and sound. A ship anchored nearby, the *Providence* from Antigua, had not fared as well—one passenger had died on the way to

*New York City as Hosack would have seen it from
the* Mohawk *in the summer of 1794*

New York, and two more would die in town in a few days. The chairman of the municipal health committee, John Broome, observed that the victims of the *Providence* had evidently fallen "sacrifice to common Typhus fevers, not attended to, perhaps, so early or so vigorously as is generally requisite in those disorders." The *Providence*, in short, had had neither a Captain Allen at the helm nor a newly minted Dr. Hosack among its passengers.

Almost as soon as the anchor splashed into the harbor, Captain Allen found himself besieged from all sides. Merchants looking for their shipments clambered onto the ship just as he was trying to assist the passengers eager to disembark. Captain Allen was anxious to deliver his cache of letters, not least John Jay's, to the post office, but it took him so long to handle the flurry of requests for his attention that the office was closed by the time he could even think of going ashore. He would sleep on the *Mohawk* that night and deliver the letters the next day. Meanwhile, the passengers stepped with sea legs onto Crane's Wharf. Shouts and curses rang out in different tongues as servants, sailors, merchants, and enslaved humans circumnavigated mountains of stacked crates and coiled ropes. The waterfront was a notoriously pungent place, where a crush of unwashed bodies in sweat-drenched clothing added spice to the brackish notes of the water stagnating beneath the wharves. Sometimes the crowd of men and horses was so dense it was hard for pedestrians to squeeze between them.

The next day, Captain Allen left the ship and walked to the post office at 30 Wall Street to deliver the letters. Within hours, Sarah Jay was tearing one open and devouring its contents. She replied immediately. "My d[ea]r Mr. Jay, It is impossible to describe what were my sensations upon beholding again the handwriting of my husband and my son." That same day, the first-class passengers, Hosack among them, placed a public letter to Captain Allen in the *New-York Daily Gazette*. "Having observed your very liberal and humane treatment of the steerage passengers, and having ourselves experienced every attention, we cannot refrain from expressing to you our unanimous sentiments of your conduct, and from conveying to you our cordial wishes for your prosperity."

Hosack was overjoyed to be home, as he wrote Rush soon afterward. Making his way across Water Street into the heart of the city, he headed for his parents' house on William Street, where Kitty was living, along with Hosack's younger siblings. He had been absent for two years, twice as long as he had spent with Kitty before leaving for Britain. They were practically strangers to each other now, and they had also lost a son. Whatever Hosack's emotions on reuniting with Kitty, he would soon be directing his grief over Alexander's death into his medical practice. In a letter Hosack wrote to one of his Edinburgh professors not long after he returned to New York, he described in harrowing detail the morning he was called to the side of a dying newborn—the week-old son of lawyer Nathaniel Pendleton and his wife. As Hosack tried to count the faint pulse, the baby turned a "dark livid blueish colour" and repeatedly fell into "fits of screaming" and "convulsions." Hosack, after exhausting every other remedy he could think of—including camphor oil, laudanum (a narcotic derived from poppies), and a mustard plaster applied to the whole body—decided to submit the baby to a hot bath of Peruvian bark, spirits, and smelling salts. After repeated bathings, the baby's pulse quickened, his skin returned to normal, and his eyes brightened to "their natural expression." Soon he was nursing contentedly, out of danger.

Hosack may have lost his own son while he was in Britain, but he could now hope to rescue other children. The venerable old cities of Edinburgh and London, with their universities and gardens, their anatomists and botanists, had given him such generous gifts. Above all, one dazzling idea had lodged itself in his mind. So much of his home continent was unexplored. There must be undiscovered medicines growing out there in those vast, verdant lands.

Chapter 4

"HE IS AS GOOD AS
THE THEATRE"

COMPARED WITH THE MAJESTY AND SWEEP OF LONDON, NEW YORK looked newly humble to Hosack. He had set sail from the world's largest city and its one million inhabitants, and now he had washed up among fewer than sixty thousand. During his British sojourn, the abodes of kings and lairds had taken up residence in his memory. Kensington Palace, Edinburgh Castle, even the jewel-box Brodie Castle dwarfed the mansions of New York's bankers and merchants. The city's houses, churches, and shops were still crammed into the pointy little end of Manhattan Island. All around him, goats ran through kitchen gardens and cows liberated themselves from their sheds to saunter down the streets. Pigs rooted in the dirt by the Columbia College fence. A world away in Soho Square, Banks and Smith continued their sparkling conversations, and he was not there to listen. On this backward little island, there was no Royal Society, no Linnean Society, no Brompton Botanic Garden. Hosack turned the situation over in his mind and decided it suited him perfectly.

Soon he received exciting news from London—the Linnean Society had held its election while he was at sea. He was now a full-fledged Fellow, the first foreigner to be so designated. Henceforth he had the right to sign his name "Dr. David Hosack, F.L.S." Perhaps it was this good news that encouraged him to dream that the Fellows of the Royal Society might one day permit him to trail their honors at the end of his name like the tail of a brilliant comet: F.R.S. First, however, he would have to prove himself worthy of sharing such a distinction with the most revered of all American scientists, Benjamin Franklin, who had been elected in 1756. Not even Benjamin Rush was a Fellow of the Royal Society.

Hosack would need to tread a careful path through the political minefields that had been laid across the island in his absence. Both the city and the nation had entered precarious new times. Even with the national government in Philadelphia, at a safe remove from New York, local partisans of Hamilton and Jefferson had become ever more starkly polarized in the early 1790s. At the warehouses, wainwrights' shops, and ropewalks, the men who rolled sloshing barrels of Madeira off the ships, who wove the ropes that held firm when huge sails billowed in the trade winds—these have-nots watched with growing rancor as speculators lined their pockets. New York's stock traders, one Antifederalist charged, were "minions of despotism, who are living on the vitals of our citizens." Meanwhile, many New Yorkers were not allowed to vote because they didn't meet the minimum land ownership or rental thresholds—or because they were women or enslaved (or both). Some of the city's disenfranchised men organized themselves into the General Society of Mechanics and Tradesmen and made common cause with the Clintonians in attacking Hamilton and his financial policies. When Jefferson and Madison arrived in New York City to begin their botanical tour of the Northeast in May 1791, they had quietly met with Governor Clinton's allies, who now included members of the powerful Livingston family and the shrewd Revolutionary War veteran Aaron Burr.

Emotions might not have run so high so fast had the Antifederalist camp not been swept along by a tidal wave of political inspiration that rolled from France across the Atlantic. From Secretary of State Jefferson on down to New York's dockworkers, Antifederalists—now more often called Democratic-Republicans—had for the previous two years been mesmerized by the fast-moving events of the French Revolution. Jefferson himself had been in Paris on July 14, 1789, when a crowd stormed the Bastille and came away with the head of the prison director bobbing on a pike. Even as he acknowledged the Revolution's violent excesses, Jefferson greeted the overthrow of the old order with cheerful optimism, observing to his friend General Lafayette that "we are not to expect to be translated from despotism to liberty in a feather-bed." After departing for the United States that autumn, Jefferson kept an avid watch on the news from France.

As did New Yorkers. They couldn't have ignored the Revolution if they had tried, since there were days when it seemed to be unfolding in the streets of Manhattan, where Democratic-Republicans reveled publicly and loudly in the news that the most humble Parisians had marched a despotic monarch out of his own distant palace and back to their city. By mid-1793, after Louis XVI had been beheaded and Britain had gone to war against France, Democratic-Republicans and pro-British Federalists were skirting one another in the New York streets with their hackles raised. The threat of physical violence pulsed through the city.

It erupted into the open that summer. In June, a group of French officers from the frigate *L'Embuscade*, anchored in New York Harbor, were publicly insulted by British sympathizers. Pro-French New Yorkers retaliated by hoisting one of the Revolution's most recognizable symbols at the brand-new Tontine Coffee House on the corner of Wall and Water Streets, where a group of brokers and dealers had been meeting since early in 1793. It was a red "liberty cap," the sort worn by working-class French Revolutionaries. Word quickly circulated through New York that anyone foolhardy enough to remove the cap would earn the "scorn and hatred" of all true American patriots.

Meanwhile, out in the streets, crowds of Democratic-Republicans paraded along singing the *Marseillaise* and sporting liberty caps of their own. The captain of the British frigate *Boston*, also anchored in the harbor, challenged the captain of *L'Embuscade* to a naval battle. Wagers were laid on the outcome and fistfights broke out. On August 1, 1793, just around the time Hosack was beginning his course of botanizing in the tranquil English countryside with William Curtis, the French and the British went to war in New York Harbor. Hosack's younger brothers may well have been among the thousands of New

Yorkers who thronged the tree-lined park at the Battery to egg on the combatants. The *Embuscade* trounced the *Boston*, but that hardly eased the friction. All that next year, while Hosack was away in London, the electric charge of French Revolutionary politics crackled in the New York air. In May 1794, when John Jay had set sail from lower Manhattan for his negotiations with the British government, his cheering well-wishers were countered by Governor Clinton, Chancellor Livingston, and the French consul, who boarded a French naval ship in the harbor and sang French anthems together.

In France, this was the season of the Reign of Terror, when aristocrats and even moderate Revolutionaries were fleeing the guillotine in herds. Many landed in Manhattan, where they became hairdressers, dancing masters, schoolteachers, farmers, or dinner-party fixtures. "The city is so full of French," a British visitor observed in 1794, "that they appear to constitute a considerable part of the population." The French consul himself lived just a few doors from Hosack's parents on William Street. Culturally and intellectually, Hosack leaned decidedly British; to accomplish what he was envisioning for his city and his nation, he would have to keep his political opinions to himself.

Hosack confided his hopes to Rush in a letter just after the *Mohawk* reached New York in late August 1794. He first relayed greetings from a number of his Edinburgh teachers and inquired whether the so-called Whiskey Rebellion in Pennsylvania had been put down. (It hadn't.) He was asking, Hosack told Rush, not because of "the immediate evils" threatened by the uprising but because he hated the British to have any cause for smugness about the fortunes of the United States. Then Hosack turned to his grand plans, telling Rush that Banks, Smith, and Curtis had introduced him to botanical gardens, sparking his new love of natural history in general and botany in particular. He was now ready to hurl himself into what he called the "almost untrodden field" of American natural history. He planned to study zoology in the winters and botany in the summers, with mineralogy stuffed into the chinks of his schedule. He also wanted to get his hands on every possible genus of animal, because he hoped to dissect them and prove American scientists misguided in their focus on superficial nomenclatural quarrels when they should really be investigating internal animal structures to discover similarities and differences (an astute insight in the pre-Darwinian era).

Even as he wrote these lines, Hosack realized that he sounded ambitious in an age when overt ambition was frowned upon, and he tried to preempt Rush's reaction: "You will say this is a great undertaking for an individual." He conceded that his plan would take a whole lifetime but insisted that "on the score of Industry & inclination I feel myself secure." The twenty-five-year-old Hosack knew by now that he was most cheerful and energetic when mentally absorbed in efforts to improve the health and comfort of people around him.

IF HOSACK HAD CHOSEN to settle nearer to Rush that autumn, he would have had an easier time pursuing his scientific dreams. It was not simply that Rush had become a father figure who encouraged Hosack through gentle sparring matches and liberal praise. It was that in Philadelphia, of all cities in the United States, natural history reigned supreme. Philadelphians thought they inhabited the Athens of America, to the annoyance of many New Yorkers. The credit for Philadelphia's stature was due largely to Franklin, Jefferson, and Rush, but their friend Charles Willson Peale, a talented painter and naturalist who would loom large in Hosack's life, had plenty to do with it, too. After fighting in the Revolutionary War, Peale had founded a museum of art and natural history—known simply as Peale's Museum—in his large house at the corner of Third and Lombard Streets. He believed in the uplifting effects of scientific education on American men, women, and children alike and saw his museum as a way to help fashion an educated citizenry. Peale was a bouncing optimist who would never in his long life lose his childlike wonder at the natural world. In a speech he gave in the 1790s, he insisted that natural history "ought to become a NATIONAL CONCERN, since it is a NATIONAL GOOD."

Peale inspired strong support for his museum from his circle of Philadelphia friends. Franklin was one of the first contributors to the collection, donating the body of an angora cat he had brought back (alive) from his tour of duty as the American minister in Paris. Over the next few years, Peale assembled the most comprehensive array of plant and animal specimens in the United States—hundreds of mounted mammals, birds, fish, and reptiles, labeled according to their Linnaean classifications and arranged in naturalistic settings. He put live animals on display, too, and sometimes he stuffed them after they died.

Despite the undeniable draw of exotic specimens, Peale tried to prioritize education over shock value or novelty. "I neglect many little contrivances which might serve to catch the Eye of the gaping multitude," he confided to a friend. He showcased the most humdrum American species alongside the unfamiliar ones. He found a natural ally in Jefferson, with whom he shared a passion for all things scientific and whose politics he found congenial as well. Jefferson began serving as the president of the museum board in February 1792. Hamilton joined the board, too, setting aside partisan politics in favor of the national scientific interest and running at least one meeting when Jefferson couldn't make it.

Peale's Museum contributed to Philadelphia's reputation as the center of scientific knowledge in the United States. He added even more luster to the city with his own paintings, which many people considered among the best works of art produced in the country up to then. Famous Americans and illustrious visitors from abroad clamored to have their portraits done. In 1791, shortly before Jefferson and Hamilton joined his museum board, Peale painted portraits of each of them; he had first painted George Washington in 1772 and eventually completed seven portraits of him from life. Peale was the patriarch of a fantastically talented family, and his son Rembrandt likewise became a frequent painter of Washington and Jefferson. Some of Rembrandt's brothers—Rubens, Raphaelle, Titian, and Benjamin Franklin Peale—also inherited their father's outsize talents in painting and science. Charles Willson Peale's daughters, several of whom were named for female painters, also received training in the arts, but they were less free than their brothers to pursue their talents outside the domestic sphere. Hosack would come to know many of the Peale children well, and Rembrandt would paint Hosack's portrait more than once.

In March 1794, Peale named his latest child Charles Linnaeus Peale. "The first Linnaeus perhaps so named in America was born this morning," he wrote happily to a friend. "May he be a light to this New World, like him of Sweden whose percevering labours in Natural History hath eluminated the Old World." (Peale had creative spelling.) The baby's first name was soon dropped from use. He also acquired a nickname: Lin.

———— ❧ ————

BUT TO HOSACK, New York felt most like home, and he intended to stay there. After he returned from Britain, he and Kitty moved into a house at 60 Maiden Lane, a few blocks from his parents' house (and just steps from the house at 57 Maiden Lane where Jefferson had held his dinner party for Madison and Hamilton in 1790). Hosack began setting up his medical office in rooms at home, the customary arrangement for a private practice. He needed surgical equipment such as drills, amputation saws, trocars, lancets, syringes, catheters, and pliers for pulling teeth. He also needed pillboxes, ointment pots, stoppered bottles, and vials, all of which he bought at Joel and Jotham Post's medical supply shop at the corner of Wall and William Streets.

Some of his medicines he prepared from scratch, but others he had no choice but to purchase from the shops. At Philips & Clark, a druggists' shop just up Maiden Lane from his house, he bought spermaceti, which served as an ointment base for rubbing medicines into patients' itchy or oozing skin. He also bought smelling salts and mercury. Hosack felt deeply ambivalent about mercury, because it exhausted patients and made their gums so raw and soft that their teeth often fell out. He reluctantly decided to keep it on hand (in the form of calomel pills), but he much preferred remedies that came from the plant kingdom. He bought myrrh, made from a tropical tree resin into reddish drops that could be dissolved in water and prescribed to induce a mild sweat. Another sudorific Hosack bought at Philips & Clark was camphor, produced from the tropical *Laurus camphora* tree (today it is called *Cinnamomum camphora*). It is used externally as an oil or salve today, but eighteenth-century doctors prescribed camphor pills or powder for internal use. In high doses, ingested camphor could provoke more vomiting than desired, in which case the *Edinburgh New*

Dispensatory recommended prescribing opium pills—which Hosack also purchased.

His home office readied, he circulated word that he was prepared to receive patients. In the tradition of the era, he was also seeking private pupils who wished to supplement their courses at Columbia College by apprenticing with a practicing physician. New York was still chronically short of doctors, and in that first year Hosack earned about $1,500 from his practice (about $30,000 today). It was a respectable showing for a new doctor, but when a professorship in botany opened up at Columbia in the spring of 1795, he put his name forward.

Columbia's outgoing professor of botany, Samuel Latham Mitchill, helped Hosack secure the position. Mitchill, who was not much older than Hosack, had decided to become a doctor after watching his five-year-old brother die of croup; he later told Hosack that as a grown man

Hosack would have owned a medicine chest much like this one,
which belonged to his friend Rufus King

he could still hear his brother's "distressful croaking" and "suffocating anguish." Mitchill had trained at Edinburgh before Hosack, and by the time Hosack returned to New York in 1794, Mitchill was already one of the most prominent local physicians. He was the clear heir to the revered John Charlton and also to Samuel Bard, the doctor who had operated on President Washington's thigh in June 1789.

A staunch Democratic-Republican, Mitchill idolized Jefferson and would later observe that when it came to the encouragement of the natural sciences in the United States, there could be no doubt that Jefferson had done more "than all the Presidents of the United States severally and jointly." Mitchill loved the American wilderness so much that he once tried to prove that the Garden of Eden must have been located in upstate New York, just outside of Syracuse at Onondaga Lake. He spent his days pursuing polymathic studies in chemistry, botany, medicine, history, literature, moral philosophy, and political theory. Yet Mitchill's chief interest was chemistry, and he had decided to shed his botanical responsibilities at Columbia. His timing could not have been better for Hosack, who apparently sat down with him at some point that winter to discuss his recent studies at the Brompton Botanic Garden and the Linnean Society. In February 1795, Mitchill wrote a letter to Columbia's president saying that the twenty-five-year-old Dr. Hosack was a person of "zeal, Industry and Talents." Hosack

Samuel Latham Mitchill, Hosack's Columbia colleague, friend, and rival

was hired. From that moment forward, Mitchill would be a constant presence in Hosack's life, their relationship vacillating from warm collegiality to cutthroat competition and back again.

While Hosack waited for his Columbia appointment to be announced publicly, he sorted through the specimens he had brought back from London. He had decided to donate a little herbarium collection of foreign grasses—about sixty in all—to a local group called the New-York State Society for the Promotion of Agriculture, Arts, and Manufactures. The society had been founded in 1791 by Mitchill and other New York men hoping to foster American independence through agricultural improvement. Hosack thought there might be specimens in his collection that New York farmers could profitably cultivate.

The New-York State Society for the Promotion of Agriculture, Arts, and Manufactures had a strong practical bent. Its charter explicitly discouraged members from "performances intended to display literature and erudition." The very first query issued to the public was a call for farmers and estate owners to nominate the best varieties of manure. The society's members met once a month when the New York legislature was in session in the city. In a room at City Hall on Wall Street, they argued over questions like whether cast iron or wrought iron was better for making plowshares (the winner was cast iron) and how to help the state's farmers raise more barley to supply the city's brewers. At these meetings, Hosack could rub shoulders with New York's most powerful figures—men such as former governor George Clinton, Chancellor of New York Robert R. Livingston, and the chancellor's brother, Edward Livingston. (John Jay, still in London, was also a member.) Some of these men had more than a quarter century on Hosack, but he wasn't in the least intimidated. In the spring of 1795, flush with his new expertise, he decided to school the society's members on botany by means of an open letter addressed to Chancellor Livingston, the state's highest judicial officer and the man who had administered the oath of office to President Washington. When the letter was read aloud at a meeting (possibly by Hosack himself), the men in the room encountered the fervor for civic improvement that would fill Hosack's waking hours for decades to come. Hosack was proposing that every member of the society be deputized to collect specimens of all the plants currently growing in the state of New York. Together, these citizen-scientists would compile the state's first complete *hortus siccus*—literally, a "dry garden." This trove would permit systematic research and experimentation for the good of the state and the nation.

Hosack informed the men he would be happy to organize the assembled specimens according to their "Botanic Order" in the Linnaean system, adding that Smith and Curtis were standing by to lend their expertise from across the Atlantic. Thanks to his Brompton studies, Hosack knew how to collect a good specimen, and he gave the society members a detailed description of his favorite approach, which "after various trials . . . I find to be the least troublesome and the most successful." They should botanize only in fine weather in order to obtain totally dry specimens for pressing. Each specimen must include the plant's flower and leaf so it would be clearly identifiable later. Tree or shrub specimens should take the form of a small branch with "the flowers and some of the most perfect leaves." The specimens should then be placed between sheets of "soft spungy paper" and pressed in "a machine contrived for this purpose"—a wooden herbarium press. If no herbarium press was available, a stack of heavy books would suffice. Hosack warned them to change the blotting paper once a day until the specimen was completely dried; otherwise, mold might set in.

But Hosack had misdirected his enthusiasm. The New-York State Society for the Promotion of Agriculture, Arts, and Manufactures was no Linnean Society. Systematic botany, untethered as it was from the practical concerns of farming and manufacturing, failed to interest most of its members. The idea sank like a rock and disappeared from view.

HOSACK'S BOTANICAL HOPES may have been crushed, but his medical practice was soon flourishing, as he attracted patients and students alike to his private practice.

In May 1795, nine months after his return to New York, Hosack was summoned to an exciting case. The captain of an East India trading ship, an otherwise healthy man in his midforties, had been suffering for six years from a swollen scrotum. Each year, the swelling had continued to increase, the captain told Hosack, who in turn explained the possible treatments for a hydrocele. The captain elected to undergo the very procedure Hosack had learned from Dr. James Earle at St. Bartholomew's Hospital in London the previous year. With three other physicians gathered around him to watch, he performed the procedure for the first time recorded on American soil. Using a trocar, Hosack withdrew a pint of foul-smelling fluid from the bulging flesh and injected a mixture of alcohol and water. He then applied a poultice of bread and water and prescribed regular spoonfuls of laudanum to ease the pain. In three weeks' time, the captain reported himself fully recovered and ready to command his next voyage to the East Indies.

Shortly after this, Columbia finalized Hosack's appointment as its new professor of botany. Botany was usually offered during the summer term to take advantage of the growing season, and Hosack would give his first lecture on June 15 at College Hall. It was still one of the most imposing structures in town, with its cupola and gables. Just outside the fence enclosing the college yard, prostitutes loitered as young men in black robes went gusting past on their way to class. That year's crop of botany students filed into a classroom and inspected their new professor.

Quite young. A bit on the portly side.

Yet Hosack turned out to be surprisingly quick on his feet and animated in his gestures. As he lectured, he drew on the oratorical skills that had won him a prize in his college days, radiating the intellectual authority of a far more seasoned teacher and scholar. He arrived in class with his notes heavily underlined so that he could make his main points with maximum force. "Dr. Hosack read his lectures, and no man was ever more emphatic, impressive and instructive," one student later remembered. Another noticed that Hosack enrolled even his "very heavy black eyebrows" in his pedagogical efforts when he found it convenient to assume "a thunder-cloud frown."

The very first sentence of Hosack's botany course seemed calculated to intimidate the novice. "Natural History considering the term in its greatest latitude may be said to comprehend a knowledge of every part of the universe," he declared, before skating across the fields of

meteorology, hydrography, geology, and mineralogy. Finally he settled down to the topic at hand: the study of "the vegetables which
clothe and adorn" the planet. He reviewed the definitions of plants
proffered over the centuries by a starry roster of scientists, including
Linnaeus, Herman Boerhaave, and Joseph Pitton de Tournefort. He
couldn't resist mentioning his recent access to Linnaeus's manuscripts,
which he "had an opportunity of observing . . . in the possession of my
friend and preceptor Dr. Smith of London."

Over the next weeks, Hosack channeled Linnaeus, Smith, and
Curtis as he led his young charges on a self-assured tour of the botanical world. Laying out plant specimens like bodies on a dissecting table,
he revealed their inner structures with his knife—epidermis, cortex,
lignum, medulla—and then he sliced open minuscule seeds to show
them the cotyledons and radicles that held new life. He taught them
the critical differences among grasses, ferns, fungi, and mosses. He
explained how leaves functioned "in the vegetable economy as organs
of respiration." He told them that plants are nature's "instruments by
which she corrects the impurities of the air." He called their attention
to the intricate rhythms of the natural world, where the life cycles of
insects, birds, and animals align perfectly with those of the plants that
sustain them. He pointed out that if these creatures "were produced at
a time when those plants did not flower, [they] would necessarily perish for want of food."

He then turned their attention to a topic of some delicacy: sex. It
was now known that many plants resembled animals in their manner
of reproduction, Hosack told his students. Thanks to the dogged work
of many botanists, especially Linnaeus, "at this day I believe there are
none who study the vegetable kingdom in any part of the world who
do not admit the universality of the sexual machine."

———— ✿✿✿ ————

THAT SUMMER OF 1795 was hotter, one New Yorker observed, than
the West Indies, but without the relief of the islands' evening breezes.
At his house on Maiden Lane, Hosack checked a thermometer several times a day and wondered at the consistently high temperatures.

Hosack's former anatomy teacher, Bayley, meditated at his home on Broadway on how unbearably humid the summer was turning out to be. People kept complaining to him that everything in their houses had sprouted a rind of mold—books, shoes, floors, wallpaper.

Suddenly the city erupted like a swollen boil. In July, Hamilton was assaulted in the street. It happened on an overcast day in the middle of the month, near the corner of Broad and Wall Streets, where Hamilton, who had moved back to New York earlier that year, had gone to confront a crowd gathering to protest the terms of the Jay Treaty. Although Jay had signed the treaty in London all the way back in November 1794, the United States Senate had only just ratified it (minus one article) in June 1795. The American public had learned of its exact contents at the beginning of July, just as Jay was being sworn in as the new governor of New York. Enraged anti-British Americans—otherwise known as bloodthirsty Jacobins, said one Federalist newspaper—immediately burned Governor Jay in effigy up and down the Eastern Seaboard and denounced the treaty as "humiliating."

On July 18, opponents of the treaty converged by arrangement on City Hall before marching through the streets and burning a copy of the treaty and, for good measure, a portrait of the governor himself. Hamilton was helping to spearhead the counterdemonstration that day, and as he strained to make his defense of the treaty heard over the crowd, some people responded by throwing rocks at him. At least one struck him in the head. Hamilton retreated indignantly, in the process almost committing himself to two duels. (In the end, both conflicts were settled off the dueling ground.)

The city's soaring political fevers soon collided with the real thing. Around July 19, the British ship *Zephyr* arrived at New York from Port-au-Prince and unloaded most of its cargo at the foot of William Street before sailing out into the East River to dump twenty-two barrels of spoiled coffee. A few weeks later, a brief, painful notice appeared in a New England newspaper. "The Yellow Fever prevails in New York." The son of a prominent New York politician had been stricken: "Mr. Amiel Jenkins . . . is in the last stage of the Yellow Fever—His tongue lies clot[t]ed in blood, and his physicians consequently despair of his recovery."

New York had largely been spared the devastation visited on Phila-delphia during the terrible epidemics of 1793 and 1794. While Hosack had been rambling through meadows with Curtis in the late-summer sunshine, yellow fever weeded out thousands of people in Philadel-phia. Not every individual who contracted yellow fever died, but the course of the illness in those who did took one terrifying turn after another. First came chills, intense back pain, and yellowing skin; then came a black crust on the tongue, delirium, and a diarrhea that resem-bled molasses. In the last hours of life, victims often lay with their knees drawn up toward their chins, as if to shield themselves from an attacker. Expiring bodies purged their contents so violently that the dying seemed to be retching up chunks of their own stomachs in vomit that had the look and feel of coffee grounds.

Benjamin Rush, like many others in Philadelphia, believed that the illness had originated from a damaged shipment of coffee that had been unloaded, rotting, on a wharf. For weeks, he trudged up staircase after staircase to the sickrooms of the dying. Mapping the stages of yellow fever in his notebooks, he set down the stark details of human misery. "I found Mr. Cochran sitting on the side of his bed, with a pot in his hand, into which he was discharging black bile from his stomach, on the 6th day of the disorder. He died on the next day." By the end of the epidemic, more than four thousand putrefy-ing bodies had been trundled to gravesites and potters' fields, past the deserted townhouses of those who had escaped to the countryside or were already dead.

Rush insisted that in the treatment of yellow fever, bloodletting, along with liberal doses of mercury, was of paramount importance. The loss of blood, he claimed, reduced inflammation, lowered the heart rate, and alleviated pain. In a report he published a few years later, he noted that every patient he had saved in 1793 had been bled, and he therefore resolved to use the lancet even more vigorously dur-ing the yellow-fever epidemic that hit Philadelphia the following year. He sliced through the tender skin of a girl named Sally Eyre nine times, spilling eighty ounces of her blood into a waiting bowl. He relieved a man named John Madge of almost double that amount in twelve ses-sions. For all the horror he witnessed, Rush found a sinister beauty in the treatments. The blood of some patients left "a beautiful scarlet coloured sediment in the bottom of the bowl."

Others abhorred bloodletting. In early September 1793, Alexander Hamilton had fallen ill at his country residence outside Philadelphia, and Eliza Hamilton a few days later. (Jefferson refused to believe that Hamilton really did have yellow fever and wrote James Madison a mean-spirited letter saying so.) Alexander and Eliza were speedily cured by Edward Stevens, an old acquaintance (possibly a relative) of Hamilton from the West Indies, who eschewed bloodletting in favor of cold baths, Peruvian bark, Madeira, and brandy. On September 11, Hamilton published an open letter to Philadelphia's College of Physicians touting Stevens's approach, and within a week Stevens himself had published an account of his methods. Rush was livid. He thought Stevens's treatment ineffective and believed Hamilton's implicit criticism of bloodletting was politically motivated; everyone knew Rush was a staunch supporter of Jefferson. When the news of the epidemic had reached Hosack in London, however, he heard people praising Rush's bold methods.

Now, in the summer of 1795, as signs of illness began to make their appearance at the houses closest to the East River wharves, Hosack held out hope that the disease was not, in fact, yellow fever. Like many other people, he thought that yellow fever was contagious. If the tendrils of rumor were permitted to reach out across the green fields of Long Island and New Jersey, up toward Boston and down toward Philadelphia, the city would face economic disaster. Ships' captains would sail their vessels toward healthier ports, unloading their cargo on the wharves of other cities. Farmers would stop loading their produce and livestock onto ferries bound for the markets of lower Manhattan. Men of all classes, along with their families—merchants, artisans, and day laborers alike—would be ruined. Malnutrition would strengthen the destructive force of the epidemic.

On a Saturday evening in mid-August 1795, the city's physicians— Hosack likely among them—converged on City Hall for an emergency meeting. At their head was Mitchill, who had set aside his pioneering research on the anesthetic effects of nitrous oxide (later known as laughing gas) to help stem the rising tide of anxiety in the city. The physicians issued a terse statement to the press: "It was agreed to, as the opinion of this meeting, that no case of yellow fever exists within the circle of practice of any person, now present; and that the same be declared to their fellow citizens, with a view to calm their apprehensions, on the present occasion." Within a few days, though, Mitchill decamped to his country house on Long Island. Confusion reigned in New York. Was it, or wasn't it, yellow fever?

On September 3, Hosack wrote to Rush. "You will no doubt expect a line upon the subject of fever as prevailing here." The three patients Hosack had treated thus far, he told Rush, appeared to have contracted typhus, and after his experience on the *Mohawk* the previous year, Hosack believed typhus less deadly than yellow fever. The approximately fifty cases of fever he had heard about were mainly along the East River, where laborers and sailors lived in the kind of filthy quarters favorable to the spread of typhus. Hosack informed Rush that Columbia's medical professors were treating so many patients in that part of town that "it has actually become proverbial that the faculty is always to be found in Water Street."

All that autumn, the body count climbed and terrifying rumors whirled through the city. It was said that a dockworker had stuck his arm into a damaged bale of imported cotton, and when he withdrew it, it had turned yellow from the contagion festering inside. Along the Battery, soldiers unleashed cannonades in a vain effort to disperse the sickly air rising from stagnant water around the wharves and in the island's marshy lowlands. Panicked letters were loaded onto mail stages up and down the coast. Richard Bayley noted sadly that the terror brought out the worst in people. In early September, a group of neighbors on Water Street attempted to have a sixteen-year-old girl forcibly deported to the countryside. She had apparently come down with a cold, and she recovered in two days.

The advice coming in from Philadelphia, a city still licking its wounds from the 1793 and 1794 epidemics, was continued vigilance. "Let not the good people of New York and the other towns be too sanguine," one newspaper recommended. "Rather let them be cautious. It is better to fear too far than to trust too far." In New York, the Federalist newspaper *American Minerva*, edited by Noah Webster, mocked that advice: "The alarm of the Philadelphians at the Yellow Fever resembles the alarm of our Jacobins about the treaty. Raw head and bloody bones, ghosts, goblins, slavery, famine, pestilence, and the Lord knows what evils will befall us, if we have a treaty with Great Britain." Some New Yorkers simply persisted in maintaining that all was well. An anonymous "Inhabitant of Peck's Slip"—the East River wharf at the epicenter of the outbreak—swore in a newspaper in early September that "there has been not an individual, except two, in the whole slip, been ailing, and they had trifling colds, and they are now perfectly well."

But over the next weeks, as the dead began to emerge from their houses to be spirited away for burial in churchyards and potters' fields, many New Yorkers, including Hosack, conceded that yellow fever had seized the city. Eleven people had reportedly died on August 25 alone, and by September 8, John Broome, the chairman of the city health committee, had recorded nearly a hundred deaths. The hired men who carried the bodies away sometimes showed a shocking lack of respect for both the living and the dead; in a letter, the mother of one of Hosack's students referred to these men as "the hearse monsters."

The epidemic threw the whole Eastern Seaboard into turmoil. Governor Thomas Mifflin of Pennsylvania issued a proclamation prohibiting all contact between New York City and Philadelphia by land or water, on pain of a $300 fine. When Mayor Richard Varick of New York presented this news to the city council members, they sent word back to Pennsylvania that "a much greater Degree of general Health prevails in this City at present than is usual at this Season of the Year." Many people agreed with the city council that Governor Mifflin was overreacting. An anonymous letter "To the Citizens of Philadelphia" that ran in an Albany paper near the middle of September ridiculed the hysteria as akin to the Federalists' needless anxiety about the French Revolution. "Cousins, we are all popping off here like rotten sheep," the letter went. "Two hundred carcasses have been burned on the Battery—500 hanged for fear of catching the Yellow Fever, and about 35 or 40 guillotined—all the windows in town are broken by the firing of cannon—several of our seven story houses have fallen down *slam bang* of their own accord—Federall hall has got a case of the fidgets, and two yearling pigs have died of the measles—pray send us about 100,000 dollars, to stop the contagion."

But the bravado masked tragedy. As the death toll mounted during the month of September, New Yorkers closed their shops and traded horror stories. Elihu Hubbard Smith heard a story about an Irishman who was found dead and alone on the floor of his house after his family had abandoned him in his last hours of agony. "This destructive Terror, this malady of the mind," Smith wrote in his diary, is "a thousand times more dreadful & pernicious than all corporeal evils." Still, no one knew how the disease was transmitted or how to save those who fell ill. Some doctors, Smith among them, chose to follow Rush's methods. At the hospital, Hosack saw the blood of the city's poor running down their arms and dripping off their fingers into bowls. He hesitated to pick up a lancet himself, afraid he would drain his patients of their last energies.

It wasn't that he was completely opposed to bloodletting. But when a body was already wracked by illness, Hosack's instincts told him that blood loss would give more power to the fever, not less. Better to keep flesh, bone, and blood safely encased in skin. The trick, he thought, would be to stimulate the circulation and help the patient sweat out the fever. In sickrooms around the city, Hosack began washing limbs and torsos with cold vinegar and water, then wrapping them tightly in blankets. He held cups of pungent liquid to cracked, puffy lips. The burnt bitterness of *spiritus mindereri*, made from acetic acid and ammonia. Tangy tamarind water from the tamarind tree (*Tamarindus indica*) of the African tropics. A touch of Virginia snakeroot (*Aristolochia serpentaria*), poisonous in high doses but a safe and powerful diaphoretic—sweat-inducer—if diluted. Gradually, the fever released its hold on Hosack's patients. The tongue would not hatch the telltale black crust, the eyes would not go glassy and blank. Hosack was jubilant. "When I find my patient sweating within a few hours after the attack of the disease, I congratulate him as secure from danger." There is still no cure for yellow fever today. Hosack's approach was probably most beneficial for what it did not do—weaken a patient with blood loss just as their internal organs were threatened with collapse.

As autumn wore on, the deaths began tapering off. Shopkeepers opened their doors and the streets once again teemed with carriages and horsecarts. One of Hosack's students, Alexander Anderson, who was ministering to patients out in the countryside, received a cheerful note from his mother in the city reporting that "some of our acquaintance had just returned to town & by way of thankfullness to their Maker—for theirs & our preservation from the great calamity as they called it—they had got themselves compleatly Drunk."

THE WORST HAD PASSED for the city, but not for Hosack. On a clear, mild day the following February, Kitty died in childbirth. The baby died, too, leaving Hosack all alone. It must have struck him as a bleak, wintry defeat. He had been out of the country when Alexander died, but even his presence and his clinical skills had not been able to save Kitty and her second baby. Hosack channeled his helplessness into a new professional focus on obstetrics, painting vivid scenes in his Columbia lectures of the overwhelming chaos of childbirth. He told his students that before they entered the delivery room, they must possess complete mental mastery of the birthing process. Once there, they would have "no time for much reflexion much less for consulting books." He told them it was a matter of life and death that they memorize everything about the reproductive organs—"not merely the soft parts" but also "the structure and configuration of the bones of the pelvis." He argued that childbirth so fully engaged the whole female body "by means of its membranes, nerves, and blood vessels" that giving birth was "almost one of the vital functions," like breathing and circulation. He told his students that the physician must radiate confidence both "to yourself" and "to your patient or her friends" because the least uncertainty "manifested by you is observed by them." They should speak to the mother in a calm, respectful tone while at the same time taking charge.

Hosack believed that every woman, rich or poor, deserved good obstetric care. He began to think about founding a "lying-in" hospital, devoted to helping women give birth safely even if they could not afford to have a doctor making regular house calls. If a doctor's own wife and baby couldn't survive childbirth, what was happening in the almshouse, in the boardinghouses, and in all the miserable shacks at the edges of the city?

Only two weeks after Kitty's death, Hosack was nominated to be the new professor of *materia medica* at Columbia. He would now be teaching courses on both *materia medica* and botany, which he thought was a perfect combination, given the importance of botany to medicine. He also joined the staff of the New-York Hospital. He threw himself into his work at the college, at the hospital, and in his private practice. Following Rush's example, he avoided liquor, chose his diet carefully, rose early, and worked late—all to double the hours of his usefulness. He would later remark that people who drank too much and ate a meat-heavy diet "dig their graves with their teeth."

But no matter how hard Hosack worked, he was only one man. Mothers across the city continued to bleed to death in their birthing beds; entire families died of contagion at the almshouse five minutes' walk from Columbia. He needed an army of doctors, and he was going to muster and drill them in College Hall. Every restless student in Hosack's classroom who stilled himself, suddenly alert, to ponder the theories of Rush on yellow fever, of John Hunter on digestion—there was a potential new recruit. Hosack would unleash these young men, generation by generation, on the enemies: injury, disease, premature death. "If we except the art of war, there is certainly no other profession which calls for the same promptness in the exercise of the understanding," he told them.

Saliva, bile, semen. Tears, milk, blood. Hosack initiated his students into the animal fluids and the suspended, stewing mass of organs that depended on their healthy circulation and secretion. The students memorized the tree-branch tracery of the blood vessels, learned the heft and texture of heart, intestine, and bone. They discovered that their teacher had an insatiable curiosity about human bodies, including his own. "Dr. H.," a student jotted down in his notebook, "remarks that frequently after a hearty dinner, when he has found his mind heavy and dull, and disinclined to an effort, he has lost a few ounces of blood from the arm and experienced immediately a happy elasticity of mind."

Teenage boys do not naturally favor the folding of exuberant limbs into hard-angled wooden desks. To rally them to his cause, Hosack joked, charmed, and seduced. He made himself master of the double entendre, of the vivid personal anecdote. The dullest fact somehow became fascinating as it crossed his lips. Young men from all over the United States began flocking to Columbia, where they "were ever anxious for the hour of his lecture to arrive," one student observed. They struggled to keep up with his nimble mind as they dashed his words across the pages of their notebooks. "Dr. H. thinks that Brown and Dr. Danvin were fools. Dr. H. is of the opinion that the system of induction pursued by Lord Bacon and Sir Isaac Newton, has contributed more to the advancement of knowledge than any other method pursued. Physicians have a greater variety of facts to record than any other persons."

His students came to admire, even revere him. One student who spent many hours in his classroom later summed up Hosack's gifts. "Let it be sufficient to remark that, distinguished beyond all his competitors in the healing art, for a long series of years, he was acknowledged by every hearer, to have been the most eloquent and impressive teacher of scientific medicine and clinical practice that this country has produced." Another student described Hosack's "fine, manly, and commanding voice" and his "graceful, powerful" gestures. "No reader can imagine them. He must have heard, and have seen the man, to understand what he was in the lecture room." Privately, his students were just as effusive. A young man named Moses Champion reflected on Hosack's appeal in a letter home. "His lectures on Obstetrics are both instructing and very amusing. When he undertakes to please he is as good as the theatre."

Hosack exhorted these future physicians to cultivate a gentle bedside manner. "Make frequent visits to the sick," he told them. "Record their age, their sex, their occupation, their temperament, their general external appearances. Patiently listen to the history of their complaints." He railed against drinking, gambling, and attending horse races as "altogether incompatible" with the dignity of a physician. He humiliated any student who arrived after the start of class, ordering him front and center for a dressing-down, because a tardy physician could kill a patient. One student later recalled, "The punishment was terrible, yet all loved him." Medicine was a sacred profession that demanded complete dedication and discipline. "From a want of proper guidance and instruction," he told a group of students early in his teaching career, "I have observed some among you betrayed into habits of inattention." Indeed, one young scholar who was scribbling away as Hosack spoke about allergic reactions missed a key word when he wrote down in his notebook that "a Lady, with whom Dr. H. was in company, suddenly became faintish, insisting there were some *Poppies* (Puppies?) roasted in the room." Or maybe it was one of Hosack's frequent joking asides.

Hosack taught his young men to respect the past and present masters of medicine—Hippocrates, Paracelsus, Boerhaave, Rush, and others—but also urged them to take up every treatise with wits sharpened and skepticism at the ready. "Remember that period of superstition, in which it was supposed that diseases were inflicted by evil spirits, and were only to be cured by charms and incantations." They must always be on their guard, he warned them, for "there is a sort of fashion in medicine as well as in dress." It was on the topic of choosing medicines that he grew most heated. He suspected that many of his fellow doctors were drawn to the grand gesture, with sometimes fatal results for their patients. Hosack would never completely reject the prevailing humoral approach, yet he advocated ever greater precision in diagnosis, arguing that the proper treatment was not a drastic assault on the whole bodily system via bleeding or purging, but gentler medicines aimed at the primary ailment.

Those medicines, he told his students, could often be found in the plant world. One student later recalled that when Hosack spoke on "the beautiful science of botany," he added "charm to a discourse already beaming with observations of the highest import to humanity." He offered them the example of agaric, a tree fungus that would staunch the bleeding in the case of a hemorrhage. He spoke to them of Peruvian bark, of Carolina pinkroot, of sassafras, of butternut—each had its medicinal uses, he explained, and there were hundreds, maybe thousands more. All known botanical remedies, he would tell his students year after year, must be committed to memory and tested against the living, breathing, and dying evidence of one's patients. When he insisted to them that plants "cannot be thrown aside as [a] mere matter of speculative inquiry," he was clearly remembering an objection he had heard before. Perhaps his proposal to the very practical gentlemen of the New-York State Society for the Promotion of Agriculture, Arts, and Manufactures had been dismissed with precisely this sentiment.

Beyond the city's border, the riches of a whole continent were flourishing and dying each year, uncatalogued and unused. Hosack whipped up the students with talk of the "immense treasures of America" and the "great reputation" awaiting the man who devoted himself to the discovery of new medicines. He showed them colored engravings of plants and lent them his botanical books, but he also echoed William Curtis by urging them to leave the classroom for the meadows and woodlands surrounding New York. Here poisonous plants mingled with the benign and the medicinal. The doctor who could not distinguish among them toyed with the lives of his patients, he told them, using phrases he had heard from Curtis. "How degrading," he said, for "the Physician not to know his food from his poison, [to] mistake a *Hemlock* for a *Parsley*, or the leaves of *Foxglove* for those of *Mullein*!" Hosack meant to spare his students embarrassment. More to the point, though, he meant to spare their future patients improper treatment or outright poisoning. The United States deserved doctors just as knowledgeable and skilled as Europe's.

But his task was Herculean. He had no botanical garden.

Chapter 5

"THE GRASS IS THREE FEET
HIGH IN THE STREETS"

T HERE *WAS* AN AMERICAN GARDEN OUT THERE. HOSACK KNEW
that perfectly well. On a pretty, sloping riverside farm to the west of
Philadelphia, a man named William Bartram presided over what many
people considered the most important collection of plants in North
America.

Bartram lived for plants. He was a very shy man who felt most at
ease working in his garden or botanizing in the wild. He spent years
collecting in the American South, broke his leg in a twenty-foot fall
while gathering seeds from a cypress tree, and at the behest of a curious medical student gamely swallowed some pills made from blood-
root (*Sanguinaria canadensis*). He got a headache and vomited twice,
but it was entirely worth the trouble in the service of medical botany.

The garden on the Schuylkill River that William Bartram ran with
help from one of his brothers, John Bartram Jr., had been founded by
their father in 1728. In the 1730s, the senior John Bartram had joined
forces with a plant-worshipping merchant in England named Peter
Collinson. Their arrangement had John Bartram traveling in pursuit
of plants all over the American colonies, then returning to his farm
on the Schuylkill to cultivate specimens and prepare them for ship-

ping to Collinson, who sold them to collectors around Europe. Visitors to John Bartram's nursery outside Philadelphia had to time their visits carefully to catch him at home, but if they did, they would find him tending to "his Idol Flowers" and "his darling productions," in the words of his friend Alexander Garden (for whom the gardenia is named).* In Europe, Linnaeus himself eagerly awaited shipments from Bartram. While men like Collinson lusted after the showiest and rarest American specimens that Bartram could find, systematic botanists—Linnaeus foremost—wanted every plant, no matter how humble or plain. Bartram's own interests likewise encompassed the entirety of the plant world. Not only the most striking blooms but also agricultural crops and medicinal plants enthralled him.

In 1738, Bartram had made an electrifying discovery: an American species of ginseng (*Panax quinquefolius*) growing near the Susquehanna River. Chinese ginseng was well known to Asian and European doctors as one of the most versatile and potent medicinal plants in the world. Benjamin Franklin was so excited that Bartram had found ginseng growing wild in America that he published an announcement in the *Pennsylvania Gazette*. Bartram's discovery gave fuel to Franklin's belief that the "Mountains and Swamps in America" were filled with plants "whose Virtues and proper Uses are yet unknown to Physicians." Franklin tried to raise subscription funds for Bartram to go in search of these plants. When this failed, he and Bartram decided to focus instead on creating an American edition of a British guide to medicinal plants: Thomas Short's *Medicina Britannica*. For their new edition, which Franklin published in 1751, Bartram wrote little editorials regarding which of the plants in Short's lists could be found growing wild in North America. He wrote, for example, that a decoction made from the roots of horse weed (*Collinsonia canadensis*) "is much commended for Womens After-pains," and that Native Americans drank a juice made from the roots of an orange-flowered plant called pleurisy root (*Asclepias tuberosa*) to fight dysentery. (Also known as butterfly milkweed, *Asclepias tuberosa* can be highly toxic.)

---

* Bartram's beautiful house and garden can still be visited today.

John Bartram sparked a passion for medicinal plants in his children. Two sons, Isaac and Moses, became apothecaries and ran a shop together in Philadelphia, where they dispensed herbal medicines made from such plants as saffron, valerian, and juniper berries, alongside chemical and mineral remedies derived from mercury, saltpeter, and other such substances. John Bartram also tried hard to persuade his son William to become a physician or apothecary, but it was probably his own fault that William refused to leave the garden and coop himself up in a shop or clinic, because when William was a teenager his father had taken him on botanizing trips up and down the Eastern Seaboard. William loved botanizing so much, in fact, that in 1773 he embarked on a four-year plant-gathering expedition of his own. It would result in one of the most influential pieces of nature writing ever to come out of North America.

It took him until 1791 to bring the story of his travels to the reading public. When readers finally opened the long-awaited volume—begun in a British colony and published in a newly independent America—they discovered that the title alone traversed a lengthy distance down the page: *Travels through North & South Carolina, Georgia, East & West Florida, the Cherokee Country, the Extensive Territories of the Muscogulges, or Creek Confederacy, and the Country of the Chactaws*; containing *An Account of the Soil and Natural Productions of those Regions, together with Observations on the Manners of the Indians.* In the pages of his *Travels*, Bartram brought to gripping life the perils and wonders of the American natural world. He described, for example, the moment two large alligators tried to overturn his canoe, "roaring terribly and belching floods of water over me" and snapping their powerful jaws together "as almost to stun me" with the din. As he flailed at the beasts with a club, Bartram "expected every moment to be dragged out of the boat and instantly devoured," but they sank back into the water. As he made his way back to his garden on the Schuylkill, Bartram carried a prayer book in his pack, the gift of a Charleston friend named Mary Lamboll Thomas; he later told her the book had brought him comfort "on those dangerous & dubious scenes in my Travels."

Yet Bartram had also encountered landscapes of serene botanical beauty and was just as adept at capturing these for his readers. "Observe these green meadows how they are decorated," he wrote. "They seem enamelled with the beds of flowers." After Bartram's *Travels* appeared, a young Philadelphia physician named Benjamin Smith Barton gushed that "no man in America is so constantly employed in reflecting on the beauties and the wonder of Creation." Barton especially appreciated the fact that Bartram valued medicinal plants, and the two men corresponded about some of the botanical remedies Bartram had seen in use among Native American peoples—for example, a holly tree (*Ilex cassine*) that contained caffeine and was "the most powerful & efficacious vegetable Diuretick yet known" because "its effects are almost instantaneous."

Benjamin Smith Barton, who had made the pilgrimage to study medicine in Edinburgh just a few years before Hosack, believed that "the man who discovers one valuable new medicine is a more important Benefactor to his species than Alexander, Caesar, or an hundred other conquerors." Although William Bartram did not incline to such bombast, he did agree that plant discoveries were critical to the progress of American medicine. Around 1796, following in his father's footsteps and likely working with his brothers Moses and Isaac, Bartram prepared a small *pharmacopoeia*, a catalogue of medicines with instructions for preparation and dosage. This new catalogue took its inspiration from the sorts of volumes Hosack had just spent two years lugging around Great Britain, especially the *Edinburgh New Dispensatory* and William Cullen's *Treatise on the Materia Medica*. Bartram's pharmacopoeia followed Cullen's innovative decision to organize medicines not alphabetically but by their effects on the body: diuretics, diaphoretics, purgatives, and so on.

Among the plants featured in Bartram's new Philadelphia pharmacopoeia were many that were now known to be growing in North America—thanks in large part to the Bartrams themselves. This project of identifying native medicinal plants was no mere exercise in systematic botany. It was a matter of life and death, for while Benjamin Smith Barton and William Bartram corresponded about *Ilex cassine*, and Moses and Isaac Bartram boiled medicinal plants at their apothecary shop, Americans were being felled by fevers, tumors, infections, and a host of other disorders and disasters whose cures might be growing in the swamps of Florida, in the mountains of North Carolina, or even in the woods just outside of Philadelphia. Hosack knew how much the Bartrams had done for botany, especially medical botany. But almost half a century had elapsed since Franklin had published John Bartram's 1751 update of Thomas Short's *Medicina Britannica*, and it was still among the most comprehensive works on American medicinal plants. What the nation needed, Hosack thought, was a new kind of garden—a botany classroom, chemical laboratory, apothecary shop, plant nursery, horticulture school, and lovely landscape all rolled into one. The kind of garden that was already pushing up its first pale shoots in his mind.

⎯⎯⎯⎯✺⎯⎯⎯⎯

HOSACK SPENT MORE AND MORE HOURS in 1795 dreaming of his garden, but he was an unknown, impecunious young doctor. He needed public goodwill and financial support to bring it to life. Then, at the end of the year, an unexpected invitation cleared a path for Hosack straight into the heart of national politics and the most fashionable houses in New York. It came from Samuel Bard, President Washington's former physician.

Bard, now in his fifties, had a jutting lower lip and deep lines that etched permanent parentheses around his nose and mouth. He was

Samuel Bard,
Hosack's mentor and
medical partner

big-hearted and genial, a sturdy pillar of New York medicine. During medical studies in Edinburgh in the 1760s, Bard had won a botany medal and written a thesis on the medical effects of opium, after testing it on his roommate and himself. In 1767, he had helped found New York City's first medical school, at King's College, and then the New-York Hospital a few years later. In clinical practice, Bard prized both compassion and honesty, telling his students to "remember always that your Patient is the Object of the tenderest Affection, to some one," but also warning them that they should "never buoy up a dying Man with groundless Expectations of Recovery."

Bard was a founding member of the New-York State Society for the Promotion of Agriculture, Arts, and Manufactures, and probably one of the few people who had mentally rejoiced at Hosack's proposal for a *hortus siccus*. Bard knew this was not the first effort at cataloguing New York's flora, for as a boy he had once been sent to recuperate from a childhood illness in the fresh country air at an estate named Coldengham, where a talented young botanist named Jane Colden had taught him Linnaean botany. Jane's father was the wealthy and powerful Cadwallader Colden, a passionate plantsman who would later become lieutenant governor of the province of New York, serving until the outbreak of the American Revolution. Linnaeus once called him "*Summus Perfectus*" and also named a flower for him. Colden encouraged his daughter's interest in botany, and at Coldengham she met John Bartram Sr. and William Bartram, as well as Linnaeus's student Pehr Kalm. In the 1740s, Jane produced ink drawings and descriptions of more than three hundred plants growing in the vicinity of her father's estate, up the Hudson River near Poughkeepsie. In 1756, the British merchant Peter Collinson wrote Linnaeus that Jane Colden "is perhaps the first lady that has so perfectly studied your system. She deserves to be celebrated." Her New York flora manuscript was never published, but Bard knew about it from his time studying with her, and he had remained deeply interested in botany ever since.

When Hosack joined Columbia in 1794, he had become Bard's colleague on the medical faculty. As they talked about medicine and botany, the older man came to see Hosack as a possible professional heir. Bard dreamed of retiring to his splendid estate up the Hudson at Hyde

Park, where he could enjoy his family and his views of the Catskill Mountains. Toward the close of 1795, Bard invited Hosack to join him as the junior partner in his private practice. Bard soon became deeply attached to his diligent new protégé, viewing him as a surrogate son just as Benjamin Rush did. Paternal warmth suffuses a Christmas note Bard later sent Hosack from Hyde Park. "My dear Hosack," he wrote, "I cannot express my wishes for your happiness at this season of festivity and congratulation, in stronger terms, than that, at my time of life you may sit down to a festive family board with feelings of equal enjoyment, and sentiments of equal gratitude for similar blessings."

For his part, Hosack found that his new association with Bard brought its own blessings, even beyond the gift of being mentored by an experienced and affectionate elder. He now gained access to Bard's many wealthy and powerful patients, and he strove to honor the great trust placed him. Hosack always labored under the cracking whip of his own conscience, which he once described as "an internal monitor" that hounded him if he ever ignored any chance to improve himself as a physician. The pangs that resulted from his own disappointment in himself were "more painful" than any "bodily sufferings" he could ever face. This inner voice had goaded him in Edinburgh and London, and now it drove him to excel at his new duties in Bard's practice. When Bard returned from his trips to Hyde Park, he heard glowing reports about Hosack from his patients.

Only two years after disembarking from the *Mohawk*, Hosack had established himself as one of the city's most respected physicians. A witty, gregarious man with an ample supply of stories about Sir Joseph Banks and other British luminaries, Hosack was also increasingly sought after as a guest at New York social gatherings. His new status was confirmed in the winter of 1797, when he received an invitation to a very exclusive dinner party. On February 28 Aaron Burr had written his thirteen-year-old daughter, Theo, from Philadelphia, where he was serving in the Senate, asking her to show hospitality to a distinguished visitor named Joseph Brant, who would be traveling to New York after having his portrait painted by Charles Willson Peale. Her mother, Theodosia (for whom Theo was named), had died in 1794, and Theo was now the mistress of the household. Burr asked Theo to find some gifts for the man's daughters; he thought some earrings might do nicely. Theo threw a dinner party in the visitor's honor and invited several of the most distinguished New Yorkers she could think of. Hosack was among the guests that night. Joseph Brant, Chief of the Six Nations and of the Mohawk people, turned out to have "simple, easy, polished, and even court-like manners"—which made sense, because he had already been received at the court of King George III, more than once.

When they weren't staying in the city, the Burrs lived in a rented mansion just north of New York called Richmond Hill, built in the 1760s and nestled in acres of gardens, meadows, and woods overlooking the Hudson. The mansion had served as George Washington's headquarters for part of 1776, and, after the war, Vice President John Adams and his wife, Abigail, had taken up residence there, hosting many dignitaries before the government moved to Philadelphia—among them Jefferson, newly returned from France and resplendent in "red waistcoat and breeches, the fashion of Versailles," as another guest later remembered. Abigail Adams had loved Richmond Hill, as "delicious" a spot as any she had ever seen. To the north she saw the island's endless green fields, to the south the spires of New York, and across the Hudson "the fertile country of the Jerseys, covered with a golden harvest." Burr leased Richmond Hill in the early 1790s, and surviving records of his purchases from a local plantsman suggest that he indulged his love of horticulture. He bought hyacinths, daisies, polyanthus, and also dozens of wood planks, perhaps to build hotbeds for his gardens, or the icehouse he would advertise some years later when he was trying to sublet the estate.

While Congress was in session, Burr was forced to leave his gardens and his wife and daughter to go to Philadelphia, but Theo wrote him frequently during these absences, relaying household news and messages from her mother, who was ill. "Ma begs you will omit the thoughts of leaving Congress," Theo wrote on one occasion. Unlike many of his peers, Burr believed in thoroughly educating the new nation's daughters, and he seized every chance to instruct and challenge Theo. Medicine and botany both figured significantly in these efforts, and even as Burr queried Theo about her mother's health, he pushed her to learn some medical botany. He corrected her spelling of *laudanum*, and he instructed her to "be able, upon my arrival, to tell me the difference between an *infusion* and *decoction*; and the history, the virtues, and the *botanical* or medical name of the bark." He also promised to send Theo "a most beautiful assortment of flower-seeds and flowering shrubs." A decade later, when Burr was considering making an offer on a spectacular estate in northern Manhattan, Theo would urge him to do so because of his love of ornamental gardens. "How many delightful walks can be made on one hundred and thirty acres!" she wrote her father. "How much of your taste displayed! In ten or twenty years hence, one hundred and thirty acres on New-York island will be a principality."*

In late 1793, Burr, stuck in Philadelphia and worried about his wife, consulted Benjamin Rush for a second opinion about her illness. On Christmas Eve, he wrote her to say that Rush had stopped by his lodgings earlier in the evening and was recommending that she take a tiny dose of hemlock, which "has the narcotic powers of Opium superadded to other qualities." Rush had told Burr that he had seen hemlock "work wonderful cures." If Theodosia followed Rush's prescription, however, it didn't seem to improve her health. The following month Burr wrote with more advice from Rush, this time directing it to ten-year-old Theo, who was helping care for her mother in New York. "Doctor Rush thinks that [Peruvian] bark would not be amiss." Over the next months, Burr came to rely on Theo, addressing her affectionately as "my dear little girl." He wrote her constantly, sometimes even from the Senate chamber.

In the spring of 1794, Burr's domestic world was shattered. On May 19, he received a week-old letter saying that his wife was feeling much better. Just one hour later, an express delivery informed him that she had died on the eighteenth. Burr would later observe that the loss "dealt me more pain than all sorrows combined." He now devoted himself to parenting the motherless Theo, who became the center of

---

* Burr ended up not buying the estate, but he would move there thirty years later after marrying Eliza Jumel, a widow whose late husband had bought the estate in 1810. Today it is open to the public and is known as the Morris-Jumel Mansion.

his emotional life. One winter night in 1795, when Theo was away on a visit, Burr rattled around in the Richmond Hill mansion, missing her terribly. He finally sat down just before midnight and wrote her a tender note. "You are now in the arms of Somnus, or ought to be. . . . Dream on."

HOSACK, IN ADDITION TO JOINING the Richmond Hill social circle, now began serving as one of Burr's and Theo's regular physicians. By the spring of 1796, thanks to Hosack's association with Bard, he had also begun helping care for Alexander Hamilton's family. The Hamiltons lived in a house full of children at 26 Broadway, close to the leafy park that spread its shade over the Battery. Alexander Hamilton was just entering his forties; Eliza was about two years younger. Their son Philip, fourteen in January 1796, was the oldest child. According to a surviving portrait, Philip was a beautiful boy with an elfin face, all angular cheekbones and large eyes—a perfect synthesis of the most fetching features of his attractive parents. Next came their daughter Angelica, named for Eliza's sister Angelica Schuyler Church, and there were three younger boys in the Hamilton family by this time: Alexander, James, and John.

Hamilton's law practice kept him constantly on the move as he raced to meetings and trials around the city and up and down the East Coast. "When will the time come," he lamented to Eliza, "that I shall be exempt from the necessity of leaving my dear family?" Political affairs also made persistent assaults on Hamilton's time, especially after the tumultuous reception of the Jay Treaty. Although he had left public office in 1795, he was as attentive as ever to national politics and was laboring that spring over a momentous document for the president, who had decided not to run for a third term. Washington wished to take his leave with a formal written address to the nation and had enlisted Hamilton's eloquent pen for the purpose. Throughout the summer of 1796, their drafts flew back and forth between New York and Philadelphia, and on September 19, Washington's Farewell Address was published in the papers. It was a solemn and beautiful meditation that moved people to tears. "Interwoven as is the love of liberty with every ligament of your hearts," Washington told his fellow Americans, he felt assured that they would honor and protect the institutions that guaranteed that liberty. Hamilton was deeply distressed, however, when it became clear that the second president would be John Adams, a man he thought too egotistical to govern wisely. Adams returned the sentiment. In a crude insult grounded in humoral theory, Adams would later tell Rush that Hamilton suffered from excess semen, which turned into "choler" and "ascended to the brain"—making Hamilton's military and economic policies the product of "a superabundance of secretions which he could not find Whores enough to draw off."

After Washington's departure, Hamilton continued to advise Federalist comrades on domestic and international affairs from behind the scenes, fretting in April 1797 to his friend Oliver Wolcott Jr., "The situation of our Country My Dear Sir, is singularly Critical." A joyful family event that summer, the August 4 birth of a healthy baby boy, found Hamilton deeply mired in national politics. Then, in September, Philip fell deathly ill with a mysterious fever. Neither Bard nor Hosack was in attendance. Bard may have been away at his Hyde Park estate, and perhaps in this case Hosack was initially deemed too inexperienced. Whatever the reason, another physician was summoned to the house at 26 Broadway—John Charlton. Charlton was older than Bard and his equal in professional standing. He had begun as a physician in the court of King George III, settling in New York after a tour of duty in the French and Indian War. By the time Charlton was called to Philip's bedside, he had logged many decades of clinical practice.

Even with his son in such good hands, Hamilton faced a wrenching decision. He had urgent legal business in Connecticut. He felt awful about leaving Philip, yet he tore himself away. Before he had even crossed from New York into Connecticut, he wrote to Eliza. "I am . . . very anxious about my Dear Philip. I pray heaven to restore him and in every event to support you." Hamilton urged her to make sure the doctor gave Philip cold baths if his fever remained elevated. Charlton had undoubtedly attempted this—it was a common treatment for fevers of all kinds, especially in the earlier stages—and at last he had exhausted every remedy he knew. He thought of Hosack. The latter was relatively untested as a physician, but his youth might be an advantage precisely because he had studied so recently with some of the world's greatest medical minds. He also possessed excellent instincts and a strong stomach for experimentation. Charlton sent for him.

It was a short walk from 60 Maiden Lane to 26 Broadway. Hosack arrived at the house to find Philip near death. A courier was dispatched to bring Hamilton home immediately, and seeing that Eliza was "overwhelmed with distress," Hosack sent her out of Philip's room. Her husband was a war veteran who was accustomed to the sight of dying young men, but Hosack saw no reason for Eliza to watch her son endure a painful end. It was at this terrible moment, Hosack later recalled, that an idea occurred to him. He probably couldn't save Philip, but he might be able to keep him alive until Hamilton could reach his bedside. Now, just as Hosack had done in the yellow-fever epidemic two years earlier, he decided to place his trust in heat, amplifying its stimulating effects by mixing smelling salts, alcohol, and a strong dose of Peruvian bark directly into the steaming water. Then he lowered Philip in.

As Hosack hovered and waited, Hamilton rushed down the dark country roads toward the city. The previous month had been one of the most distressing of his life. On August 4, the same day Eliza had given birth to their son William in Albany, Hamilton had posted a letter to James Monroe hinting at the possibility of a duel. He was furious with Monroe, whom he suspected of playing a central role in exposing a scandal about to engulf his family. The events now under renewed scrutiny had unfolded six years earlier. In 1791, a beautiful woman named Maria Reynolds had arrived at Hamilton's house in Philadelphia saying that her husband had abandoned her. Gallantly offering his assistance, Hamilton discovered that "other than pecuniary consolation would be acceptable," as he delicately phrased it, and he had embarked on an affair with Reynolds, soon finding himself the subject of ongoing blackmail by her husband.

By the summer of 1797, both affair and blackmail had long since ceased, but Hamilton's political opponents felt it expedient to keep the rumors of wrongdoing alive. Now the story had resurfaced in the press, putting him on the defensive again. On August 25, Hamilton had published a detailed statement in which he described his affair with Maria Reynolds and his dealings with her husband, appending so many supporting documents that it reached the length of a short book. Hamilton's aim was to clear his name of charges that he had compromised his position as secretary of the treasury by passing inside financial information to his lover's husband. The chief result, however, was public shame for Eliza and scathing ridicule for her husband. When Philip fell desperately ill a few weeks later, his parents were struggling to recover their dignity and privacy in the harsh glare of the national press.

As Hosack had hurried to 26 Broadway to examine Philip, he was acutely aware of what he later described as the "great distress then existing in your family." He was desperate to spare them a further, unfathomable misery. When Hamilton finally arrived home and learned that Hosack's ministrations had saved Philip, Hosack felt to his marrow how satisfying and significant his life's chosen work could be. Back at Columbia, he reworked his lectures on fevers to include the moment he had opened his eyes to find Alexander Hamilton, hero of the American Revolution, kneeling in grateful prayer before him. Hosack could be boastful, but this wasn't just vanity. It was the most inspiring story he could tell a classroom full of teenage American boys considering a career in medicine.

———— ∞∞∞ ————

ONE MONTH AFTER HOSACK's night at the Hamilton home, his friend Dr. Elihu Hubbard Smith dined with Hosack and a mutual friend, Dr. Thomas Parke of Philadelphia. The three men fell into a heated argument about yellow fever. Hosack insisted the fever was not of local origin but imported each year to the United States—he just wasn't sure how. (Hosack was wrong about the disease being contagious, but he was correct about its importation. It was the *Aedes aegypti* mosquito, originally native to Africa, that carried yellow fever to the northeastern United States on ships sailing from South America and from Southern ports such as Charleston and Wilmington.) Smith argued that yellow fever was of local origin, and later that night he wrote in his diary that he was "satisfied that it is impossible for Hosack, (as he expects) to prove it to have been imported."* As for treating fever victims, Smith allied himself with Parke—"Parke & I are bleeders"—but Hosack defended his gentler approach of using botanical remedies.

Less than a week later, the medical board of New York recommended that the chief bleeder himself, Benjamin Rush, be hired at Columbia. Rush was longing to move to New York, in part because the controversy in Philadelphia about his treatment of yellow fever had scared off many patients; he knew all too well that his name was now "mentioned with horror in some companies." But one powerful Columbia trustee was adamantly opposed to the idea of hiring him: Hamilton, with whom Rush had clashed so publicly during the 1793 epidemic. Hamilton managed to block the appointment, and Rush gave up, telling a friend wryly that he felt honored that "the opposition to my appointment has come from that gentleman."

When Hosack learned that his old mentor was being thwarted by his glamorous new friend, he must have felt personally conflicted, but his principle of political neutrality helped him navigate the situation with both relationships intact. He and Rush continued their frequent exchange of warm letters, with Rush calling himself Hosack's "friend and brother in the republic of medicine," while Hosack's connection with the Hamiltons was only deepening. As Hosack observed Hamilton at home with Eliza and the children, he was struck by the tenderness they inspired in this remorseful husband and father.

Hosack also began to experience firsthand what he called Hamilton's "highly endowed and cultivated mind." He had likely read some of Hamilton's political works before meeting the man himself, but now he learned how brilliant Hamilton was in conversation and what a range of subjects he had at his command, medicine among them. At some point, Hosack confided in Hamilton about his idea for a botanical garden. Hosack had been thinking and talking about it for months,

---

* It was not until 1900 that United States Army Major Walter Reed and colleagues would prove through experiments conducted in Cuba that yellow fever was carried by the *Aedes aegypti* mosquito.

even buttonholing one of his students in the street to ask him to procure specimens from an uncle who ran a botanical garden in the West Indies. But it wasn't until after saving Philip Hamilton's life that Hosack finally took action. Only weeks later, he approached Columbia's president and trustees. The current situation, he informed them, had been causing him "great regret." He could not teach his medical students botany without a garden. He had tried using books and engravings, but they were unhelpfully abstract and they bored the students. "The obvious and only effectual remedy would be the establishment of a Botanical Garden," Hosack told the trustees.

Three dimensions rather than two. It was being done in the surgical theater, where professors and students sliced into human cadavers, so why not in the equally urgent study of medicinal plants? The president of the college was intrigued, and he convened a committee to consider the idea. They quickly concluded that Hosack should receive three hundred pounds annually for the new enterprise. He was elated at the news. He would have to wait for the funds to materialize, and he hoped it wouldn't be long.

IN THE MEANTIME, Hosack got his own house in order. He had been a widower for a year and a half, but shortly before Christmas, Elihu Hubbard Smith told a mutual friend, "Hosack has been some time at Phila., where he is marrying. When he returns, there will be feasting &c." In January 1798, Smith called to take tea with Hosack and his new wife, Mary Eddy Hosack, a dark-haired, slender woman of twenty-eight—the same age as Hosack. They had probably first met during Hosack's student days in Philadelphia, where they had moved in overlapping, rarefied social circles. Orphaned as a young girl, she had been adopted into the learned, well-connected Wistar family; from time to time, she had sat across a chessboard from Benjamin Franklin. Her foster brother, Caspar Wistar, a prominent physician and friend of Hosack, was later to achieve botanical immortality when the botanist Thomas Nuttall christened a luxurious, purple-clustered vine in his honor: *Wisteria*. Just a month before the wedding of Hosack and

Mary Eddy, her foster sister, Catharine Wistar, had married Franklin's grandson William Bache. One of Hosack's botany students in New York later described Mary Eddy Hosack as "the only lady here with whom I can converse on scientific subjects."

After their wedding in December 1797, Hosack and Mary settled into a townhouse at 65 Broadway, on the most fashionable street in the city. Members of the powerful Livingston clan lived both next door and across the way. Dr. Bayley lived at 44 Broadway, Dr. Charlton at 34 Broadway, and the Hamiltons at 26 Broadway. Hosack's medical practice was thriving, and Mary may have brought still more money to the marriage. Their new household soon included servants, a coachman, horses, and at least one milk cow.

Their household also included enslaved people.

Hosack had grown up with slaves. When the first-ever United States census counted the members of Hosack's father's household in 1790, the latter had owned two of the city's approximately two thousand enslaved men and women. Some of the most powerful men in New York were agitating to end slavery, but Hosack seems not to have joined these efforts, in spite of the fact that he counted antislavery advocates among his friends and knew Rush's strong opinions on the issue. As a physician, naturalist, and civic leader, Hosack would spend most of his life racing ahead of his fellow Americans, but when it came to slavery, he apparently lacked compassion and imagination. The 1800 census, conducted three years after his marriage to Mary, shows he had five enslaved people in his household at 65 Broadway; as of 1810, he would still have one. Like other botanical and horticultural enthusiasts of the early Republic—including Washington, Jefferson, and Madison—Hosack may have put enslaved people to work in his garden. No record survives of whether or not he did so.

ON A WARM, CLEAR EVENING in the middle of September 1798, Elihu Hubbard Smith called in at 65 Broadway to check on the Hosacks. Mary was pregnant with her first child, and just when Hosack should have been marshaling his energies and his medical skills to make sure she didn't die in childbirth as Kitty had, he had fallen seriously ill with yellow fever.

All over the city, New Yorkers were again feeling fatigued and feverish. Dozens had died in the previous three weeks. Smith felt so listless he could barely bring himself to go out. His eyes and nose gushed with what felt like "perpetual tears, & perpetual drizzling," and his nights were "tormented with musquitoes & incongruous dreams." For weeks, the papers had maintained a conspiratorial silence about the insistent signs that yellow fever had returned to New York. Then they had turned to ridicule. "Our country friends," the *New-York Gazette* had written in late August, are claiming the city is so eerily empty that "the grass is *three* feet high in the streets." These naïve alarmists, the paper said, had all but recommended that the authorities burn the whole city down and "then introduce the North [Hudson] and East River through the Streets" to flush out the fever. Yet many New Yorkers had in fact already hastened up the Boston Post Road into the countryside, and frightened farmers no longer dared to bring their produce to the city's markets. Thomas Greenleaf, the editor of *Greenleaf's New York Journal*, complained that he was having a terrible time getting his paper published, because four of his apprentices had fled. "These boys are hereby charged to return immediately," he warned in his paper. "If they do not do it, we shall . . ."—here he trailed off ominously.

Greenleaf died ten days later, on September 14, the day that Smith stopped in to see the Hosacks. On September 18, Charles Willson Peale's son Titian died, having been in town while his father painted portraits of well-to-do New Yorkers. Peale was inconsolable, confessing to a friend almost a year later that he was still crying such "floods of tears" he feared that "this *cankerworm*, Grief, will prey on my Vitals, and shorten my days." (In true Peale fashion, Titian's brother Rembrandt wrote a poem about him, while his brother Raphaelle painted a miniature of him in watercolor on ivory.) Hosack's friend Elihu Hubbard Smith died the day after Titian. A few days later, the Common Council finally acknowledged the fever's return with a series of emergency measures, including an increase in the number of city watchmen to protect deserted properties from robbers.

Until Hosack contracted the fever himself, he had been ministering to his wealthier patients in their homes and to the poor at the hospital and the almshouse. Just as he had done during the 1795 fever season, he was again shying away from inflicting mercury and bleeding on his patients. Hosack noted that some doctors thought of mercury as the "Sampson of the *materia medica*" for its great purgative power, but he was doubtful about the claim. "In the yellow fever it has truly proved a Sampson, for I verily believe it has slain its thousands." Instead of mercury, therefore, Hosack prescribed milder purgatives such as castor oil, magnesia, and his favorite of all, *Eupatorium perfoliatum*, a plant that grew wild on the island of Manhattan. Local country people called the plant *boneset*, and they often drank a potent infusion made from its stem and leaves to provoke vomiting, an effect thought to be salutary in calming a fevered body. Hosack was using it for his yellow-fever patients with good results, and above all without the poisonous aftereffects of mercury.

In mid-September, after weeks of exhausting clinical rounds, the symptoms of the fever finally invaded Hosack's own body: chills, aching muscles, the sickly yellow tint to the skin. Although he was right in thinking that yellow fever was not domestic in origin, he also still thought, incorrectly, that it was contagious and that he might infect Mary and her unborn baby. He made himself choke down bitter infusions of boneset until he vomited, and then he wrapped himself in blankets and sipped milder herb teas made from catmint, sage, and snakeroot, until he felt sure he had sweated out the fever. A few blocks away on Broad Street, Samuel Bard was following the same regimen. When Bard had first learned of the epidemic, he had been at his Hyde Park estate and had rushed down to the city to lend help, but soon he, too, had contracted yellow fever. He dispatched a pitiful letter to his wife, "without whose aid I can neither bear sorrow nor sickness." When she arrived from Hyde Park at the Broad Street house, she reportedly found her husband lying on a sofa nearly unconscious, his skin yellow. Now Bard, like Hosack, was spending his days wrapped in heated blankets and swallowing infusions of boneset. Both men made complete recoveries.

Hosack emerged from the 1798 epidemic more certain than ever that New York and the nation desperately needed a botanical garden. If a humble native plant like boneset could prove its merit over mercury, he was convinced that other medicinal remedies were every minute being "trodden under foot as unworthy of regard." Although he and Bard had survived the epidemic and managed to save at least some of their patients, hundreds of other New Yorkers had perished for lack of adequate clinical care and medicines.

As cooler weather arrived, the number of fever deaths waned. Mary was safely delivered of a baby boy in October. They named him Samuel Bard Hosack and commenced the vigil over his delicate infancy. Thirteen months later, in November 1799, Mary sent a fretful letter to her foster sister, Catharine Wistar Bache, in Philadelphia. "My little boy has not been very well for a few days," she wrote, praying that his indisposition was simply a side effect of teething. Elsewhere in Philadelphia, Charles Willson Peale and his wife had just welcomed a baby boy over whom they, too, would be watching with anxiety. They named him Titian, for his dead brother.*

Alexander and Eliza Hamilton also celebrated the birth of a child that fall, a daughter whom they called Eliza. Unlike Hosack and Peale, Hamilton had not seen any of his children perish, but Philip's brush with death in the fall of 1797 had been a frightening reminder of the constant threat. Even the most robust of men could be swiftly felled by illness. Three weeks after Eliza's birth, Hamilton's mentor and friend George Washington succumbed to a throat infection he had contracted after inspecting his Mount Vernon estate in wintry weather. He died after his doctors had bled him half-dry. "It is our painful duty, first to announce to our Country and the world, the death of their illustrious benefactor," wrote the *Alexandria Times* two days later. As Washington's body was prepared for burial in the family crypt at Mount Vernon, the newspapers began weeks of somber adieux and odes. "From Vernon's Mount behold the HERO rise! Resplendent forms attend him thro' the skies!"†

---

* The second Titian would grow up, almost inevitably, to be an extraordinary painter and naturalist, and eventually a pioneer in the nascent art of photography.

† There were botanical tributes, too; Hosack would describe to a visiting Scottish botanist named David Douglas a "very delicious" plum whose color lay "somewhat between cream and sulphur." Douglas noted in his diary that the plum was named for President Washington—as "every product in the United States that is great or good is called."

In New York during the week immediately following Washington's death, the air echoed daily with the tolling of the city's church bells, muffled to sound mournful. On December 31, Hamilton, Hosack, and hundreds of other New Yorkers joined a memorial procession through lower Manhattan to St. Paul's Chapel on Broadway. The focal point of the parade was a symbolic, empty coffin surmounted by a four-foot sculpture of an American eagle with "extended but drooping wings." The coffin was carried by a group of Revolutionary War veterans, among them the incumbent mayor of New York, Richard Varick. The procession moved up Broadway between crowds of hushed onlookers as cannon were fired on the Battery.

Of all the disconsolate New Yorkers who turned out that day to honor the departed president, none grieved more than Alexander Hamilton. "My imagination is gloomy, my heart sad."

⸎

AT EVERY TURN, Hosack saw premature death, fever, imperiled infancies—and still the Columbia trustees left him without the funds he needed to begin on the botanical garden. Columbia was struggling financially, with the state capital having moved to Albany in 1797 and the legislature's annual grant to the college terminated. More than two years had passed since the trustees had greeted Hosack's proposal with enthusiasm, but their finances had not permitted them to make good on their promise. Early in 1800, he finally gave up on Columbia and turned to his old college friend DeWitt Clinton, who had gone to work after graduation for his uncle, Governor George Clinton. When the latter had decided not to run for reelection in 1795, DeWitt Clinton had gone back to Columbia briefly to indulge his passion for natural history, studying chemistry and zoology with Samuel Latham Mitchill, and botany with Columbia's newest professor—his friend Hosack.

More recently, however, Clinton had begun to make a name for himself in city and state politics. After marrying a wealthy and beautiful young woman named Maria Franklin in 1796, he was elected first to the State Assembly and then, in 1798, to the State Senate. Even as Clinton's political fortunes rose, he retained his firm conviction that government should offer unstinting support to scientific inquiry. When he opened Hosack's letter in Albany in February 1800, therefore, he was predisposed to find the contents intriguing. Did Clinton not agree, Hosack asked, that New York should have the honor of founding the first botanical garden in the new United States? The garden would be a powerful weapon against what Hosack called the "formidable and dangerous diseases" that plagued the nation each year, and it would also facilitate the introduction of useful new agricultural crops. Hosack told Clinton he had just submitted a written appeal to the legislature to fund his garden and was hoping that Clinton would "cheerfully lend your exertions in its favor."

As it happened, DeWitt Clinton never got the chance to rise to his feet, stretch himself to his considerable height, and urge his fellow senators to fund the proposed garden, because Hosack's petition was routed to the lower legislative house—the State Assembly. Nevertheless, Hosack's idea for a garden grabbed legislative attention on its own merits, and a committee of three assemblymen recommended to their colleagues that he receive the funds he sought. So Hosack had done well to turn from Columbia to the state legislature. Now he would simply need to wait for a bill to be introduced and the money to be appropriated.

DeWitt Clinton, Hosack's closest friend, as an elder statesman

Then, nothing. In the crush of a thousand matters from every county upstate and down, the assemblymen dropped the topic. The machinery of the state ground on. No further attention was paid to Hosack or his garden, and soon all eyes were on the impending end of President Adams's first term. With New York State's electoral votes likely to play a decisive role in the upcoming national election, Hamilton and Burr worked to shore up their respective political positions. Burr, hoping to improve the chances of a strong Democratic-Republican showing, forged a sweeping alliance with the Clintons and the Livingstons. On the evening of April 17, 1800, members of this coalition gathered at a house on William Street and agreed on a slate of candidates for the State Assembly that included George Clinton and Brockholst Livingston. They also endorsed Samuel Latham Mitchill, Hosack's Columbia colleague, as their candidate for New York City's congressional seat. Hamilton had seen his own slate of Federalist candidates approved at a meeting two nights earlier. Over the next ten days or so, Burr and Hamilton and their supporters hurried around town trying to mobilize voters. In the newspapers and in the streets, the two sides hurled insults back and forth. Federalists cautioned that a Republican victory would result in a Jefferson presidency—in other words, a complete catastrophe, since "Mr. Jefferson, and of course Jacobins at large, wish to destroy the Constitution of the United States." Democratic-Republicans, meanwhile, piously urged all citizens to report Federalists for any instances of bribery or fraud, a directive that struck one Federalist as about as fair as "a prostitute revil[ing] an honest woman as a w——."

The elections at the end of April proved an unmitigated triumph for Burr and a humiliation for Hamilton, an outcome at least partially attributable to the end of the French Revolution, long a useful scare-mongering weapon for the Federalists. The Democratic-Republicans took every single State Assembly seat, and their congressional candidate, Mitchill, would be going to Washington. Everyone knew that in the upcoming presidential election Jefferson would be the most powerful Democratic-Republican, and now Burr's resounding success in New York put him in the running for the presidency as well. Mean-

while, Hamilton was lobbying hard for his friend Charles Cotesworth
Pinckney of South Carolina to get a place on the Federalist ticket
alongside President Adams, with whom Hamilton was at increasingly
bitter odds that summer.

In August, Hamilton brought his frenzy of political activity to a
standstill just long enough to transact some personal business. He was
planning to create a quiet retreat for his family—away from the city
and its fevers, yellow and political alike—and he now purchased a
piece of land toward the northern end of Manhattan, past the village
of Harlem. On a rolling spread of thirty-five acres, the architect John
McComb Jr. began designing an elegant mansion for Hamilton.

That fall, the presidential election resulted in an electoral-college
tie between Jefferson and Burr. With his friend Pinckney out of the
running, Hamilton threw his support behind his longtime rival Jefferson, regarding Burr as a slippery character who was "wicked enough
to scruple nothing." It took until February 1801, and thirty-six rounds
of balloting in the House of Representatives, before Jefferson was
named the victor. Burr would become vice president, knowing full
well that Hamilton had blocked his path to the presidency. On March
4, 1801, Jefferson donned a simple suit and walked from his boarding-
house to the Capitol for his inauguration. For Hosack, who played his
political opinions so close to the vest, the most significant fact about
the contest between Burr and Jefferson was that it had featured two
powerful Americans who adored botany and horticulture. Now one of
them was president.

Chapter 6

"DOCTOR, I DESPAIR"

Although Hosack had spared Hamilton the loss of his son, once again he was unable to spare himself. Two-year-old Samuel Bard Hosack died of scarlet fever on March 22, 1801. He was the third child Hosack had lost in less than a decade. The only mercy was that Samuel's death had not utterly emptied the house of children, because Mary had given birth to a girl—also called Mary—a year earlier. But it was still a terrible blow, and Hosack seemed raw and irritable that spring. Not long after Samuel's death, he lashed out at a fellow naturalist, Charles Willson Peale—a man grappling with his own share of paternal sorrow after the death of his son Titian from yellow fever.

The spat between Hosack and Peale was ignited by news that had arrived from upstate New York the previous fall. A farmer near the town of Newburgh had discovered the buried bones of "an animal of uncommon magnitude." When word of the find made its way south, the new president of the United States was among the happiest of all Americans. Since the 1780s, Jefferson had been on a near-constant lookout for physical proof to counter European claims of New World "degeneracy." During his time as minister in Paris, he had spent untold hours trying to procure a giant moose carcass from correspondents back home. Even after he had managed to do so, European naturalists continued to belittle North American fauna, including its humans, as smaller and weaker. The discovery of giant bones in upstate New York might now lend the United States powerful new scientific ammunition.

Philadelphia rushed past New York to seize the opportunity. On June 5, 1801, Charles Willson Peale set out by stagecoach for the farm of John Masten, about seventy miles up the Hudson from New York City, near the Shawangunk Mountains. He hoped to find there the remains of a mastodon—known to Peale and his contemporaries as a mammoth. If he did, he might in one fell swoop strike a blow against degeneracy claims *and* stun the public with an utterly novel exhibit for his natural history museum in Philadelphia.

American stagecoaches, the French traveler Jacques-Pierre Brissot de Warville had observed in the 1780s, were "true political carriages" that could throw a congressman together with "the shoemaker who elected him." When Peale climbed into the crowded coach, he found that on this trip politics would be taking the form of an obnoxious Englishman who passed the hours slandering President Jefferson and railing against Napoleon as a Corsican "boy" who should not be in charge of a country. Peale steamed in silence as he listened to the man "slobber his venum" and "pison the minds" of the other travelers. Finally he lost his temper, arguing with the Englishman most of the way to the next inn. The following day, Peale observed with satisfaction that the man was as silent and "sullen as any hog to this end of our Journey"—New York City. After a stop here with his wealthy in-laws to secure financial backing for his adventure, he arrived at the Masten farm in late June.

Peale immediately recognized the significance of the bones but tried to squelch his excitement and feign indifference. After successful negotiations, he bundled the available bones into hogshead barrels, making plans to return later in the summer to retrieve the rest of the bones from the swamp where they lay buried. He carted the barrels down the Hudson to New York City, where, as he wrote in his journal, the news of his booty "flew like wild fire." Dozens of people crowded into his lodgings on his first night in town, hoping to see the bones. "Every body seemed rejoiced that the bones had fallen into my hands . . . with me they would be preserved, and saved to this Country." Among the visitors were many prominent New Yorkers, including Vice President Burr.

Hosack was as curious as anyone else about Peale's find. After inspecting the bones that evening, he invited Peale to dine with him the next day, but Peale arrived to find his host in a bad mood. Hosack failed even to greet him, snapping instead that he rued the day Peale had gotten hold of the bones. Peale was flabbergasted. Hosack was stricken with a potent bout of professional jealousy and regional pride. The bones, he told Peale, should not leave the state of New York. Peale defended himself hotly. "Do you know any man that would put these bones togather if they had them, but myself?" Hosack had to concede that he did not, but he remained deeply annoyed. Before leaving New York, Peale recounted the tense scene in a letter he sent to a friend in Newburgh, and he also confided it to his journal. Hosack was the only person in all of New York who had "uttered a contrary sentiment." In a celebratory letter to President Jefferson the same day, Peale did not mention Hosack at all.

In August 1801, Peale traveled back up to Newburgh with his son Rembrandt, who helped him design a giant contraption to scoop water out of the morass as they dug for more bones. They enlisted local boys to run inside a huge wooden wheel that powered a system of ropes and buckets. With the help of two dozen men, the Peales retrieved as many of the remaining bones of the skeleton as possible and then moved to other bone sites to try to supplement the gaps. They ended up with nearly enough material to put together a second skeleton. Back in Philadelphia that fall, the Peales assembled the Masten mammoth—which was in fact a mastodon (*Mammut americanum*)—and also cobbled together the second skeleton out of the remaining bones and a set of wooden pieces that Rembrandt carved to mimic the missing parts. When Peale was ready to welcome visitors, he whipped up a hyperbolic handbill. "Ten thousand moons ago . . . a race of animals were in being, huge as the frowning Precipice, cruel as the bloody Panther, swift as the descending Eagle, and terrible as the Angel of the Night. The Pines crashed beneath their feet; and the Lake shrunk when they slaked their thirst." The public ate it up, even at fifty cents a visit, double the price of regular entry to the rest of

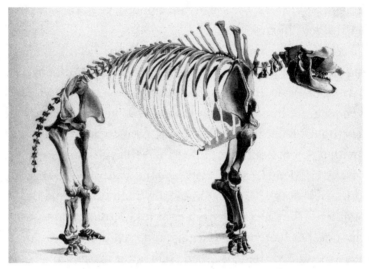

The Peales' mastodon, drawn by Titian Peale

the museum. Peale made plans for his sons Rembrandt and Rubens to take the second skeleton on a European tour. They would sail from New York in the spring of 1802.

MEANWHILE, HOSACK STEWED. He had been waiting more than three years for someone with sufficient power and money to back his proposal for a botanical garden. After his testy exchange with Peale, he reached the end of his tether. It had been exquisitely painful watching New Yorkers fall over themselves to praise a Philadelphia naturalist for whisking their own scientific treasure out of the state. Would it be foolhardy of him, or simply enterprising, if he set out alone on his garden plans? Hosack decided on the latter. He felt certain that once his benighted fellow citizens understood what a botanical garden actually *was*, they would come to their senses and support him. For now, though, they could not seem to grasp what they had never seen for themselves.

He would need a suitable piece of land. He turned his attention to the frenzied Manhattan real estate market—booming for more than a decade, ever since the New York state legislature had passed a law

in 1784 facilitating the sale of the vast spreads confiscated from hastily departed Loyalists to the British crown. Among the first properties that had been snapped up by speculators was a ninety-acre estate
stretching from the Bowery all the way to the East River. It had once
been home to James De Lancey, the exiled son of a former British lieutenant governor of New York.* On the west side of the island, too, the
city was edging inexorably beyond its old border, toward the village
of Greenwich and Burr's Richmond Hill estate. The Common Council
had hastened all this growth by laying down new streets and selling
the Hudson riverfront to merchants who built new piers and markets.
Enterprising landowners in the area sold off lots, built and rented out
new houses, and moved their own families to the country, far from the
scourge of city filth and yellow fever. Absentee landlords began their
reign over the city's shopkeepers and artisans.

Not everyone smiled upon these changes in the landscape. In the
1780s, Chancellor Robert R. Livingston had angrily called upon
Mayor James Duane to "put a stop to your improvements (as they
are falsely called) upon the north [Hudson] river." Manhattan Island,
Livingston predicted, "contains a sufficient quantity of ground for a
much larger city than New York will ever be." In the summer of 1801,
though, the city did not yet extend much past Chambers Street. As
Hosack rode north from 65 Broadway in search of property for sale, he
passed the familiar landmarks of his busy days—Trinity Church, St.
Paul's, Columbia, then the almshouse, where he now worked part-time
as a physician to the poor. A few blocks farther, he passed the New-
York Hospital, and then he crossed the lovely stretch of land known
as Lispenard's Meadow. On its north side Hosack forded a creek and
passed the towering, grass-tufted bluff New Yorkers called Bayard's
Mount. And then he rode off the city map.

Pastoral Manhattan. "Nothing is more magnificent than the situation of this town—between two majestic rivers, the north and the
east," Brissot had marveled in the 1780s. The deep forests of Manhattan had largely disappeared, but here and there the fields of crops

---

* Today, Delancey Street runs right across the old De Lancey property as it heads from the
Bowery to the East River, where it funnels drivers onto the Williamsburg Bridge.

were punctuated by little groves of woodland trees and shrubs. At the edges of the island, where fresh breezes and river views prevailed, lay the estates of wealthy New York families who maintained residences far from the dense, dirty web of city streets. As Hosack rode deeper into the countryside, Broadway became the Bloomingdale Road. From here he turned right onto the Boston Post Road, which shuttled him over to the origin point of a lane that ran up the spine of the island. This was the Middle Road. Farmhouses, barns, ponds, fields rolled past. Then he found what he was looking for, nestled among the surrounding farms.

The city government—the Corporation of the City of New York—owned the land. Hosack bought it in several parcels with his own money, beginning on September 1, 1801, the day after his thirty-second birthday.* That whole month was filled with glorious blue skies, and from a rocky bluff on the western edge of his new property, Hosack reveled in his sweeping views. When he turned and stood to look up the island, the East River lay to his right, with skinny Blackwell's Island (today's Roosevelt Island) pointing southwest toward the harbor and northeast toward Connecticut. Across the East River he could see "the fruitful fields of Long-Island," while to his left lay the wide Hudson, dotted with the sloops and packet boats that carried politicians back and forth to Albany and farm produce down to the city. The cliffs known as the Palisades rose directly across the Hudson from where he stood. Atop them sat the village of Weehawken.

Hosack loved that he could see both rivers from his property, but it was the earth beneath his feet that thrilled him most: his Manhattan in miniature, with its glacial rocks, its green fields, its moist bottomlands. It was already home to some of the island's native species, such as viburnum shrubs (*Viburnum prunifolium*) and hollow-leaved violets (*Viola cucullata*). The hills to the north were thick with mountain laurel (*Kalmia latifolia*). Because Hosack's new land already contained a variety of soils, it would be well suited for a garden where he planned

---

* The final deed to Hosack's twenty-acre property would be executed in 1804 by Mayor DeWitt Clinton. The cost was about $4,800 (about $100,000 today), plus a yearly "quit rent" of sixteen bushels of grain.

to cultivate every single species he possibly could. Later on, after he had struggled to get all the rocks removed, he grumpily admitted that the tract had been "exceedingly rough" when he bought it.

Two weeks after Hosack purchased the first parcel of land, he received word that the brig *Rambler* had arrived from Edinburgh. He made his way down to the harbor in the brilliant late-summer sunshine, signed a bill of lading proffered by a port officer, and took possession of his goods—plants, or perhaps books? His library of medical and botanical volumes at 65 Broadway had been growing as rapidly as a coddled hothouse specimen. In 1794, the year he had returned to New York, he had taken out a subscription to Curtis's *Botanical Magazine*, and he now owned every volume that Curtis had managed to publish before his death in 1799. Sometime in 1801, Samuel Bard asked Hosack if he could borrow these volumes of the *Botanical Magazine*, along with six volumes of James Edward Smith's *English Botany*. Hosack kept careful track of these treasured tomes. It wasn't only that they were written by his two most important British mentors—he needed to consult them as he planned his new garden. Bard returned them all before the year was out.

———◦◦◦◦———

ON A CHILLY, OVERCAST AFTERNOON toward the end of November 1801, Alexander Hamilton appeared on the doorstep at 65 Broadway. He had learned that his son Philip had rowed across the Hudson to Paulus Hook to fight a duel, and he had rushed to fetch Hosack, who was in fact already on his way to help. Hamilton was so distraught that he reportedly fainted at Hosack's house.

The affair had begun three nights earlier at the Park Theatre on Chatham (now Park) Row. Evenings at the Park Theatre, which seated two thousand, often treated audiences to a lengthy comedy or drama set off by music-infused one-acts and pantomimes. Among the theatergoers, as Washington Irving described them in a satirical piece the following year, were "the beaus of the present day, who meet here to lounge away an idle hour, and to play off their little impertinences for the entertainment of the public." The advertisement in the *Mercantile*

Advertiser for the performance of Friday, November 20, carried the usual warning that theatergoers were not to "carry a lighted Segar into any part of the Theatre, or to attempt to renew the dangerous practice of smoking, either in the Lobbies or the presence of the Audience." That night, words alone proved combustible enough to set young men ablaze.

Nineteen-year-old Philip Hamilton was justifiably proud of his father and bristled at criticism of his policies as secretary of the treasury. In 1801, with Jefferson in the White House and local Republicans puffed up by their recent political gains, Philip had frequent chances to read attacks on Hamilton and his fellow Federalists and hear them verbally abused in person. One scathing treatment proved too provocative for him to ignore. In June 1801, the *American Citizen and General Advertiser* had announced that New York's upcoming Fourth of July celebrations would feature a patriotic speech by "George I. Eacker, Esq., one of our Republican Young Men." In his speech, Eacker lionized Jefferson as "the man of the people" and exulted over his election to the White House. He concluded by proclaiming that "no more shall personal virulence, disguised under the protecting mantle of Federalism, destroy the happiness of the persecuted patriot." For months afterward, Philip boiled with indignation at Eacker's many pointed attacks on his father's politics. When he crossed paths with Eacker at the Park Theatre on the night of November 20, the two men exchanged angry words. Three days later, they faced each other down on a dueling ground directly across the Hudson from lower Manhattan. Dueling was illegal in both New York and New Jersey but less stringently prosecuted in New Jersey.

Eacker fired first, piercing Philip in his right side. Philip fired, then collapsed. Eacker was unharmed. Philip's companions rowed him across the river and carried him to the home of his aunt and uncle, Angelica and John Barker Church. Meanwhile, Alexander Hamilton arrived at 65 Broadway too late to find Hosack—he had already gone to the Churches' house. Four years after saving Philip from fever, Hosack now saw there was nothing he could do. First Alexander and then Eliza joined him at the bedside as Philip writhed in pain and

Philip Hamilton

delirium. Hosack later described the heartrending scene to another of
Hamilton's sons. "As soon as your Father ascertained the direction of
the wound . . . and felt the pulse of your brother, he instantly turned
from the bed, and taking me by the hand, which he pressed with all
the agony of grief, he exclaimed in tones and manner that can never
be effaced from my memory, 'Doctor, I despair.'" Eliza and her husband lay down on the bed, one on each side of their son. Philip died
the next morning.

Alexander and Eliza Hamilton had lost the bright light of their lives.
At the funeral, one mourner reported afterward to a friend, Hamilton
"was with difficulty supported to the grave of his hopes!" On December 5, Hamilton's friend Robert Troup wrote to Rufus King, the American ambassador to the British royal court, that "for twelve days past
the city has been much agitated" over the duel. As friends and acquaintances in other cities learned of the loss, condolences poured in. From
Philadelphia, Benjamin Rush sent a mournful letter to Hamilton, beg-
ging him to "permit a whole family to mingle their tears with yours
upon the late distressing event. . . . You do not weep alone. Many,
many tears have been Shed in our city upon your Account." When
Hamilton replied to Rush after a long delay, he conveyed the depth of
his loss in simple, searing terms by asking for a copy of a letter Philip
had written to the Rushes after a visit to their house. "You will easily

imagine that every memorial of the goodness of his heart must be precious to me." Hamilton closed his letter with thanks from Eliza, who "has drank deeply of the cup of sorrow."

Hamilton struggled to regain his footing, but he eventually returned to the affairs of his law practice. His friend Troup observed in early December that "Hamilton is more composed and is able again to attend to business." Grieving fathers had little choice.

Hosack, anguished but helpless, turned back to his teaching and his land. It was an awful practice, dueling. He strove to remain aloof from the vicious quarrels that his friends and acquaintances seemed to enter with such gusto. His medical experience gave him too vivid a display of the potential consequences, and his success as a physician depended on his neutrality. Now the success of the garden did, too.

Chapter 7

"THERE ARE NO INFORMED
PEOPLE HERE"

Hosack, having chosen a site for the garden, finally permitted his pent-up enthusiasm to spill over. He went into a frenzy of spending on livestock, tools, and building materials. He bought a pair of oxen, a horse, a dog, two sows, a wheelbarrow, a plow, a sledge, a harrow, a horsecart, a millstone, spades, rakes, hoes, flowerpots, hay, potatoes, corn, clover, piles of lumber, hundreds of bricks, and endless shipments of dung.

Nothing in his thirty-two years had prepared Hosack for the physical intensity of launching the garden. Thus far, he had spent most of his days hunched over medical and botanical volumes, examining patients, and attending and delivering lectures. His genteel botanizing excursions with Curtis in the fields near London hardly counted as life on the farm. Now he began to encounter the hardscrabble rural lives that kept the city clothed and fed. "Wanted: A Sober, industrious MAN, to manage a Farm on the island of New-York," ran a job offer in the local press. "An Amer'can Farmer, who has no family, will have a preference. Application to be made at the last house on the Bloomingdale Road." The island was swarming with men and women desperate for work, both American-born and recent immigrants. Some of them had come from Ireland, fleeing the aftermath of the failed 1798 rebellion inspired by the French Revolution. Many more had crossed from war-torn England and France to a city bubbling with opportunity but also marred by stark inequality.

Recent immigrants and out-of-work native New Yorkers—these were the men who turned up at the property to ask Hosack for jobs. He hired them gratefully, writing out their last names—Power, Duffie, Flynn, Kelly, Doyle, McGregor, and others—in a bulky leather-bound notebook, along with details about their contracts. At night, the men would have trudged off to sleep in rented rooms at nearby farms or made the long trek down the Middle Road back to boardinghouses in the city. What did they think of the young doctor who was spending so much of his own income on a garden? These laborers likely sympathized with the anti-elitist causes of New York's Democratic-Republicans. Whatever their private opinions about Hosack and his project, they returned each day to help him turn his soggy, rock-strewn patch of Manhattan into an American Eden.

Hosack at the helm meant long hours, as his private medical students could have told the new hired hands. Each year, these students had to sign their names beneath a stringent set of regulations before he would agree to instruct them. They were required to report to his office by nine o'clock each morning, and there they remained for the next twelve hours—unless he released them to attend medical lectures at the college or took them on his rounds at the hospital. Idle chatter among the young men attending his medical practice bothered Hosack, and he explicitly forbade "conversation to be held upon any other than medical subjects." Now he took these exacting standards up the Middle Road to the new garden property. When he found any of his laborers wanting, he docked their pay or sacked them outright.

The men began to bring tentative signs of order to the property. They unloaded fifteen casks of gunpowder and blew up the boulders strewn across the land. They hacked at stumps, drove oxen and horses onward before plows, and spread manure over newly cleared fields. They pulled down an old barn and put up a new one for the livestock, and they dug several wells from which to haul buckets full of water for the fields, the animals, and their own parched throats.

A stone wall more than seven feet high and two feet thick rose up to girdle the grounds. The men set two gates into this wall where it ran along the west side of the Middle Road—one at the southern end of the property and one at the northern end. During his year in London Hosack had passed through the Brompton Botanic Garden's formal entrance dozens of times, and now he ordered his men to construct a snug little porter's lodge much like Curtis's at each of the two new gates. When he was further along with the landscaping and planting and New Yorkers began riding up from town to tour the garden, he would be able to receive them properly.

IF HOSACK HAD HAD A CHANCE to read the epitaph engraved on Curtis's tombstone across the sea, he would have found heartening words for the new chapter in his own life.

While common herbs shall spring profusely wild,
Or gardens cherish all that's blithe and gay,
So long thy works shall please, dear Nature's child,
So long thy mem'ry suffer no decay.

Hosack could have used the encouragement. Aside from Bard, there were few people with whom he could converse in any detail about his garden plans. Burr was probably the best candidate, but he was a busy vice president, and if Hosack did speak with him about the garden at this early stage, he left no record of it. Hamilton knew a good deal about plant-based medicines but was not very knowledgeable about horticulture. Hosack's friend DeWitt Clinton, who had briefly studied botany with Hosack at Columbia, had been in Washington since February 1802 as a senator. Samuel Latham Mitchill knew plenty about botany, but he, too, was serving in Congress. And the truth was that even if Clinton and Mitchill had been around for Hosack to talk to, neither of them nor anyone else in New York was even remotely as expert in Linnaean botany as Curtis, Smith, and Banks. No one in the United States was, except William Bartram in Philadelphia.

Then, in the spring of 1802, on the island of Manhattan, Hosack came face-to-face with European botanical royalty. François André Michaux was one year younger than Hosack. He was very striking, with a Gallic nose and impressive cheekbones. His receding hairline was gradually leaving him with more than his fair share of forehead, but he had been partly compensated for this abandonment by his mass of fluffy curls. He even bore the scars of his devotion to botany; one of his eyes had been injured during a botanizing excursion a decade earlier when he accidentally stepped into a partridge hunter's line of fire.

François André's father, André Michaux, was one of France's most famous botanists. In 1785, Louis XVI had sent the elder Michaux to the wilds of North America. France had been sapped by repeated crop failures as well as by the recent fighting on the American side in the Revolutionary War, and the king had ordered André Michaux to establish a nursery for the collection and shipment of American plants and trees to France. "I imagine myself setting out to the conquest of the new world," Michaux crowed to a friend. He took his teenage son, François André, with him on the voyage. They landed at New York City, bought an English dictionary for a dollar, and went out to botanize. Dismayed by the complete lack of botanical knowledge he encountered, the senior Michaux complained in a letter home, "There are no informed people here, not even amateurs."

François André
Michaux

André Michaux concentrated on sharing his love of plants with his young son. As they hiked the woods and meadows collecting for the French crown, they both fell in love with the "majestic" and "magnificent" trees of the American forests, especially the great oaks. They went south to visit the Bartram family's garden near Philadelphia, where both father and son became friends of William Bartram. Within months of arriving in New York, the elder Michaux sent hundreds of tree specimens to the Jardin du Roi, mostly white cedar (*Chamaecyparis thyoides*). He also sent samples of a fruit he called "Cramberrie," which the locals used for jam. André Michaux purchased twenty-nine acres of land on the New Jersey side of the Hudson, near the Hackensack River, and established a nursery for the French crown there. A local paper described Michaux's new property as "naturally wild and romantic," but a visitor from Connecticut found it "awful, gloomy, lonely, miserable." Michaux seems to have agreed with the latter. In 1786 he left a French assistant in charge and moved with François André to sunny Charleston, South Carolina, where he established another garden that became his home base as he traipsed through the South collecting specimens for the French government.

Their comfortable new arrangement was suddenly upended when, in June 1791, the king himself embarked on an expedition of sorts. Fearful of the French Revolution, Louis XVI fled in disguise with his family toward the border; within hours, however, they had been recognized, arrested, and returned to Paris. Royal interest in botany declined precipitously, as did Michaux's funds. He soon returned to France, where he began several years of such feverish work that by 1800, when he sailed off to botanize for Napoleon in the South Seas, he had largely completed two pioneering books based on his American travels: the *History of the American Oaks*, in French, and the *Flora Boreali-Americana*—the first large-scale catalogue of American plants—in Latin. After Michaux left France on this new voyage, his son was dispatched by Napoleon's minister of the interior with instructions to settle the affairs of the two American gardens his father had founded. The New Jersey garden in particular presented problems. Both Michaux men felt that the French gardener

who had been left in charge for ten years had treated it as his own
private farm. The gardener protested that he had acted patriotically
when he refused multiple offers for its purchase, including one from
Vice President Burr. François André set sail to appraise the situation
for the French government.

When André Michaux had arrived in New York City with the teenage François André in the 1780s, he had pronounced it a complete
botanical backwater. But now, in the spring of 1802, when François
André returned as a grown man, someone directed him to Hosack,
who was "held in the highest reputation as a professor of botany,"
as François André noted in his journal. Their meeting was momentous for Hosack. Michaux was up to date on the latest developments
in European botany, and he had also traversed hundreds of miles of
American wilderness studying the native flora Hosack was so eager
to learn about. Best of all, Michaux was enthusiastic about his garden
project. Hosack soon took his new French friend up the Middle Road
to see the work underway there.

The spirit of William Curtis accompanied the two men as they
toured the land together. Hosack had mined his happy memories
of Brompton in drawing up his plans. The most important plot, of
course, would be the one devoted to medicinal plants. Although it
would include foreign species, he was especially interested in medicinals that grew locally, such as boneset, Virginia snakeroot, and others yet undiscovered. Once the plants were mature, Hosack would
bring his medical students to the garden and show them, as Curtis had once shown *him*, how to connect the botanical remedies
described in their medical textbooks with the plants as they appeared
under cultivation. Hosack also intended to organize several plots so
that they would illustrate the differences between Linnaeus's sexual
system and Jussieu's natural orders. Then the principles that tripped
up his students in the classroom would lodge easily in their minds
as they leaned over the beds of plants reading the Latin and English
labels. They would see the links between species inscribed right on
the land around them. Together he and his students could investigate the medicinal virtues of plants through structural and chemical
analyses, and he would teach them how to prepare pills, powders,

decoctions, and infusions from the plants. They would finally grasp the most important truth he had learned at Curtis's side—that gardens restore both bodies and souls.

Medical botany was Hosack's first love, but he also dreamed of improving the agricultural fortunes of his state and nation. Not far from the medicinal plot, he planned to tinker with crops such as wheat, oats, barley, and sunflowers. He hoped to pinpoint the perfect conditions of sunlight, water, and soil for each species, and also to determine which new foreign species might usefully be introduced to American farms. He had lots of space for these crops, for the rural spaciousness of Manhattan had afforded him a property at least twice as large as the city-hemmed Brompton Botanic Garden. Hosack would be able to conduct agricultural experiments to his heart's content. He was especially eager "to naturalize as soon as possible to our climate the productions of the southern states and of the tropics." His edible crops might one day help farmers feed the United States. He also wanted to help clothe his fellow citizens. He had just planted a species of cotton (*Gossypium herbaceum*) he had acquired from the West Indies; it was originally native to Africa. If this species turned out to thrive on Manhattan, it might well lessen the northeastern states' dependence for raw cotton on the West Indies and, increasingly, on the American South— and perhaps even lessen their dependence on Great Britain for finished cotton textiles.

Hosack's tributes to Brompton continued elsewhere on the property. Curtis had cultivated a healthy respect for poisonous plants, and Hosack frequently repeated Curtis's remarks about how important it was that doctors and farmers be able to recognize these species in the wild. Hosack, too, planned to devote an instructional plot to plants that could poison humans and livestock, such as wolfsbane (*Aconitum napellus*) and nightshade (*Nicandra physalodes*). Eventually he would bring his herbarium up from the city, with the Linnaean duplicates from James Edward Smith as the crown jewels of the collection. Students inspired by the garden's living plants would then be able to pore over specimens that the world's most famous botanist had mounted and labeled with his own hands.

Despite the enormous size of his property, Hosack was working to

encircle it with forest trees and shrubs. This was another nod to Brompton, and of all the features of Hosack's design, François André Michaux found this one the most personally exciting. When the younger
Michaux published the most important work of his life, *North American
Sylva*, a decade later, he recalled how intrigued he had been to find
Hosack raising a pretty little tree that generally preferred a milder climate: loblolly bay (*Gordonia lasianthus*), which Michaux loved for "the
luxuriance of its vegetation, the beauty of its flowers, and the rich-
ness of its evergreen foliage." In the introduction to his book, Michaux
issued one of the earliest pleas for American forests, helping to launch
the forest conservation movement in this country; Thoreau would
consult it as he wrote *Walden*.

Michaux seconded Hosack's enthusiasm about the garden
property—it was indeed well suited to growing a great variety of species. Hosack, for his part, enjoyed his new friend's company so much
that he took him to see several of the other establishments to which
he was devoting his energy. Michaux found the New-York Hospital
spacious and clean but thought the thin beds looked horribly uncom-

Loblolly bay from Michaux's
North American Sylva

fortable. They also toured the state prison on the banks of the Hudson near the village of Greenwich, just north of New York, where Hosack somehow found time in his packed schedule to serve as an attending physician. Michaux's impression here was of a tranquil and shipshape establishment; he thought the prisoners seemed content to perform the various jobs assigned to them, such as making nails. (The occasional prison break suggests Michaux was off the mark.) Hosack bought some of the prison nails for his men to use on the new barn they were building at the garden.

IN THE SPRING OF 1802, just around the time Hosack and Michaux met, Rembrandt Peale arrived in New York with the second set of mastodon bones the Peales had retrieved. He set them up for public exhibition practically on Hosack's doorstep. A short walk up Broadway and a right on Wall Street took Hosack to the exhibition in the old assembly room at City Hall. The mastodon was finally back where Hosack thought it had always belonged, although he was surely irritated all over again that the Peales were basking in glory and fortune that should have gone to a New Yorker. No one else seemed to mind. Within two days, Rembrandt had raked in $340. Some visitors did complain about the number of wooden replacement bones in the incomplete skeleton, and reports began to circulate that the entire thing was a fraud. Rembrandt quickly published an ad to dispel the profit-killing rumor, quipping that the presence of a few wooden bones "can only be objectionable to those whose dullness of intellect cannot comprehend that this animal must have resembled all others in one circumstance at least, viz, *that one half of him was the exact counterpart to the other.*"

As crowds continued to visit the mastodon, Rembrandt Peale kept postponing his departure for Europe with his brother Rubens. Finally, in June, he announced to the public that they had booked passage on a ship bound for England. By autumn the brothers were breakfasting with Hosack's old mentor Sir Joseph Banks. Meanwhile, in Philadelphia, their father, Charles Willson Peale, continued to collect specimens and mount them for display in his museum. That September, he took out a lost-and-found ad for a "Red and Blue Macaw that flew from the State House Garden last week." Freedom appeared preferable to stuffed immortality. Peale announced that whoever captured his macaw and brought it back would "be entitled to a sight of that wonder of the western world; the skeleton of the Mammoth."

It was galling to Hosack that, at the late date of 1802, New Yorkers were still so dependent on Philadelphia for America's scientific enlightenment and artistic achievement. When it came to creating charitable institutions, however, New York was probably busier than any other city in the United States. There was, for example, the Society for the Information and Assistance of Persons Emigrating from Foreign Countries, which had given aid to some of the *Mohawk* passengers in the autumn of 1794. Also in the works was a charity for the sailors who had spent their lives ferrying such travelers back and forth across the oceans. It had been the pet project of a retired seaman named Robert Richard Randall, the son of a well-to-do Caribbean pirate. In June 1801, weeks before his death, Randall had hired Alexander Hamilton to draw up a will that left a valuable bequest to fund a home for "aged, decrepit and worn out sailors." People praising New York's generous spirit could also point to the hospital, the almshouse, the Society for the Relief of Distressed Debtors, and the public pharmaceutical dispensary—all organizations in which Hosack was active. The counterfeiting of medicines, false claims about the efficacy of patent remedies, and highway-robbery prices were such a problem that the city's doctors had banded together in 1791 to found a public dispensary to "keep the poor from being preyed on by merciless and unfeeling quacks." At the dispensary, poor New Yorkers could buy physician-certified medicines at reasonable prices so they would not be forced to seek emergency help at the hospital or the almshouse.

For women who could not afford a physician to deliver their babies at home, Hosack had recently helped found an obstetrics clinic—the Lying-In Hospital. He was also involved with a fever-quarantine hospital that had recently been created in a converted mansion on an East River estate called Bellevue. Yet for all this frenetic compassion, New

York still lacked most of the cultural and scientific institutions befitting a great city. To be sure, there was the Park Theatre on Park Row (formerly Chatham Row), and up the Middle Road Hosack was building his own answer to the Brompton Botanic Garden and the Linnean Society combined. But New York still had no equivalent to the Royal Society—or rather nothing like Philadelphia's more suitably democratic American Philosophical Society. New York also had no museum of art, while Philadelphia, of course, had Peale's Museum, where he displayed paintings along with his natural history specimens.

Hosack belonged to a generation of New Yorkers attempting to give the city more to be proud of. In the summer of 1802, he was invited by friends to help found the city's first museum of fine arts. The Livingston family was leading the effort. Robert R. Livingston, Jefferson's minister to France, was joined there in 1803 by James Monroe, a presidential envoy sent to aid Livingston in negotiations over French-owned territories in North America. Even with the glories of Paris at his feet and a weighty ministerial portfolio on his desk, New York was never far from Livingston's mind. When he was taken to see an astonishing collection of Greek and Roman sculptures looted from Italy by Napoleon, Livingston hatched an idea. He wrote to the mayor of New York City—his younger brother Edward. What if he were to ship home plaster casts of the most impressive sculptures in Napoleon's collections? Then New Yorkers could see the world's greatest art without having to spend six weeks at sea to do it. His brother was delighted with the idea, and in a public announcement of the plan for the new museum, Mayor Livingston slyly invoked the degeneracy debate that had sent Jefferson looking for a giant moose. The American "climate will not be less favorable" to the growth of the fine arts, the mayor said, "than that in which they have heretofore flourished"—Europe.

It was exactly the kind of project Hosack loved. He signed up as a founding member, as did DeWitt Clinton and many other men in Hosack's circle. Meanwhile, in Paris, Minister Livingston arranged for plaster casts to be made of great sculptures including the *Laocoön*, the *Apollo Belvedere*, and the *Dying Gaul*. These copies set sail from France and were soon installed in a room in City Hall. In a show of gratitude, the members elected Napoleon to honorary membership in the New-York Academy of Fine Arts. The emperor sent them twenty-four lavish volumes on Italian art. But Minister Livingston sent Hosack something from Paris even more exciting than art. He sent seeds.

———— ⬚⬚ ————

FIVE MILES UP the Bloomingdale Road from Hosack's garden, Alexander Hamilton nursed his broken heart. In May 1802, he had recorded a grim entry in his account book. "Expence (Philips funeral &c) 266.11." The next month, Eliza gave birth to a boy, their eighth child. They named him Philip.

The Hamiltons' new country house, the Grange, was completed that summer. By fall, when the goldenrod glowed in the island's meadows, Hamilton was distracting himself with plans for a garden at the Grange. The architect of the national banking system pored with delight over tomes on trees and soil. He asked some of the farmers who worked the nearby land for practical advice. As he listened to them, Hamilton developed a deep respect for their hard-won intimacy with the island, passed down to them in Dutch, English, and then American accents. He told Eliza, "I may yet live twenty years, please God, and I will one day build for them a chapel in this grove."

Hamilton also consulted old political comrades whose horticultural knowledge he admired. Writing to the Philadelphia lawyer and farmer Richard Peters, he half-joked, "A disappointed politician . . . is very apt to take refuge in a Garden." Peters teased in reply that Hamilton was too political an animal to disappear into his garden forever: "A strong Passion for horticultural or rural Persuits has sometimes lulled, but has never yet eradicated the stronger Propensities for political Operations." Hamilton begged Peters for advice about what to plant and how to keep it alive, quipping that as a novice gardener at the Grange, he was in a position "for which I am as little fitted as Jefferson to guide the helm of the UStates."

Hamilton was spending as much time at the Grange as he could steal away from his legal practice. His route out of the city and up the Bloomingdale Road took him near the western edge of Hosack's garden. At about the halfway point of his journey, Hamilton sometimes reined his horse to a stop and went in search of Hosack. The garden was not very far along, but Hosack loved explaining his plans to anyone who would listen, and he must have been especially delighted to show Hamilton around the property. The family physician now began to double as a horticultural adviser. Hosack gave Hamilton bulbs and cuttings for the Grange. Around this time, Hamilton drew a sketch for Eliza of ornamental plantings inspired by decorative beds Hosack had planted at the botanical garden. At the very center of the Grange's flower garden, Hamilton indicated, he wished to have a smaller circular garden planted. It should be eighteen feet in diameter and contain nine tulips, nine lilies, nine hyacinths, and nine of whatever other flower the Grange's gardener thought most suitable. Wild roses just outside the circle would furnish a soft border, setting off the formality within.

Perhaps it was these visits with Hosack that gave Hamilton his growing measure of horticultural confidence. He issued detailed orders—for example, that "a few waggon loads" of compost should be spread on the grounds and a ditch should "be dug along the fruit garden and grove about four feet wide." This ditch was a *ha-ha*, an ingenious landscaping trick that kept animals out of the garden and at the same time preserved an unbroken view across the land. Hamilton also decided that "a few dogwood trees, not large, scattered along the margin of the grove would be very pleasant." But he didn't want this work to "interfere with the hot bed."

Hamilton also wrote to his friend Charles Cotesworth Pinckney of South Carolina, the Federalist he had favored for the 1800 presidency. Pinckney was an avid plantsman, and Hamilton asked him to send watermelon and muskmelon seeds, which Pinckney did in March 1803 via the brig *Charleston Packet*. The seeds, he told Hamilton, were the same kind he had previously sent to Martha Washington at Mount Vernon. He admitted they hadn't done well there and speculated that they were adapted for a warmer climate than Virginia's, which didn't bode well for their fate in New York. Pinckney told Hamilton that he planned to send several sets of flower seeds by another ship, settling on species that would produce bold splashes of color. He chose coral bean

(*Erythrina herbacea*), for example, a plant native to the Deep South that lures butterflies and hummingbirds in the spring and early summer with its long red blossoms. If it managed to flower at the Grange, the Hamilton children would need to be warned about its tempting red seeds, because they were poisonous. Pinckney also promised seeds of a "beautiful purple convolvulus" (*Ipomoea purpurea*)—morning glory. With any luck, the landscapes at the Grange would be vivid with color for many months each year.

Even with his mind on flowers and hotbeds, Hamilton never stopped thinking about politics. This inveterate writer of political pamphlets couldn't resist seeing the land as another blank page. He sketched the nation's founding history right there in the front yard of the Grange. He planted thirteen sweetgum trees, one for each of the original states. Hamilton was modest about his gardening skills, but he knew just as well as Jefferson and Burr what wondrous powers lay hidden in American plants.

The Grange as it appeared in the late nineteenth century,
with Hamilton's sweetgums in the foreground

IT WAS HOSACK'S OWN FAULT that so many people and organizations were clamoring for his attention these days. "Flora does not
receive from me those attentions which are due to her," he lamented
to a Danish-Norwegian botanist named Martin Vahl. Vahl, a former
student of Linnaeus, had heard about the new American garden and
shipped Hosack a small collection of dried plants. Hosack felt honored by the gesture. He promised Vahl that he would soon send over
some specimens from "this almost unexplored country." He confided
to this brother botanist that he longed to devote himself to his garden.
Instead, he was penned indoors in the lecture hall, the sickroom, the
surgical clinic. He needed more pairs of hands.

He found them. Two of Mary's nephews, Caspar Wistar Eddy and
John Eddy, came to work with him at the garden. John, who was nine-
teen, already lived in New York, while Caspar, John's twelve-year-old
cousin, came from Philadelphia. Hosack trained the boys in Linnaean
botany, thus passing on to these young Americans what the British heirs
of Linnaeus had taught him in London. He made sure they understood
that no species was too humble for their regard and that they should
write down exactly when and where they found each plant. He taught
them to collect as much of the plant as possible—roots, stem, leaves,
flowers, seeds—instead of tearing at them like heedless children. He
showed them how to carry the specimens back to the garden safely. He
explained which ones they should set aside for planting and which they
should press, dry, and mount to go into the herbarium.

Then he sent his nephews out into the sunlit world. Patches of green
old Mannahatta still sprouted between the boardinghouses and shops,
behind the mansions and the shacks, on the beaches below the prison
and the piers. Caspar did most of the collecting, while John excelled at
preparing and mounting the plants. Hosack lectured at Columbia and
walked the halls of the hospital and the prison. The boys wound their
way among raspberry brambles and wild roses. As they scouted for
specimens around the city, they drew upon the stored-up wisdom of
their uncle. They brought him back a sample of boneset—his prized

fever remedy—and another of a pretty yellow buttercup (*Ranunculus acris*) with medicinal powers that Curtis had also been growing at Brompton. Caspar collected branches of white oak (*Quercus alba*), scarlet oak (*Quercus coccinea*), and dogwood (*Cornus sericea*). He found a fern that Linnaeus had named *Pteris aquilina*, eagle fern, for its winglike fronds. He brought Hosack a branch of a shrub called inkberry (*Prinos glaber*), whose dark fruits were sometimes used to dye cloth. And in some corner of the city where the boys' boots could still sink in boggy earth, a place not yet hardened over by streets and sidewalks, Caspar found the damp-loving *Saururus cernuus*, known familiarly as nodding lizard's tail.

The boys ranged farther and farther from the city. Near the village of Harlem, Caspar harvested rabbit's-foot clover (*Trifolium arvense*). The boys ventured across the Hudson to Hoboken and found primrose-leaved violets (*Viola primulifolia*). Near Weehawken, it was wild grapes (*Vitis labrusca* and *Vitis vulpina*) and viburnum (*Viburnum dentatum* and *Viburnum lantanoides*), and on the Paulus Hook beach, Caspar collected a beautiful flowering vine called sea pea (*Pisum maritimum*). It was probably on the Palisades that he stumbled on a carnivorous plant called side-saddle flower (*Sarracenia purpurea*). They went east to Long Island, and near the town of Brooklyn, they found sea lavender (*Statice limonium*).

John Eddy was deaf. He had lost his hearing at the age of twelve, during an attack of scarlet fever. He could not hear the waves beating against the Paulus Hook ferryboat, the whisk of dry grass around his ankles as he crossed a field, or the snap of a branch between his hands. Hosack was teaching him botany by means of "an artificial alphabet formed by the fingers," as he wrote Martin Vahl. He was illustrating John's lessons with the volumes he had brought home from London, and at Hosack's side, John was discovering a world that eluded most New Yorkers. The city's thick green backdrop divided itself into the tidy compartments of his uncle's mind. Herbaceous, woody; perennial, annual; poison, remedy. Hosack encouraged John to assemble a little herbarium of native plants. They could use the specimen sheets prepared by Linnaeus as models, and then they would send John's handiwork to Vahl in Denmark.

John chose plants that captured the whole range of New York's flora, from tiny woodland flowers to snippets of great trees. He dried and pressed dozens of specimens, then positioned them on clean sheets. The crisp, starry flowers of wood anemone (*Anemone quinquefolia*), the faded pink of Carolina roses (*Rosa carolina*), a twig of white oak (*Quercus alba*), the floppy white blooms of a laurel that Hosack knew as rock rose (*Rhododendron maximum*). Hosack wrote a friendly letter to Vahl and enclosed it with the collection. He said both his nephews showed great promise as naturalists, but John had made "very astonishing progress." He wanted to spread the word of John's accomplishments to botanists around the world, because his story might inspire others. Hosack was touched that John took such "delight in Botanic pursuits that it very much diminishes his misfortune in being deprived of the many sources of pleasure which pass thro the channel of hearing."

In turn, Hosack was reaping the fruits of John's newfound passion—and of Caspar's, too. The boys circled back to him from their ramblings with their stacks of native specimens to be planted or dried. Hosack took out his leather-bound memorandum book and wrote down the Latin names of the ones he could identify. When he needed help, he scoured the illustrations in Curtis's *Botanical Magazine* and the descriptions in Linnaeus's *Systema Naturae*. The clay pots he had set out in the garden began to fill up with plants. He bought more.

Hosack was smitten with botany. When Mary gave birth to a son that winter, their first since Samuel's death, he named the baby James Edward Smith Hosack, in a nod to his London friend, founder of the Linnean Society. He went back up the Middle Road and walked his land until his feet knew its shifting textures: dense and damp, or skittering away in dust and pebbles. He learned the angle of the light across his fields as the sun sank toward the Hudson. The kitchen garden and the fruit nursery would go here; there, the medicinal plot. He would put a path up the hill. He made adjustments to his plans, checking his Brompton memories and his botanical volumes against the plants growing in their beds and pots—or *not* growing, not as they should. Hosack was still learning from Curtis, was coming to know in his bones what Curtis had already lived every day for decades when they had first met: the pleasure and exhaustion of raising a huge garden out of the bare earth.

For five years, Hosack had been trying to convey in words an imagined landscape of scents, colors, and healing powers. Now it was finally coming into being, outside his solitary mind. At first, to travelers taking the stagecoach up the Boston Post Road, the land that jostled past in the distance, with its tiny figures walking behind tiny plows and oxen, looked like any of the surrounding farms. Then the walls of an enormous building began to rise at the northwest corner of Hosack's property.

IN JUNE 1803, Hosack gave a public lecture at Columbia. With his rich voice he painted a seductive vision of the garden he was building. He explained that he would be training the nation's new doctors there, and he spoke about the thousands of plants he was collecting from around the world—medicinal, agricultural, commercial, and ornamental. He told his listeners he intended to collect every known species native to the continent, and, as explorers discovered new species, he would safeguard specimens of each one at the garden. He reminded his audience of the critical medicines and crops the nation was forced to import each year from "distant quarters of the globe." These very plants, Hosack promised, would soon be growing less than four miles north of where they now sat.

The speech was a success. A positive review ran in several newspapers, including Hamilton's *New-York Evening Post*, which praised Dr. Hosack's garden for being the "first attempt of the kind in this country." In Europe, noted the article, people understood the importance of botanical gardens. It was time to support this "public spirited" American man, and later, when the garden was ready, it should be purchased by the federal government. Hosack couldn't have gotten better press if he had written it himself.

At the garden, his men were nearly finished with the new building. Few New Yorkers had ever seen anything like it. More than sixty feet long and twenty feet tall, it sat atop the highest point on his property, facing south. Its white façade had seven graceful arches, each stoppered with glass windows. Inside the building, rows of graduated shelves—known as *stages*—would accommodate dozens of potted plants. Under these stages, Hosack's men were laying hidden flues that would carry heat from a stove tucked out of sight. Wide walkways would run the length of the greenhouse, allowing access to the plants.

Curtis had had a greenhouse at Brompton, of course, but Hosack also had a model much closer to home. To the west of Philadelphia, a wealthy private collector named William Hamilton had built just such a greenhouse to shelter his extraordinary collection of exotic plants. By the late eighteenth century, his estate, The Woodlands, was nearly as famous among plant-lovers as the Bartrams' garden. Hamilton hosted many distinguished guests at The Woodlands, including Thomas Jefferson, who thought it was the best garden outside Britain. André Michaux and the teenage François André Michaux had also visited The Woodlands. It was after one of these visits, in fact, that François André had been shot in the eye by a partridge hunter in 1789.

Hosack knew the estate well—probably from his year studying with Benjamin Rush. In July 1803, Hosack wrote his friend Thomas Parke in Philadelphia to say that his greenhouse was nearly identical to the one at The Woodlands. There, the plants were arranged on stages rising to either side of visitors, so that as they strolled they seemed to be traversing a valley flanked by green slopes. Even the view from outside the greenhouse was carefully curated. Seen from the front, the tallest plants stood at the center of the windows, with the others placed in descending order, thus forming "a miniature hill clothed with choice vegetation," one visitor wrote. As Hosack's men completed his own new greenhouse, he enlisted Parke's help. "My collection of plants is yet small," he told Parke. "I have written to my friends in Europe and in the East and West-Indies for their plants. I will also collect the native productions of North and South America." He asked Parke to contact William Bartram: "Request him to send me a catalogue." Hosack's friends and correspondents responded by burying him in specimens. Down at New York Harbor over the next months, he wrote his signature on bills of lading and then took possession of boxes and crates filled with seeds, cuttings, and dried specimens. He hauled China,

Egypt, India, and the Cape of Good Hope up the Middle Road to the garden. He and his men planted seeds and arranged clay pots on the stages in the greenhouse, and Hosack worried over his young plants as though they were children.

It was around this time that he finally chose a name for the garden. Britain had inspired him to create it, so he settled on a name that recalled his Scottish ancestry and the serene seascape he had discovered during his trip to northern Scotland in 1793. He would call it the Elgin Botanic Garden.

Hosack was not the only one changing the face of the island these days. When he drove from his house to the garden, he passed crews working on a construction site that had recently been slashed into the earth on the east side of Broadway. An elegant new City Hall was slowly going up. The design was by Hamilton's architect for the Grange, John McComb Jr., who, together with a Frenchman named Joseph-François Mangin, had won the city's architectural competition. Once complete, the new City Hall would have a two-tiered marble façade with a handsome portico out front and a cupola high above. By design, the rear of the building was left bare of marble. No one but farmers and laborers would think of living to the north, except for wealthy families on their country estates, and they would be too far from City Hall to mind the raw look of its back elevation.

By the end of 1803, New York had a new mayor: DeWitt Clinton. Clinton had sorely missed New York while living in Washington, but his return to the city was also a piece of political strategy. He had come to the conclusion that he could wield greater influence—even on the national scene—as mayor of New York than as a senator. Since Clinton would not be able to move into the future City Hall anytime soon, he set up his office in the old City Hall on Wall Street. Hosack was fortunate. Having his best friend in the mayor's office might well prove useful for the garden.

⁓⸗⸗⸗⸗⸗

NEW PLANTS WERE ARRIVING all the time. In response to Hosack's written requests, his correspondents continued sending him specimens they had collected themselves on their world travels, as well as others they had received at their own botanical gardens in Europe and South America. Some sent paper packets that spilled out mysterious seeds when Hosack opened them. Others sent wooden cases lined with earth in which seedlings grew—except when they arrived dead, casualties of the ocean voyage.

Even as he accumulated these plants from around the world, however, most of Hosack's own continent lay tantalizingly beyond the reach of his eager pen. He knew from the works of naturalists like John and William Bartram that the native peoples of North America were adept in the use of medicinal plants, and he suspected—just as Franklin had, half a century earlier—that there must be thousands of species unknown to European settlers. In his recent public lecture at Columbia, Hosack had spoken of his dream of amassing and studying the native plants of North America. But he was a doctor, not an explorer. Privately, he was forming another plan. In Philadelphia, his mentor Benjamin Rush had just been meeting with a young captain named Meriwether Lewis. At President Jefferson's request, Rush was helping Lewis prepare for a journey across the continent.

This wasn't the first expedition Jefferson had backed. In 1793, he had helped the American Philosophical Society organize a western expedition for André Michaux, but then Jefferson himself had accidentally sabotaged it by involving the French minister to the United States, Edmond-Charles Genêt, who wanted to use Michaux as a French spy. Michaux's expedition was abandoned. Then, in 1802, Jefferson read an account of a recent British expedition to the Pacific Northwest that made him more anxious than ever to secure a route to the Pacific Ocean for the United States—preferably an all-water route via rivers and lakes, for ease of future travel and transport.

This time, Jefferson placed an American in charge of his expedition, which would be called the Corps of Discovery: Meriwether Lewis, who as a young man had lobbied Jefferson unsuccessfully to allow him to join what had become André Michaux's aborted trip in 1793. Since those days, Lewis had risen to the rank of captain in the United States Army before moving to Washington in April 1801 to take up a post as Jefferson's personal secretary. The two men lived and worked side by side in the cavernous President's House. Lewis worshiped Jefferson, telling a friend how thrilled he was to be serving a man "whose virtue and talents I have ever adored, and always conceived second to none." For his part, Jefferson began to feel certain that Lewis was the right man to assemble a corps of explorers bound for the Pacific, and also to oversee the complicated preparations for their expedition.

So it was that in the fall of 1802, the days became a blur of maps, treatises, debates, meetings, and errands, as Lewis and Jefferson tried to prepare for a voyage whose precise route and duration were unpredictable. They knew that Lewis would depart with his team from St. Louis and travel up the Missouri River to the villages of the Mandan people in what is today North Dakota. After that, the map went blank. They could only speculate about the hardships, the threats, and the friends and foes Lewis and his men might encounter. Lewis did his best to calculate which items to purchase and in what quantities, and how to pack them for transport on the boats—or on the backs of his horses and his men, should an all-water route prove elusive. They would need enormous stores of food, of course, and weapons, ammunition, camp supplies, and gifts for the native peoples he and his men would meet as they floated along the Missouri and hunted on its banks. He also stockpiled pens, ink, pencils, and paper, because Jefferson wanted him to describe and draw the animal, vegetable, and mineral curiosities he encountered on the voyage. They agreed that Lewis would bring back what specimens he could manage to preserve and carry, so that Jefferson and his fellow naturalists back east could see for themselves at least some of the treasures of the American West.

Although Lewis had learned a good deal about agriculture while managing his family's Virginia plantation, he had little to no formal training in botany. Jefferson, by contrast, had amassed a very considerable stock of botanical knowledge over his decades as a devoted gardener and scholar of natural history. In his *Notes on the State of Virginia*, written twenty years earlier, he had reflected at erudite length on the medicinal, edible, ornamental, and commercially useful plants of his state. Of the pecan nut, for example, Jefferson had written, "Were I to venture to describe this, speaking of the fruit from memory, and of the leaf from plants of two years growth, I should specify it as the Juglans alba, foliolis lanceolatis, acuminatis, serratis, tomentosis, fructu minore, ovato, compresso, vix insculpto, dulci, putamine, tenerrimo. It grows on the Illinois, Wabash, Ohio, and Mississippi." Now, in the halls of the President's House and the gardens of Monticello, Jefferson taught Lewis the fundamentals of Linnaeus's succinct system of naming and classification, and he urged him to keep careful track on his travels of "the dates at which particular plants put forth or lose their flower, or leaf."

In May 1803, Jefferson sent Lewis to Philadelphia for advanced scientific study with members of the American Philosophical Society. It would likely have surprised Jefferson to learn that New York was home to one of the finest young Linnaean botanists in the United States; at any rate, Philadelphia's concentration of scientific expertise across all fields was unmatched. Jefferson directed Lewis to Benjamin Smith Barton, who had just published the nation's first botany textbook. Barton further acquainted Lewis with botanical nomenclature and showed him how to prepare and label plant and animal specimens. As they talked, Barton became so excited about the expedition that he invited himself along, although at the last minute he changed his mind and stayed home. He was a sickly man and not at all cut out for a wilderness trek. But between them, Barton and Jefferson produced a fine botanist in Lewis.

Rush met with Lewis to give him medical advice on how to treat fever and other ailments. He also sold him six hundred of his purgative "Thunderclapper" pills (which contained mercury) and hundreds of doses of various other medications. Rush gave Lewis a list of questions about the medical and cultural practices of native peoples and implored him to keep track of what he learned. Finally, before leaving Philadelphia, Lewis sought out another doctor whom Jefferson had recommended: Caspar Wistar, Hosack's brother-in-law. Wistar was a professor of anatomy at the University of Pennsylvania, but he was also the country's leading expert on fossils. He and Jefferson enjoyed an affectionate friendship anchored in their shared fascination with the fossil evidence of the continent's extinct—or perhaps simply undiscovered?—creatures. In January 1802, Wistar had sent Jefferson a minute description of the mastodon that Peale had just put on display at his museum, to which Jefferson replied with his own reflections on what Peale had recently told him about the condition of the skeleton. It was Wistar, in fact, who had alerted Jefferson to the publication of the British Pacific expeditionary account, thus spurring the president to organize the American expedition led by Lewis. When Lewis met with Wistar in May 1803, they discussed fossils and mastodons.

By late June, Lewis had settled on a fellow officer for the expedition: William Clark, a trusted old US Army friend. In a great stroke of luck, Jefferson received word in early July of Livingston's and Monroe's successful conclusion of the Louisiana Purchase in France, which meant that for a good stretch of their journey westward, Lewis and Clark would be traversing territory newly in the possession of the United States. After several more months of advance work, Lewis finally met up with Clark in the Indiana Territory in October. They traveled together to St. Louis and wintered there with their men. On May 22, 1804, they all pushed off in their canoes, bound for the Pacific Ocean. Two volumes on Linnaean botany, stowed in Lewis's luggage, went floating along the Missouri with them. In Washington, Jefferson began the long wait for news and specimens, as did Barton, Rush, Wistar, and the other members of the American Philosophical Society in Philadelphia. In New York, Hosack was waiting, too.

Chapter 8

"H–K IS ENOUGH, AND EVEN
THAT UNNECESSARY"

In MAY 1804, A LIVE OSTRICH LANDED ON MANHATTAN ISLAND. It was a young male more than six feet tall, with enormous eyes, long lashes, and taloned legs that might kill a man with one kick. An émigré Frenchman named Monsieur Delacoste (his first name has not survived) had gone to great expense to procure the ostrich from Africa. He was displaying it at 38 William Street. For a small fee, New Yorkers could inspect the ostrich, which could itself probably inspect some uncomfortably familiar feathers swaying on the hats of wealthier ladies.

The ostrich exhibit was part of a bigger plan. Delacoste was founding the city's first scientific museum of natural history. Aside from the ostrich, his animal specimens were dead and stuffed, but many had shock value nonetheless. They were alien creatures hinting of wild lands far from the tame island of Manhattan: a jaguar, a seven-foot-long viper, a toucan, an anteater, a three-toed sloth with its baby. Like Peale, however, Delacoste was a completist when it came to collecting from nature and was therefore displaying items such as a "hair-ball found in the stomach of an Ox." As committed to plants as to animals, Delacoste had assembled exotic botanical specimens most New Yorkers had never seen before—banana, cacao, guava, papaya, and dozens more. Perhaps it took a Hosack to delight in the extensive collection of varieties of wood.

Hosack warmly welcomed Delacoste's initiative. In London, Hosack had frequented the Leverian Museum, a breathtaking private collection of animals, plants, and gemstones. Thus far, the closest thing the United States boasted was Peale's Museum, but with Delacoste's help New Yorkers could now dream of surpassing Philadelphia in natural history. On June 4, along with two Columbia colleagues, Hosack signed an open letter in support of Delacoste's museum, the Cabinet of Natural History. "We cheerfully express our approbation of the establishment which you contemplate," they wrote. "It is the first institution of *this nature* which has been established in this city." Delacoste paid for the letter to run regularly in the local papers for weeks to come.

FOR THE MOMENT, HOWEVER, Peale's reputation as the premier American curator was as tightly intact as a sealed display case. In fact, the day Hosack and his colleagues signed the Delacoste letter in New York, Peale was attending a jolly meal in Washington with President Jefferson and the world's greatest explorer. Baron Alexander von Humboldt was the very model of the dashing adventurer—brilliant, vivacious, handsome, and rich. He was thirty-four years old, two weeks younger than Hosack. After five years exploring South America, the celebrated Prussian naturalist had arrived in Philadelphia in May 1804 with two friends in tow: a handsome young man from Quito (in today's Ecuador) named Carlos Montúfar and a French botanist named Aimé Bonpland. The men quickly made the acquaintance of Hosack's brother-in-law, Caspar Wistar, and they also found their way to Peale's Museum.

Peale was smitten. Humboldt was "without exception the most extraordinary traveller" he had ever met. The man could hold forth with ease in multiple languages on astronomy, botany, mineralogy, natural history, philosophy, and more. Peale struggled to find adequate praise for his new friend: "He is the fountain of knowledge which flows in copious streams—to drop this metaphor to take another, he is a great luminary defusing light on every branch of science." Peale went with Humboldt to Washington to meet Jefferson. They left Philadelphia on May 29, and as the stagecoach bounced south Humboldt entertained Peale, Bonpland, and Montúfar in rapid-fire, heavily accented English, Spanish, and French. He told Peale about a miraculous warmth-inducing plant that saved men's lives in the snow-capped South American mountains (later Peale could not recall the plant's name). "The native Indians when traveling and benighted, take the leaves and make a bed & covering themselves with some bowes [boughs] they sleep comfortably warm amidst the Snows."

Peale was excited to take Humboldt up to the cupola of the grand new Capitol in Washington, where they could look out over the growing city. When they met with President Jefferson, Peale gave him a novel gadget that he had lugged along on the stagecoach. This was a polygraph, invented by a business partner of Peale's named John Hawkins. A person holding a pen attached to this contraption could write out a document while another pen moved in concert, thus producing a faithful copy. Peale had been collaborating with Hawkins on improving its mechanics, and he now presented Jefferson with the new prototype. The president loved it. "I only lament that it had not been invented 30 years sooner," he wrote Peale afterward.

A few days later, the party of travelers gathered at the White House for dinner. Jefferson and Humboldt traded stories about their explorations of natural history and human cultures. The president had dispatched the Corps of Discovery into the North American continent only weeks earlier. As Lewis and Clark were paddling their canoes away from the only civilization they knew, the returned explorer leaned over Jefferson's table and described the miraculous sights he had seen in South America—electric eels, steaming volcanoes, ancient monuments with intricate calendars carved into their surfaces, and more. Peale was delighted that the dinner guests offered a toast not to party politics but to natural history.

The next night Peale and his party joined Secretary of State

Madison and his wife, Dolley, for a "sumtious" dinner paired with wonderful wines. During the meal Peale's dentures broke, so while Humboldt entranced the Madisons with his stories, Peale raced to a gunsmith and got them repaired. He was back at the table in less than half an hour. Rattled by the mishap, he took to keeping a spare set in his pocket and often "changed my Teeth while at Table without the Company Observing what I was about by holding a handkerchief before my mouth." He was under the impression that the process looked to other people like he was simply "picking something from my teeth that might be troublesome."

After dinner, Humboldt came to Peale's room to report that he had been conversing privately with Jefferson about the museum. Humboldt said he had asked why the federal government hadn't purchased Peale's collection yet; in Europe, such institutions were funded by the state. According to Humboldt, Jefferson had said "it was his ardent wish." Peale was over the moon that a naturalist of Humboldt's stature had been privately lobbying the president on his behalf. When they returned to Philadelphia on June 21, Peale felt that the three weeks they had spent together had passed as quickly as three days. Humboldt began to plan a trip to New York. Peale felt tempted to tag along.

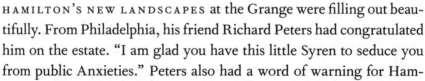

HAMILTON'S NEW LANDSCAPES at the Grange were filling out beautifully. From Philadelphia, his friend Richard Peters had congratulated him on the estate. "I am glad you have this little Syren to seduce you from public Anxieties." Peters also had a word of warning for Hamilton. "Make your little Farm your Plaything—but see that you have other Business, that you may afford to pay for the Rattle." Hamilton's law practice was humming, but he still managed to stay involved with the most mundane garden tasks. In October 1803, when the edge of autumn was tinting the hardwood forests of upstate New York, he had written to Eliza from the hamlet of Claverack, forty miles south of Albany, with detailed instructions for a new compost bed. It should "consist of 3 barrels full of the *clay* which I bought 6 barrels of *black mould* 2 waggon loads of the best clay on the Hill opposite the *Quakers*

place this side of Mrs. Verplancks (the Gardener must go for it himself) and one waggon load of pure cow-dung."

Hamilton's country refuge absorbed and delighted him. "You see I do not forget the Grange . . . nor anyone that inhabits it," he told Eliza. "Accept yourself my tenderest affection." When he came home from his trips, the house rang with laughter. Hamilton's son James later recalled that his "gentle nature rendered his house a most joyous one to his children and friends." The handsome, vivacious father sat beside his delicate daughter Angelica at the piano, accompanying her while she sang. As early summer 1804 enveloped Manhattan in clouds of greenery, Hamilton savored peaceful days at the Grange with his family. He invited friends to dine and dance at the mansion, and as he held court on his piazza, he was silhouetted against "the distant outlines of the variegated landscape of hill & dale, oceans & rivers," one of his sons would later recall. These were, his son wrote, "the last sunny days." On June 27, 1804, Aaron Burr challenged Alexander Hamilton to a duel.

The angry exchange of letters leading to the challenge had begun ten days earlier, after Burr learned that Hamilton had criticized him at an Albany dinner party held in March. At the time of the party, Burr had been running for governor of New York against Morgan Lewis, and the conversation that night turned to politics. According to a dinner guest named Charles Cooper, Hamilton had spoken scathingly of Burr. Cooper, challenged publicly by Hamilton's father-in-law, Philip Schuyler, to back up his claims, protested in print that he could "detail to you a still more despicable opinion which General Hamilton has expressed of Mr. Burr." By early summer, Burr had seen the Cooper allegations for himself, and on June 18 he asked his friend William P. Van Ness to deliver a copy to Hamilton, along with a personal note in which Burr demanded "a prompt and unqualified acknowledgment or denial" of Cooper's claim about Hamilton's abusive remarks. The exact content of those remarks remained murky to those who had not been at the dinner party. Over the next week and a half, Hamilton and Burr traded letters via Van Ness and a lawyer close to Hamilton, Nathaniel Pendleton. Pendleton also had strong ties to Hosack. Hosack's younger brother William had joined

Pendleton's law practice at 62 William Street by 1803, and in August of that year they announced that they would be moving their offices to 65 Broadway—in other words, to Hosack's townhouse.* Pendleton and Van Ness raced back and forth trying to stave off a duel, but Hamilton refused to confirm or deny the allegations.

Burr and Hamilton both concealed the escalating conflict from their loved ones. On June 24, Burr wrote a cheerful letter to Theo, who now lived in South Carolina with her husband, Joseph Alston, and their young son, Aaron Burr Alston. Burr missed Theo so much that the previous evening he had brought her portrait into the dining room at Richmond Hill and celebrated her birthday with a dinner and dancing party. He thanked her for a recent letter full of news about his grandson. "You can't think how much these little details amuse and interest me." But he also chided her affectionately for neglecting both her own education and her son's. "If you were quite mistress of natural philosophy, he would now be hourly acquiring a knowledge of various branches, particularly natural history, botany, and chymistry." By this point, Burr had already drafted words of challenge to Hamilton. Even in this moment of extreme political crisis, his other great loves—his daughter and the study of the natural world—were not eclipsed.

IN PHILADELPHIA THAT SAME DAY, June 24, Charles Willson Peale began work on a portrait of Alexander von Humboldt. He would have to paint it quickly. Humboldt had just learned that an American ship was setting sail for Bordeaux on June 28, and he would be seizing the opportunity to return to Paris, where he planned to compile and publish the voluminous records of his explorations. Peale packaged up a preserved "alegator" to send with Humboldt to a professor of zoology at the National Museum of Natural History in Paris. Peale wrote a note to go with the alligator saying he would be happy to send more speci-

---

* It is not clear how long this arrangement lasted. By the time the 1804 city directory was published, Pendleton was listed at 17 Wall Street and William Hosack was not listed at all.

mens in the future but at the moment was too "fearful of the *Rapacious paws of the War Hawks*." At any moment on the open Atlantic, naval or pirate ships could swoop in with sails flapping and abscond with precious scientific cargo.

Between portrait sittings for Peale, Humboldt drafted letters of regret and thanks to his American contacts. On Monday, June 25, he wrote to Hosack. He said he had read some of Hosack's essays and found them "so interesting." "It was one of my most pleasant plans to visit the beautiful city in which you reside and to present my respects to you," Humboldt wrote Hosack. Now that he had to leave so suddenly, he could only hope to be "honored with your correspondence." If Hosack would send seeds to him, Humboldt promised that he and his French botanist friends, among them Jussieu, would take perfect care of them. Humboldt also invited Hosack to let him know whether he could help by sending anything from France to New York. How disappointing—Hosack would not get to lead the world's most famous explorer on an expedition through the Elgin Botanic Garden. A visit from Humboldt would have drawn national attention to Hosack's work there, and a foreigner who felt comfortable urging the president of the United States to fund Peale's natural history museum surely would have urged public support for this pioneering American garden.

On June 27, Peale finished his portrait. He had depicted Humboldt gazing at the viewer with a blue-eyed intensity that is softened by full, rosy cheeks. Peale was elated. He had proved to himself "that at the age of 63 I could paint as good a portrait as I could at 50 years of age."

That same day, Humboldt composed more letters in preparation for his departure. In one, he thanked Secretary of State Madison for helping secure the proper passport for him and his luggage, which contained some sixty thousand plant specimens, as well as for information Madison had shared about a particular species of wheat. "I love seeing a Secretary of State so interested in the agriculture of the land he inhabits," he wrote. Humboldt also penned an appreciative letter to Secretary of the Treasury Gallatin, and another to Jefferson thanking him for his hospitality at the President's House. "I had the pleasure of seeing the first Magistrate of this great Republic living with the simplicity of a philosopher."

Alexander von Humboldt
in 1804, by
Charles Willson Peale

The fourth and fifth letters Humboldt drafted on June 27 conveyed his regrets to Hamilton and Burr, two other Americans he had particularly hoped to meet. He was very pained, Humboldt assured each man, at having to return to Paris "without enjoying the pleasure of meeting You in person and seeing the interesting Circle in which you live." Humboldt missed his chance with one of them forever. That same day in New York, Burr challenged Hamilton to a duel.

Hamilton now began to organize his affairs. He drew up a report on his finances and drafted a statement explaining why he had allowed himself to get into debt. The Grange had drained him financially, just as his friend Richard Peters had warned him it would, but Hamilton took comfort in the "progressive rise of property on this Island." His estate surely would appreciate over the coming years, and Eliza would be the beneficiary. On July 3, Hamilton paid Hosack for some accumulated medical bills.

Peale left Philadelphia for New York that same day. He had hoped to be accompanying Humboldt but decided to go alone—he wanted to drum up sales for his polygraph by giving demonstrations at the Tontine Coffee House. He drafted an advertisement for the New York papers. "The advantages of multiplying writings in the same moment that the original is made, are so manifest to every thinking man, that

it is unnecessary to mention them." Peale was also eager to visit Delacoste's natural history collection. He welcomed the competition to his
own museum, later observing, "Whilst we witness with horror, the
deplorable and desolating effects of European wars, we cannot but
dwell with emotions of gratification and delight, upon the effects of a
rivalry in the sciences and arts."

Peale spent the first night of his trip in New Jersey, waking up
on the morning of the Fourth of July to spectacular weather. Nature
was in "full bloom," he wrote in his diary. The trees were "cloathed
in their deep coloured Green & thickest foliage and the fields loaded
with a bountious store of either grass or grain." He saw fruit ripening along the lanes and in the orchards. Splendidly dressed soldiers,
their weapons glinting in the sun, paraded past to the sounds of fifes
and drums. The stagecoach horses took fright at the din, but Peale
enjoyed the whole scene immensely. He took a ferryboat across the
Hudson to lower Manhattan and found the city in a state of festive
chaos. He hired a porter to wheel the bulky polygraph to the City
Hotel, but they were forced to wait for a gap in the marching troops
before they could dash across Broad Street. He deposited his luggage at the hotel and walked to William Street to see the Cabinet of
Natural History.

As soon as he entered Delacoste's rooms, Peale felt the stress seep-
ing out of him. "Subjects of a well arranged Museum," he wrote to his
children in Philadelphia, "are the best means to restore tranquility to
a troubled mind." He moved contentedly among the exhibits, observing with approval how scrupulous Delacoste had been in arranging
his specimens in the glass cases. The birds had not yet been mounted,
but Delacoste had placed them on cloth-covered shelves, arranging
them in "the linian System" with "Ribons put between each Genus."
The collection contained birds and beasts Peale had never seen. Over
the previous two months, Delacoste had managed to secure dozens of
subscribers to the museum at $2 each. Hosack and Clinton had signed
up, as had Burr and Hamilton. Preparing the specimens was slow and
costly, however, and as Delacoste showed Peale around, he hinted that
a "society of Gentlemen" was on the verge of buying the collection

outright. Delacoste himself would stay on as curator. Privately, Peale was skeptical that Delacoste's scheme would succeed. He knew how hard it was to get Americans to support museums.

HAMILTON CONTINUED TO SHUFFLE his papers into order. On July 4, he wrote a note to Pendleton thanking him for "his friendly offices in this last critical scene" and said he had taken the liberty of naming him as his executor. As New Yorkers caroused in the streets outside, Hamilton composed a loving note to Eliza and then tucked it away among his papers. "This letter, my very dear Eliza, will not be delivered to you, unless I shall first have terminated my earthly career. . . ."

Three days later, Delacoste announced that he had gathered commitments from a small number of New Yorkers who agreed to purchase shares in the museum for $50 each (about $1,000 today). Hamilton, perhaps because he was immersed in the details of his debts, chose not to subscribe at this steep rate, but Burr, Hosack, and Clinton were among those who signed up.

Now it was time for Burr and Hamilton to settle the matter of the doctors. At least one attending physician was customary, and Hamilton chose Hosack. Despite Hosack's distaste for duels, he was moved by Hamilton's trust in his medical skills and his discretion. On July 9, Burr scrawled a note to Van Ness saying he was tired of waiting for the duel to be scheduled. "I should with regret pass over another Day . . . anything so we *but* get on." Burr felt a doctor's presence was unwarranted, but he agreed to Hosack, while trying not to implicate him by name in the affair: "H———k is enough, & even that unnecessary."

The duel was fixed for Wednesday, July 11. Hamilton spent the night of the tenth at his townhouse. Before dawn he arose, waking one of his sons to say that because four-year-old Eliza was sick—this was a subterfuge—he would be going to the Grange immediately with Dr. Hosack. A few blocks away, Hosack and Nathaniel Pendleton climbed into a carriage and rode together through the darkened streets. It may have been during this ride that Pendleton informed Hosack that Hamilton was planning to spare Burr by throwing away his shot. This strategy was meant to safeguard Hamilton's honor but leave no one dead.

When Hosack and Pendleton arrived at the Cedar Street house, Hamilton joined them in the carriage. They drove to a dock on the Hudson to board a waiting boat. Elsewhere along the shore, Burr and Van Ness settled into another boat. Although the day promised to be fine, Van Ness carried an umbrella.

RAMBLING ALONG THE HEIGHTS of the New Jersey Palisades on a clear summer day in 1804, a wayfarer who turned north to look up the broad river would have encountered a vista of primordial scale and shimmering beauty. Forests of chestnut, hickory, and oak stood on the cliff tops. Osprey sheered out over the water. Within a generation, the Hudson River would ignite the souls of painters—especially Thomas Cole and Asher Brown Durand—who were destined to put American art on the map of the world. It was as magnificent a river as any on earth, according to one awestruck European traveler after another.

On this mid-July morning, the oarsmen aimed the boats for the shore hugging the cliff beneath the town of Weehawken, where the river lapped at beaches trimmed in swaths of green. Hosack's nephews Caspar and John had collected plants for Elgin nearby, bringing back two kinds of wild grape, *Vitis labrusca* and *Vitis vulpina*. Here, saltmarsh cordgrass (*Spartina alterniflora*) grew in dense colonies along the shoreline, and wild pinks (*Silene caroliniana*) clung to the face of the cliffs. Deep in the woods above, a vine called Dutchman's pipe (*Aristolochia macrophylla*) draped itself languidly over the boughs of trees. Mountain laurel (*Kalmia latifolia*) and black raspberry (*Rubus occidentalis*) bloomed in sunny spots. The seaside goldenrod (*Solidago sempervirens*) would not bloom until late summer, after Burr had fled New York.

Hamilton's bargemen guided the boat to a small beach. Hamilton, Hosack, and Pendleton climbed out. The dueling ground itself was on a rocky ledge higher up the hillside, and Hamilton and Pendleton disappeared into the bushes, leaving Hosack behind. He waited. Within minutes, a shot rang out. After several tense seconds, he heard another shot and then a panicked cry. Pendleton was shouting for him. He dashed into the underbrush and up to the ledge. He was shocked by what he saw: Pendleton was crouching on the ground, cradling Hamilton in his arms. The expression on Hamilton's face instantly burned itself into Hosack's memory. Until his own dying day, he later said, he would never forget that "countenance of death." As Hosack rushed past Burr to get to Hamilton, Van Ness threw his open umbrella in front of Burr's face, and they left the dueling ground. In the event of legal proceedings, Hosack would be unable to provide an eyewitness account of Burr's presence.

Hamilton was grievously injured. He spoke to Hosack, saying, "This is a mortal wound, Doctor." Then he went limp. Ripping away the bloody clothes, Hosack saw that the bullet had pierced vital organs. He searched in vain for a pulse, finally pressing his hand to Hamilton's heart. He couldn't feel a heartbeat. He directed Pendleton to help him carry Hamilton down to the beach, where they lowered him into the bottom of the boat. As the oars plunged into the Hudson, Hosack hovered over Hamilton, rubbing his lips, face, and chest with spirit of hartshorn in the hope of jolting him awake. In desperation, he tried to pour some of the searing liquid directly into Hamilton's mouth. Hamilton stirred and opened his eyes, murmuring, "My vision is indistinct." Hosack now examined the injury more closely, but when he pressed around the bullet's entry near the right hip, he saw that he was causing intense pain, so he stopped. In the next few minutes, Hosack later reported, Hamilton "asked me once or twice, how I found his pulse; and he informed me that his lower extremities had lost all feeling." Hamilton requested that Eliza be brought from the Grange, but he also asked that his friends shield her from the seriousness of his injury. To Hosack, he confided that he would soon be dead.

As they landed, Hamilton's friend William Bayard was waiting on the dock, Bayard's servant having informed him that Hamilton had crossed the Hudson earlier that morning with Hosack and Pendleton. Bayard had immediately divined the reason and feared the worst. He lost his composure when he realized that Hamilton was still alive, while Hosack marveled at Hamilton's calm in the face of agonizing pain. As his distraught friends conveyed him to Bayard's house, the wounded man "alone appeared tranquil and composed."

Hosack took control. He summoned Dr. Wright Post, his Columbia colleague, to aid him in ministering to Hamilton, who was on the verge of fainting again from the torture of being moved. Hosack undressed Hamilton and darkened the bedroom in which he had been laid. Hosack knew that Hamilton had been suffering from a gastrointestinal ailment in recent months, so he took care to avoid administering any medicines that might increase this discomfort. As it was, Hosack thought Hamilton's suffering "almost intolerable." Hamilton endured the pain with dignity, speaking to Hosack repeatedly of his "beloved wife and children."

Eliza finally arrived, and when she understood the situation, she went wild with grief. It was a harrowing repetition of the black day in 1801 when Hosack had watched Eliza and Hamilton at Philip's deathbed. Eliza now stood at the brink of ghastly anguish, simultaneously familiar and unfathomable. Hamilton tried to soothe her. *Remember, my Eliza, you are a Christian,*" Hosack heard him say.

AT RICHMOND HILL, Burr waited for news. Accounts of his mood in the twenty-four hours following the duel varied wildly. He was utterly jubilant, or he was so remorseful he had gone home and tried to shoot himself. Sometime later that day, Burr asked Van Ness to come out to Richmond Hill. Showing his own face in town, Burr said, "would you know, not be very pleasant." By that evening, most New Yorkers knew that Hamilton lay near death. Friends, acquaintances, and admirers were keeping vigil at the Bayard house or gathering on street corners with gloomy faces.

Hosack, meanwhile, was considering whether to try to remove the bullet. It had ripped through Hamilton's abdomen and lodged in his spine. Hosack's own opinion was that the situation was hopeless. He conferred with Dr. Post, then with a group of military surgeons who had come in from the French frigates anchored off Manhattan. Hosack met with the surgeons outside the bedchamber, describing to them the angle at which the bullet had entered the torso. Then he brought just one of the surgeons into the bedroom so as not to disturb Hamilton. The surgeon agreed that the wound appeared fatal.

Hosack decided to spare his friend the agony of having the wound probed with surgical tools.

Hamilton slept fitfully that night. On the morning of July 12, he told Hosack that his pain had diminished, but he remained concerned for Eliza and the children. When the children had been assembled, Hosack watched as their father "opened his eyes, gave them one look, and closed them again, till they were taken away."

That same day, Burr sent an anxious note to Hosack. "Mr. Burr's respectful Compliments. He requests Dr. Hosack to inform him of the present state of Genl H. and of the hopes which are entertained of his recovery." In his distraction, he left out a word: "Mr. Burr begs to know at what hours of the [day] the Dr. may most probably be found at home, that he may repeat his inquiries." Hamilton died near two o'clock that afternoon.

Church bells began tolling the news across the waiting city. The flood of private and public grief was immediate and intense. "O America! veil thyself in black!" wrote one admirer. "Deep mourns the Eagle, with shattered wing, in some lone spot." Within hours of Hamilton's death, stunned New Yorkers had begun to gather in the city's coffeehouses and meeting halls. The vice president of the United States had killed a Revolutionary War hero, one of the chief architects of the nation. An anonymous open letter to Burr asked incredulously, "What do you think will be the feelings of the United States at large, and of Europe, when they shall learn the cause of his death, and the manner in which he expired?"

Hosack had to hold his grief in check. A day or two after the duel, at the request of some of Hamilton's friends, he performed an autopsy. Dr. Post and two other witnesses observed closely as Hosack sliced into the corpse to determine the path of Burr's bullet. He found that "the ball struck the second or third false rib, and fractured it about in the middle; it then passed through the liver and diaphragm." He encountered a large mass of clotted blood in the stomach. He surmised it had pooled there because the bullet had ripped through Hamilton's liver before coming to rest in one of the first two lumbar vertebrae. Under his finger, Hosack could feel the little spikes of shattered bone.

For many people, these bleak days of mourning recalled the

national depression that had gripped the country upon the death of George Washington, with whom Hamilton had been so closely linked. Many Americans had viewed Hamilton as the only man truly worthy of Washington's mantle, and his early death sharpened the sense of tragedy. His life took on an epic glow. "Our Troy has lost her Hector," someone wrote in a Boston paper.

New Yorkers awoke on July 14 to the mournful tolling of muffled bells. Across the city, people poured into the streets, leaned out upper-story windows, and clambered onto rooftops to watch the funeral procession. Men and women wept openly as the procession departed at ten o'clock from the Robinson Street home of John and Angelica Church. At its head, uniformed militiamen marched in orderly rows. Behind them, members of the Society of the Cincinnati carried their standard, which was covered in a black shroud. Next came the clergy, and then, as a trumpet sounded, soldiers of the Sixth Regiment hoisted Hamilton's casket to their shoulders and followed them. His riderless horse swayed along before members of his stricken family. Hosack probably walked behind the Hamiltons with the next group of designated mourners—the physicians. Behind them came wave after wave of people: city officials, foreign dignitaries, military officers, merchants, tradesmen, and more. "The Citizens in general" brought up the rear of the procession.

AT HIS LODGINGS, Peale was conducting an autopsy of his own. His room smelled of vinegar and the sea. By the day of the funeral, he had preserved sixteen species of fish for his museum. He was particularly excited about a live sheepshead fish he had found at the market. It was eight and a half pounds, with bold stripes wrapping around its body.

The Hamilton hysteria left Peale cold. His political allegiances lay so squarely with Jefferson that he could not bring himself to feel any grief. While people gathered by the hundreds in the streets, he spent the early morning preserving three species for his new fish collection. "Today Alexr. Hambleton will be buried," he observed in his diary. "It is to be an uncommon parade of every class of Citizens." When Peale finally left the house, he found Pearl Street jammed as far as the eye could see. He decided to go look for Delacoste and began to fight his way through the crowds up Chapel Street. On Beekman Street, however, Peale suddenly caught sight of some of his relatives, who were watching the scene from the windows of a townhouse. He gave up on his original plan of finding Delacoste and joined them, insisting to himself that he merely craved the company. Platoons of uniformed troops passed before him. A band played the funeral march from Handel's *Saul*. Peale's stiff reserve finally softened, and he conceded that the organizers of the funeral had put together a "well digested plan."

For two hours, the procession flowed like a dark glacier through the streets of lower Manhattan. As Hosack marched up Broadway toward Trinity Church, he passed his own front door. The boom of cannon fire ricocheted off the buildings around him—soldiers stationed in the park and at the Battery were firing their weapons at one-minute intervals, and he could hear the cannonade from British and French ships out in the harbor. They continued firing for forty-eight minutes straight.

The procession came to a halt in front of the church. Under its portico, the elder statesman Gouverneur Morris gathered his thoughts. He had been asked to deliver the eulogy, and he was worried that the wrong choice of words could turn Hamilton's funeral into a riot. Morris decided to divert attention from the violent circumstances of Hamilton's death by dwelling on his youth. Addressing the students in the crowd, he reminded them that Hamilton had been a Columbia student when the American Revolution broke out. "It seemed as if God had called him suddenly into existence, that he might assist to save a world." Morris held up this brief, heroic life as a shining model of wisdom and probity. When judging the conduct of one's fellow citizens, one should ask oneself, "*Would Hamilton have done this thing?*" Morris concluded his remarks. The pallbearers lifted the mahogany coffin from the bier and carried it to Hamilton's final resting place in the Trinity graveyard, around the corner from Hosack's house.

───— ⬯ ——───

AFTER THE FUNERAL, Hosack forced himself to focus on his medical practice and the garden, but there were several melancholy interruptions to his work. On August 11, before Hamilton's friends had removed the black mourning bands from their arms—and before Burr was indicted by the State of New York for "unlawfully wilfully wickedly and designedly" firing a pistol at Hamilton—Hosack stood before city authorities and swore on a Bible. He confirmed that Hamilton had died at William Bayard's house in July 1804. "I attended him as his Physician and was with him when he died." One week later, sick at heart, Hosack sat down to compose an account of what he had seen, heard, and done at Weehawken and then at Bayard's house. He closed his account with Latin lines by Horace, written on the death of a beloved friend, that translated thus:

> *When will incorruptible Faith and naked Truth*
> *Find another his equal?*
> *He has died wept by many.*

Hamilton had also died in debt. Very quietly, his friends began to pool their funds in support of Eliza and the children. Two days after the funeral, Oliver Wolcott Jr. wrote another of Hamilton's friends about the fundraising campaign. "The design is, that a select number of Gentlemen of easy Fortunes, shall, without much eclat & publicity, subscribe what may be sufficient." Four hundred certificates of $200 each were sold. John Jacob Astor purchased the first three shares, William Bayard the next ten. Hosack bought shares No. 132 and No. 133.

The fund, kept secret from Hamilton's children, helped ease his widow's financial woes, but grief dogged her every step. Six weeks after Hamilton died, an Albany newspaper reported a scene from Eliza's lonely new life. Dressed in a flowing black gown and veils, she had attended a church service in her hometown. By her side were her three youngest sons: John Church Hamilton, age twelve; seven-year-old William Stephen Hamilton; and Philip, who was just two. During the service, John suddenly collapsed, landing face-down and still as a corpse. Eliza sank to the floor by his side "uttering such heart-rending groans" that "even Burr himself" would have been moved. Affliction had become so relentless a presence in her life that she was sure her son was dead, but he had only fainted. Two men helped John to his feet as he revived, and the family left the church. On the steps, Eliza clung to her son, laying her head on his shoulder. For having unleashed such suffering, the newspaper wrote, justice surely demanded Burr be "held up to the view of future ages—[as] a MONSTER, and an ASSASSIN."

Although in the end the murder charge would be dropped, the vice president was not safe in New York. He made plans to sell his Richmond Hill house, and John Jacob Astor stepped forward to snap it up. By late July 1804, Burr was in Philadelphia; there he sent Benjamin Rush a note asking for a package of assorted medicines to take with him as he fled into the Deep South, where he had staunch friends. He left it to Rush to decide which medicines he should have on his journey, specifying only that a "lancet might be useful." When Burr arrived in Georgia later that summer, he wrote to Theo and recommended that she get hold of a copy of William Bartram's 1791 account of his travels through the South. "Procure and read it, and you will better understand what I may write you." He was running for his life, but Burr's interest in botany never flagged.

THE AFRICAN OSTRICH DIED the same summer as Hamilton. Delacoste cut out its gizzard, heart, and lungs and put them on display. Within months, the Cabinet of Natural History itself was at death's door. Peale had been right—Delacoste's private subscription fund had proved insufficient to sustain New York's first museum of natural history. Hosack and Peale exchanged letters railing against the shortsightedness of American politicians who consistently failed to support the arts and sciences. Peale moaned to Hosack, "Must one or a few individuals bear all the cost and trouble of such important undertaking[s]?" Hosack tried to persuade Peale himself to buy Delacoste's collection, but Peale replied that he could not possibly afford it. Besides, he was already worried about what would happen to his own collection. *None of us think we will die,* " Peale told Hosack, "yet we ought to be prepaired."

Chapter 9

"THIS DELICIOUS
BANQUET"

I N SEPTEMBER 1804, ONE-YEAR-OLD JAMES EDWARD SMITH HOSACK fell ill, possibly with yellow fever, which had returned to the city that summer. Hosack's lingering sadness at Hamilton's death was now folded into what Mary described to her foster sister Catharine as their "great anxiety" for the life of their only son. For weeks, she and Hosack watched James being "reduced to skin & Bone." He finally began to recover in October, and Hosack, mercifully spared, turned back to his affairs in the city.

In a way, New York felt like the capital of something again—not of the state or the nation, of course, but of the emotion-packed political saga of the duel. The newspapers didn't have enough pages to print all the poems and speeches that continued to pour forth about Hamilton's greatness. People were sculpting waxwork tableaux in his honor and drawing blueprints for monuments that would never be built.

One November evening about four months after the duel, Hosack settled into Mayor Clinton's office at City Hall for a conversation with some mutual friends about the current state of the city. The animating spirit that night was a merchant named John Pintard, who had decided it was time for New York to sit down in the attic of its accumulating artifacts and do some sifting and sorting. When they were all dead and gone, what would—what should—the next Americans know of the city's history, of its place in the nation and the world? These men had seen the crimson and flash of British soldiers in the streets, had heard the French Revolution debated in the coffeehouses, had dined in fine homes erected where meadows and thickets had once played host to songbirds and field mice. With the death of Washington in 1799, these men had begun to mark and mourn the passing of the figures who had helped create the United States. At the same time, a brilliant future seemed to beckon. Hosack and his friends could see their own efforts slowly beginning to lift New York out of its muddy ditches toward a place among the world's great cities.

The men sat talking that night under a portrait of George Washington by John Trumbull, and just steps from the balcony where Washington had taken the oath of office in 1789. Together, they decided to form a society to celebrate the history of New York and the nation— the New-York Historical Society. It was the kind of project that set Hosack's blood coursing, and he plunged in enthusiastically. Mayor Clinton promised the rent-free use of a room in City Hall for the society's meetings and its collection of art and artifacts.* One week later, the city council commissioned Trumbull to produce half-length portraits of all New York's mayors since the Revolution and a full-length portrait of Hamilton. Around this time, Hosack also seems to have commissioned Trumbull to paint a half-length portrait of Hamilton to hang in his own house.

In Washington, DC, earlier that same month, the man many people thought of as Hamilton's cold-blooded killer had slipped out of the wilds of the American South, after more than three months away, and back into the Senate chamber. On November 5, 1804, Burr wrote to Theo, "I was in Senate to Day, but only 15 members appearing, no business could be done—Being unsetled I have seen Nobody whom you know." When Samuel Latham Mitchill and his fellow senators began trickling in to take their seats for the new congressional session,

---

* Today the New-York Historical Society occupies a palatial building on Central Park West, and its vast collection of art and artifacts spans several centuries' worth of American history.

they found Burr on the dais at the front of the room. Murderer or not, he still held the post of vice president of the United States, and thus also that of president of the Senate. It was in this latter capacity that he would be presiding over the impeachment trial of a Federalist Supreme Court justice named Samuel Chase.

In March 1805, after the conclusion of the Chase trial, Burr gave an impassioned speech before the senators to mark the end of his term as vice president. The Senate, he told them, was "a citadel of law, of order, and of liberty." In that dignified chamber, "if anywhere, will resistance be made to the storms of political phrenzy and the silent arts of corruption." Mitchill still admired Burr's political talents, and he watched sadly as the vice president stepped down from the dais, walked to the door, and exited the Senate forever. Mitchill wrote to his wife the same day, saying that when the door slammed behind Burr, "the firmness and resolution of many of the senators gave way, and they burst into tears . . . weeping for perhaps five minutes." Two days later, President Jefferson was sworn in for his second term. His new vice president was DeWitt Clinton's uncle, George Clinton.

THAT SAME WEEK, the New York state senators lounged in their seats in Albany, listening, or perhaps not, as a petition was read aloud to them. It had been sent in by Dr. Hosack of New York City, who was pleading for funds to support a botanical garden. Three senators were tasked with considering Hosack's request, and they reported back ten days later that they had learned from a "respectable authority" that Hosack was a solid man who had launched his garden project "with a zeal and ability which leaves no room to doubt of his ultimate success." The three senators were so firmly convinced of the garden's future medical and agricultural significance, in fact, that they introduced a bill meant to fund it.

But then it happened again—something, or someone, gummed up the legislative works. That year's session ended without a vote on the garden bill. Hosack was deeply disappointed. He found Albany opaque and infuriating. He felt certain the Elgin Botanic Garden was destined to become an irresistible draw for young American doctors and naturalists, a place where they could study specimens of the whole planet's flora and experiment with medicines and crops. After the gratifying praise he had received from the internationally revered Alexander von Humboldt, Hosack had every reason to feel confident that other foreign dignitaries and naturalists who visited the United States would soon be flocking to this unique American institution. Once there, as they strolled past the medicinal beds conversing with Hosack and his students, they would surely share their valuable scientific and cultural experience with New York and the United States. But Hosack was teetering on a knife-edge. After the bad news from Albany, he realized that he could shake loose public funds only if he brought the garden to a state of obvious utility. To do that, he would have to spend much more of his own money, even at the risk of ruining himself and his family.

That spring, therefore, Hosack squared his shoulders and hired additional men. He also bought more plants and supplies—a thousand asparagus plants, six bushels of white sand, heaps of plaster of Paris, a pile of lead pipes, and two cows. In mid-March, he hired a head gardener, a British nurseryman named Andrew Gentle who had arrived in New York only three days earlier, probably with a recommendation from someone in England whom Hosack trusted in horticultural matters. All that spring and summer of 1805, while Hosack continued to write cajoling letters to scientific gentlemen around the world and to organize the seeds and specimens he had already received, Gentle laid out the medicinal, kitchen, and agricultural beds and oversaw the planting of the forest tree species around the perimeter of the property.

Hosack and Gentle were immersed in these projects when a distinguished visitor arrived for a tour of the garden. It was Governor Morgan Lewis, who had defeated Burr in the recent gubernatorial contest to replace George Clinton. Half of the Elgin Botanic Garden was still trapped inside Hosack's head, and the other half was a mess of clay pots and manure, but as the governor strolled along the garden paths under a summer sky, it began to dawn on him that he was in the presence of an extraordinary American. Hosack had yoked his dreams for

A page from Hosack's memorandum book, where he recorded his progress on the garden

the health and prosperity of the young Republic to a venerable Old World institution. Every bit of medical, agricultural, and botanical knowledge this young doctor had absorbed in British gardens he was now plowing into the soil of his own nation with his own funds and for the good of his fellow citizens.

Governor Lewis found there was so much to learn and absorb at the garden that he soon visited a second time. He grew especially intrigued when he learned of Hosack's plans to make the garden a bustling hub for plant exchanges between the northern and southern United States. Hosack's correspondents in the South had already begun sending him seeds of crops such as cotton and yams, which he and his gardeners were planting at Elgin to test which ones might survive the New York climate. At the same time Hosack was sending his Southern friends specimens of plants native to the Northeast, so that they could conduct their own similar experiments. Each time Hosack's packages went sailing out of the harbor and down the coast to botanists and planters in Wilmington, Charleston, and New Orleans, he was inking in New York's spot on the scientific map of the United States a little more boldly. Each time a Southern gentleman folded up a letter, sealed it, and scrawled "Dr. Hosack, New York" on the front, the city gained a little more in the way of national stature.

Governor Lewis left Hosack's garden convinced that these twenty acres would make New York shine—while at the same time benefiting the entire country. Hosack himself was a new species of American, one who devoted most of his waking hours to inventing and organizing the civic institutions that would guide his city, state, and nation through the fractious post-Revolutionary years. While he worked tirelessly on the botanical garden, Hosack was also a member of at least half a dozen other charitable, cultural, and scientific enterprises, *and* he was playing a leading role in the education of the country's next generation of doctors in his Columbia courses and his private practice. The man deserved to be showered with public praise, to see his image struck on a medal, to hear the scratch of the governor's pen as he signed a bill that protected the Elgin Botanic Garden for generations to come. When Governor Lewis traveled back to Albany, his mind was bursting with botanical terminology and visions of robust crops tended by prosperous, healthy citizens. He resolved to see what he could do for Hosack.

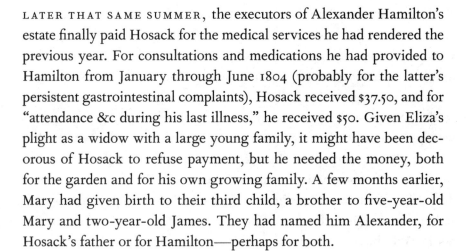

LATER THAT SAME SUMMER, the executors of Alexander Hamilton's estate finally paid Hosack for the medical services he had rendered the previous year. For consultations and medications he had provided to Hamilton from January through June 1804 (probably for the latter's persistent gastrointestinal complaints), Hosack received $37.50, and for "attendance &c during his last illness," he received $50. Given Eliza's plight as a widow with a large young family, it might have been decorous of Hosack to refuse payment, but he needed the money, both for the garden and for his own growing family. A few months earlier, Mary had given birth to their third child, a brother to five-year-old Mary and two-year-old James. They had named him Alexander, for Hosack's father or for Hamilton—perhaps for both.

IN THE FALL OF 1805, while Hosack and his medical colleagues battled another yellow-fever epidemic, he was also busy overseeing Andrew Gentle and the laborers as they prepared the Elgin grounds for winter. In December, the same month that Captain William Clark inscribed his name on a tall pine overlooking the Pacific Ocean, Hosack's men blanketed a bed of peas with manure and hay, "to be uncovered when the weather is fine," as Hosack noted in his memorandum book. He needed more help, and he began running an ad in the *New-York Gazette* for a kitchen gardener. Back home at 65 Broadway, he sat at his desk drafting letters in his crabbed script. He folded them up, sealed them with wax, and addressed them to Philadelphia, London, Paris, Copenhagen, Florence, and everywhere else he knew of someone who might send him plant specimens.

In Albany, meanwhile, Governor Lewis had not forgotten him. At the opening of the new legislative session in February 1806, the governor stood before the members of the State Senate and the State Assembly and praised the Elgin Botanic Garden to the skies. He implored the legislators to support Hosack's work, pointing out that "in a country as young as ours . . . individual fortune is not adequate to the task." Lewis told them that Hosack had already assembled one hundred fifty different species of crops and grasses.

The governor's speech moved the lawmakers to action. Within two days, several senators formed a committee to consider Lewis's plea, and just two weeks later, in mid-February, the Senate passed a bill granting Hosack an annual allowance for the garden. Now it was up to the state's assemblymen; if they too passed a bill, the garden would be safe. Hosack heard about the governor's speech and the Senate bill, but he had been disappointed twice before by Albany politics. He would try to take care of himself. He had dreamed up a financial scheme to help keep the garden afloat until his fellow citizens came to their senses. New Yorkers always craved fresh produce. If he could grow and sell enough fruit and vegetables at the garden, he might be able to offset the mounting expenses of his plant collecting and his gardeners' wages. He had already begun testing his idea by selling turnips and potatoes to a few of his friends, including the Pendletons and the Bards, and although he was chafing to launch his planned botanical and medical research, he focused instead on the more pressing problem of bankrolling the garden. In the spring of 1806, with Andrew Gentle's expert help, dozens more vegetable crops were planted.

Hosack still had much to learn about practical horticulture, but in a stroke of providential timing he could now consult the very first comprehensive garden manual in the United States, *The American Gardener's Calendar.* This month-by-month guide, published in early 1806, contained more than six hundred pages of lively instructions on everything from growing a decorative bower to heating a conservatory to preventing worms in fruit trees. The *Gardener's Calendar* was the work of Bernard McMahon, a talented Irish-born horticulturist who, with the capable aid of his Irish wife, ran a Philadelphia seed store where they welcomed customers into a clutter of gardening tools, books, baskets, barrels, sacks, and cabinets filled with seeds. The guide was greeted with glowing praise—including an anonymous review in the *Medical Repository*, a journal Mitchill had founded with the late Elihu Hubbard Smith in 1797. The review, probably by Mitchill or Hosack, praised McMahon for bringing botanical sophistication to the practical chores of gardening: "Ceres, and Flora, and Pomona, have all studied modern classification, and become acquainted with the Linnaean system." McMahon's *Gardener's Calendar* was exactly the sort of book Hamilton would have devoured happily when he was laying out his gardens at the Grange—and if he had lived long enough to do so, it would have been a point of agreement with Jefferson, who loved McMahon's book and consulted it frequently as he gardened at Monticello.

Hosack bought his own copy of McMahon's guide and set to work with Gentle. Their first order of business was to create the hotbeds that would protect the vegetable seeds from the late-winter chill as they germinated and sprouted. For these, McMahon recommended nailing together four yellow-pine planks into a rectangle about nine feet long and five feet wide, with one of the long sides of the rectangle standing taller than the other. This asymmetry would create a sloping angle for the panes of glass that would be placed atop the frame, thus keeping the rain from puddling and allowing the full warmth and light of the sun to reach the seedlings. At night, straw mats or piles of branches could be laid over the panes to insulate the plants from the cold. Once completed, the frames had to be filled with a mixture of straw, dirt, and manure, the last of which would generate heat as it decomposed. Hosack had his own steady source of manure from the cattle and pigs he was stabling at the garden—although if he followed McMahon's directions to the letter, he would have to pick his way through the piles of dung until he came across a vintage of a "lively, warm, steamy quality." McMahon warned against trying to grow seedlings directly in the manure mixture, as some garden-ers did, because it could spike to temperatures fatal for the plants. Instead, McMahon liked to plant his seeds in pots first and then plunge the pots into the soil, so he could pull them back up to the surface if the soil temperature rose too high.

For the Elgin hotbeds, Hosack and Gentle settled on workaday edi-bles that they were eager to see growing as soon as possible, such as cabbage and parsley. Then they turned to the next urgent task in the kitchen garden: preparing the outdoor vegetable plots in which they would be sowing the rest of the vegetables. It was to these outdoor plots, too, that the hotbed seedlings would be transferred when warmer weather arrived. As Hosack worked with Gentle to prepare the vege-table beds in the kitchen garden that spring, he seemed to catch gar-dening fever. He flipped his big memorandum book upside down and backward, and on the backs of the pages where he recorded the names of his private medical students, he noted each day's work at Elgin with more and more precision. On March 21, his gardeners prepared the earth for early potatoes, peas, and beans, which they planted the next day. Five days later, they sowed lettuce seeds; three days after that, it was onions, beets, and radishes. In early April, they transplanted the hotbed seedlings to the outdoor plots and then worked to get dozens more vegetable crops into the ground, among them carrots, peppers, cucumbers, squash, and additional varieties of lettuce, beans, peas, and potatoes. Inside the thick stone walls of his Manhattan garden, Hosack was leading an idyllic country life.

⊗⊗⊗

HE WAS ALSO TRYING to establish an orchard at Elgin. Several years earlier, he had bought about twenty species of fruit trees, scrawling the poetry of their names across a page of his memorandum book: bower apple, blood clingstone peach, greengage plum. By 1806 he had many dozens of species—among them nine types of cherry tree; several each of pear, apple, and mulberry; and a persimmon from Paulus Hook.

The cultivation of an orchard was an endeavor in which Bernard McMahon counseled vigilance and fortitude. Too many farmers planted their orchards and then left them to their own devices instead of pruning them regularly. It pained McMahon to see fruit trees "exhausted by moss, and injured by cattle," and "with their twigs so intimately interwoven, that a small bird can scarcely creep in among them." Late that February, just when McMahon recommended, Hosack had his fruit trees pruned. He intended to take excellent care of his orchard. He was looking forward to the time when he could set out baskets piled high with luscious fruits for sale.

But any competent Manhattan farmer with enough patience could grow apples. Or turnips and potatoes, for that matter. It was when Hosack climbed his hill and stepped through a tall archway into his greenhouse that he entered the colorful, fragrant realm where he was nurturing his own American revolution. For the previous few years, thanks to Hosack's letter-writing campaign to his British friends and to botanists he knew only by reputation, hundreds of seeds and plants had been migrating across the globe toward the garden. In cities across Europe, Hosack's correspondents had carefully labeled, packed, and shipped samples of the flora native to their home countries, along with still more specimens they had received from botanical outposts in Australia, India, the Far East, Africa, and South America. Once these specimens arrived in New York, many of them needed shielding from the harsh northeastern climate, so Hosack had planted them in his greenhouse, where they would be safe from frost. On the bitterest days, they could luxuriate in the warmth emanating from the flues he had had installed under the graduated shelves.

As Hosack moved down the long central walkway of his greenhouse that spring, he was surrounded by clay pots filled with dozens of species that few Americans had ever seen. From the East Indies, for example, Hosack had received a rare plant covered in fragrant white flowers, Arabian jasmine (*Jasminum sambac*), as well as a tall grass called Job's tears (*Coix lacryma*) for the pearly little tear-shaped grains that grew at the end of its stalks. From the Cape of Good Hope, a contact had sent him a plant with spotted, swollen leaves called tongue aloe (*Aloe lingua*) and also a spectacular evergreen native to the southern tip of Africa known as silver tree (*Protea argentea*) because of its shimmery metallic leaves. From the South Sea islands where his friend Banks had once roamed, Hosack had a striking plant with black flowers called *Lotus jacobaeus*. He was cultivating a tree heath (*Erica arborea*) from the island of Madeira and a beautiful silvery coronilla (*Coronilla argentea*) from the island of Crete.

Elsewhere in the greenhouse, Hosack was raising exquisite treasures from the Far East. One of his favorites was a sweet-scented daphne (*Daphne odora*) from China. He also loved his Japan rose (*Camellia japonica*) for its "splendid petals." He was equally proud of the many fruit trees he had acquired from China and Japan, among them species of orange and kumquat. And he was raising still other novel fruits nearby, including figs from southern Europe and avocados from South America. When the weather warmed, he would be able to move a great many of these exotic plants outdoors to soak up the fresh air and sunshine. He waited impatiently for that day, complaining in his memorandum book about how "cool and unsteady" the spring of 1806 was turning out to be. For now, the best he could do was to follow McMahon's advice to slide open the sashes along the front wall of the greenhouse around ten or eleven in the morning, and then close them in the late afternoon as the shadows lengthened and the day grew cooler.

Hosack knew that for some of his most delicate exotic specimens even the sheltering embrace of the greenhouse would be inadequate in the New York climate, so he had directed his men to erect two hothouses, one at each end of his greenhouse. These buildings were made almost entirely of glass to let in floods of light on sunny days, and they were heated by a powerful stove to near-tropical temperatures. Each of

Hosack's new hothouses was about sixty feet long, just like the green-house they flanked, thus creating a palatial conservatory complex that stretched for nearly two hundred feet across the hilltop. The hothouses jutted forward past the façade of the greenhouse a few paces, and the roofline of each hothouse sat well below that of the greenhouse, mak-ing the latter look grander still. Hosack had flower borders planted in front of the buildings, choosing species that would stipple their façades with color from early spring to summer's end: snowdrops, crocuses, violets, hyacinths, irises, asphodels.

Hothouses like Hosack's made busy diplomats of their gardeners, who were drawn into a ceaseless conflict between the excess smoke that billowed from an overstoked stove and the frigid winter air that rushed in if the smoke was vented too freely. McMahon considered it an "evil of great magnitude" to build a hothouse with a roof whose panes couldn't be louvered open to release the rank air that built up inside. Hosack seems to have agreed, because an oil painting of the garden done after the completion of the second hothouse shows roof panes open on both buildings. Inside the hothouses, Hosack and Gen-tle had two options for arranging the plants. They could place the pots on graduated stages, as they had done in the greenhouse, or they could construct hotbeds in which to put the plants. William Hamil-ton's gardeners at The Woodlands, the private estate near Philadel-phia on which Hosack had modeled his buildings, mainly used hotbeds in the hothouses. Whichever way Hosack chose to arrange his plants, McMahon recommended making sure "to sprinkle the flues occasion-ally with water, to raise a comforting steam," and to wash "dust or any sort of foulness" off the leaves with a sponge.

Hosack was soon growing many edible plants in his hothouses, including cinnamon, ginger, pineapple, mango, and even a coffee tree from Arabia. Visitors were enveloped in a swirl of spicy and sweet aro-mas that added, in Hosack's words, "zest to this delicious banquet." He also had other exotic species notable mainly for their beauty or curi-osity. His *Mimosa sensitiva* snapped its fernlike leaves tightly closed when he touched it. He had a pretty periwinkle flower from Madagas-car (*Vinca rosea*), a crape myrtle (*Lagerstroemia indica*) from the East Indies, and a prickly-stemmed Parkinsonia tree (*Parkinsonia aculeata*)

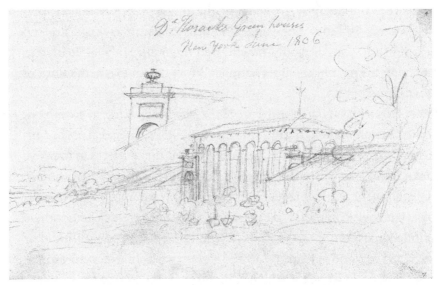

Hosack's conservatory complex,
sketched by John Trumbull in the summer of 1806

from the West Indies. One of Hosack's rarest plants was a curious shrub whose glossy green leaves looked like they had been sprinkled with gold; this was his Japanese gold-dust tree (*Aucuba japonica*). Throughout the hothouses Hosack also had splashes of red and pink, but for sheer drama, none could match his flame lily (*Gloriosa superba*), whose crimson petals, edged with yellow, leapt and curled through the air like fire.

HOSACK SURVEYED the rich plant life in his orchard, kitchen garden, greenhouse, and hothouses with mounting pride. Each species, no matter how small or plain, meant yet another entry in the living encyclopedia that his men were inscribing with their spades and rakes on the face of Manhattan. Every specimen that he acquired and successfully cultivated, or dried and labeled to add to the herbarium, meant another satisfying gift of his energy and intellect to that tenacious, globe-scattered tribe of natural historians who labored daily to identify more species. No one on earth knew how many kinds of plants were out there—growing in the jungles of Africa, on low-slung Pacific islands, alongside languid American rivers—and no one knew how long it would take to find them all. He was thrilled to be playing his part in the international effort to catalogue the flora of the whole planet.

Yet Hosack also had a sturdy toolbox of a heart. First in New York and Philadelphia, then in Edinburgh and London, he had absorbed his teachers' excitement about the promise of scientific progress to improve human lives. Ever since then, he had been alert to the smallest hint of a hidden practical advantage. His new figs, for example, might be delicious, but they could also be mashed into a hot poultice to soothe a pus-filled pocket of infected flesh. The juice from his oranges could be prescribed to ward off scurvy in the sailors who congregated around the harbor, and Hosack could also distill the orange tree's leaves into a tea to use as an antispasmodic in cases of epilepsy. The oil from his sweet bay laurel tree promised a gentle stimulant for a sluggish circulation, but if that didn't do the trick, his fiery West Indian pepper (*Capsicum annuum*) would deliver a near-electric jolt. From his cinnamon tree (*Laurus cinnamomum*), he could prepare an essential oil that was widely stocked in British medicine shops and that the *Edinburgh New Dispensatory* praised for its power to reduce "immoderate discharges from the uterus."

Hosack could read about these and a thousand other remedies in the *Edinburgh New Dispensatory*, but he was starting to outpace even the very latest edition of that gold standard of European medicine. By 1806, he was growing many plants at Elgin that were not listed in the *Dispensatory*. Hosack was learning about these new medicinals from country people on Manhattan as well as from the writings of explorers and traders who had spent time among the native peoples of North and South America. In his greenhouse, for example, he was growing Indian arrowroot (*Maranta arundinacea*), a plant he had acquired from St. Vincent that was named for its use as a poultice to remove arrow poison. He was growing a locust tree (*Hymenaea courbaril*) from the West Indies (today known as stinking toe), whose bark was used among indigenous peoples against coughs, diarrhea, and funguses. He had a tamarind tree (*Tamarindus indica*), a species native to Africa that had made its way to India and then traveled across the world to the West Indies—the source of Hosack's own specimen. Tamarind fruit could be mashed into a pulp that functioned as a gentle laxative, and he also liked to use it for fevers.

Yet these were the *known* remedies. What cure for what scourge shot silently through the veins of the plainest leaf that brushed his coat as he walked through the greenhouse? With hundreds of species now assembled at the garden, he was nearing the moment when he could focus on his medical research. For this, he would need assistants. His nephews John and Caspar had been an enormous help, and they continued to distinguish themselves as rising young botanists, but Hosack had recently taken note of a talented student in his Columbia classes.

John Wakefield Francis was a native New Yorker whose father had died in the yellow-fever epidemic that had so terrified New York City in 1795. Francis, then a small boy, had fallen so ill from the fever that he had a vivid memory of a coffin being brought into his sickroom. After his father's death, Francis's mother had managed to send him to the best New York schools and then to Columbia. Hosack recognized in the young John Francis all the attributes he valued most—intelligence, curiosity, self-discipline, and energy—and he invited Francis to study with him both in his private clinic and at the Elgin garden. Another physician later described Francis as "the prince of good fellows," praising him in the same terms often used for Hosack: a sense of humor, a talent for lecturing and storytelling, a quick intelligence, and a rich voice. Francis was also deeply racist. It is unclear whether he acquired those views before or after meeting Hosack, but in 1808 he gave a medical lecture to other physicians, probably including Hosack, on "the Bodily & Mental Inferiority of the Negro."

Francis found in Hosack a mentor and surrogate father, and he began boarding at the Hosacks' house, where he became, in Mary's words, "a member of our family." From this intimate perch, Francis now saw firsthand what enormous sums Hosack was spending on the botanical garden. Later in life, Francis would remark that Hosack

*John Francis, Hosack's
closest medical protégé*

had always thought of money solely as the means to execute his ambitious projects: "Had he the wealth of John Jacob Astor, he might have died poor."

IN LATE MAY 1806, the cool weather finally turned and Hosack decided he could safely bring out some of his exotics to bask in the Manhattan sunshine. Summer arrived, bringing pleasant, mild days and little of the poisonous humidity that always seemed to foreshadow the city's yellow-fever outbreaks.

One morning around this time, Hosack went out for a walk on the banks of the Hudson and encountered a young man collecting plants. The man apologized in French for disturbing Hosack's peace and introduced himself. He was Alire Raffeneau Delile, botanist to Emperor Napoleon. Hosack invited him home to breakfast.

Delile was nine years younger than Hosack, but he had already seen sights to entrance his host. In 1798, he had received an invitation from General Napoleon Bonaparte to join the scientific staff of his expedition to Egypt; Delile was to be in charge of collecting and cataloguing Egypt's botanical specimens. Traveling with the fleet of one hundred ninety-four ships were more than one hundred sixty scientists, scholars, and artists. On the passage over, Napoleon reportedly preferred their company to that of his officers, and in the evenings, the future emperor sat talking with them around a heavy table in a low-ceilinged room.

Napoleon intended to see the scientific and artistic treasures of the Nile Valley displayed in tidy rows on the shelves of Parisian laboratories and museums, so Delile hoisted himself onto a camel and began collecting plants. Within two days, his assistant died of the plague. In scalding heat he carried on alone, plucking fronds from the doum palm (*Hyphaene thebaica*) and marveling at the beautiful blue lotus (*Nymphaea caerulea*), sacred to the ancient Egyptians. He fell off a camel as it knelt to the sand; another camel sat on him. He named a new species of algae: *Ulva fasciata*, commonly known as sea lettuce. He took the first cast of what became known as the Rosetta Stone.* He founded the Cairo Botanical and Agricultural Garden in the backyard of an Egyptian villa, where he cultivated his recent plant discoveries and prepared specimens for his herbarium.

Then, on August 23, 1799, Napoleon abandoned Egypt for France, leaving Delile and the other scientists stranded, along with the increasingly beleaguered French forces. Delile continued to botanize until March 1801, when an army of Turks massed outside Cairo, at which point he fled to safety in the citadel. For the next six months, in a country beset by war and disease, Delile and his fellow scientists petitioned the French authorities for permission to return to France and for a vessel in which to make the voyage. They finally got a ship, but it sailed in September 1801 without Delile, because the British had attempted to confiscate his herbarium. He insisted to them that he would rather travel to England with his specimens than return to France without

---

* In 1799, when a French soldier named Bouchard spotted a stone covered with inscriptions near the city of Rosetta, Delile made a sulfur cast that would furnish the French people with their very first image of the stone. (Delile's cast was later misplaced for almost two centuries in a back room at the Montpellier Archaeological Museum; it was rediscovered in 2010.) The Rosetta Stone itself sailed to Britain, arriving at Portsmouth on the captured French frigate *L'Égyptienne* in February 1802.

them. The British released Delile and his plants, and he managed to return to Paris.

When Hosack ran into Delile in the United States a few years later, it was thanks to a new posting by Napoleon, who had sent his brave botanist to pass the time in another sweltering river delta. This time it was the Cape Fear River in Wilmington, North Carolina, an important trading port where Delile became the new French commissioner for commercial relations with the United States. Delile was delighted to learn that he would also be continuing his botanical research on behalf of the French Empire, because he much preferred roaming the Carolina cypress forests to sweating over ships' records in a claustrophobic office. The Empress Josephine, herself an avid botanist, had given Delile instructions to collect useful and pretty species from the Carolinas, and he happily obliged, on one occasion joining forces with François André Michaux to explore the mountains to the west.* Still, Delile eventually became bored with the sleepy life in Wilmington and moved north to Philadelphia, where he continued to botanize and began mingling in the sophisticated circles of the American Philosophical Society. Among his new Philadelphia friends was Hosack's brother-in-law, Caspar Wistar. Delile also made friends with other émigré Frenchmen in the city; with one of them, he conducted experiments on the best way to cook and serve the American bullfrog.

It was likely Wistar or Michaux who recommended that Delile go to New York to meet Hosack. At any rate, when Hosack ran into Delile that morning by the Hudson, he quickly realized what a treasure he had found. Hosack himself was one of the few Americans who could understand a botanical obsession so fierce that it had nearly sent Delile sailing into enemy territory rather than let the British pry his hard-won Egyptian specimens from his hands. Hosack and Delile enjoyed their breakfast so much that Delile soon joined Francis in studying medicine and botany with Hosack at both Columbia and the garden. Delile also

---

* In November 1802, François André's father, André Michaux, died in Madagascar of fever. François André, who was hiking through the forests of western North Carolina at the time, would not learn of the death for many months.

moved into 65 Broadway, and an emperor's botanist became the newest member of Hosack's household.

———— ✹ ————

WHEN HE WASN'T AT COLUMBIA or up at the garden, Hosack was seeing patients and reading and writing medical articles. The city had recently begun keeping annual mortality records, and he was trying to understand the patterns. Infants appeared briefly in the world before perishing, snuffed out by events their doctors jotted down as the flux (dysentery), fever, jaundice, teething, infanticide. Dozens more babies arrived each year already dead, and mothers were still suffering horribly and dying in labor. No wonder some doctors thought, as Benjamin Rush observed, that in a way "child-bearing is a disease." And of course women and men alike had a rich repertoire of other deaths to contemplate. In 1805 alone, according to their physicians, New Yorkers died of apoplexy (stroke), asthma, burns, cancer, consumption, convulsions, diabetes, dropsy (edema), frostbite, gout, hives, the "King's Evil" (scrofula), liver disease, smallpox, syphilis, and dozens of other illnesses and accidents.

As New York's doctors struggled to save and heal their patients, they were still working at a disadvantage compared to doctors in the largest European cities, where hospitals and medical schools provided more advanced training. New York's doctors also still relied heavily on medicines shipped to the United States from Europe. Jacob Schiffelin's apothecary shop on Pearl Street, for example, stocked anise seed, camphor oil, Peruvian bark, opium from Turkey, rhubarb from India, peppermint oil, ipecacuana, lemon essence, and many other plant-based medicinals, along with mineral-based remedies such as quicksilver, zinc, and Glauber salts. But what if Hosack could raise the foreign medicinal plants American doctors needed most? He was already growing some of the more delicate such specimens in his hothouses and greenhouse, and now he and Gentle were filling the outdoor beds with dozens of hardier species, among them many plants that William Curtis had been raising at Brompton when Hosack studied there. At Elgin, for example, Hosack had the purple-flowered common bugloss

(*Anchusa officinalis*), which could be mashed into a poultice that soothed skin inflamed by ulcers, hives, or burns; and he had sneeze-wort yarrow (*Achillea ptarmica*), the sun-loving European meadow flower whose roots could be turned into a powder to ease toothache. Nearby he was raising Roman nettle (*Urtica pilulifera*), a plant with serrated leaves and fuzzy little globes that looked charming enough for a child to play with. When brushed against the skin, the tiny hairs covering the globes released searing chemicals; Curtis had observed in the *Flora Londinensis* that "urtication, or whipping with stinging nettles, is an old practice" for stimulating a numb or rheumatic limb. Hosack planted chamomile and verbena, which made soothing teas, as did his several species of mullein (*Verbascum*). Along with his catnip (*Nepeta cataria*) and his mint plants (*Mentha*), he liked to prescribe mullein to patients suffering from dysentery.

Hosack was also cultivating and studying medicinals native to North America. He had already learned from Curtis and from the *Edinburgh New Dispensatory* about many such plants, and now in his own medicinal beds at Elgin he began planting dozens of them. He had the great blue lobelia (*Lobelia siphilitica*), for example, prized by the Cherokee people for coping with venereal disease, and also wood sorrel (*Oxalis acetosella*), whose leaves produced a delicious syrup used to reduce a high fever. He made sure to include the North American ginseng (*Panax quinquefolius*), which had so enraptured Benjamin Franklin when John Bartram had found it in the 1730s. The Iroquois, Mohegan, and other Native American peoples used American ginseng for a great variety of health conditions: as a fertility drug, as a pain-killer in case of earaches or headaches, as eyedrops for infected eyes, to help stop vomiting, as a psychiatric drug, and more. Another Native American medicinal, Virginia snakeroot (*Aristolochia serpentaria*), was a species Hosack always made sure to mention in his medical lectures at Columbia; not long after he began raising this perennial at Elgin, he had found it effective in treating anthrax. Hosack also alerted his fellow New York physicians to a report about snakeroot that he received from a doctor in Maine, who had attended in a case of lockjaw so severe that the man's mouth "would barely admit a goose-quill"—until snakeroot was injected between his clenched teeth.

Hosack was on a quest to collect every native medicinal plant he could gather locally or procure from his correspondents throughout the United States. His nephew Caspar brought him an herbaceous perennial called colicroot (*Aletris farinosa*), also sometimes called unicorn root because of the tall spike of white flowers that bloomed on its single stem between late spring and midsummer. Colicroot was a plant that John Bartram had once observed was "very plentiful in Jersey," noting that a decoction of it was good for stomach and bowel pains. Another of Hosack's native medicinals that was often used for upset bowels was skullcap (*Scutellaria lateriflora*), a pretty perennial with blue flowers that stuck out sideways. It was probably for its cough-alleviating properties that Hosack had planted his native hibiscus (*Hibiscus moscheutos*), which he called "Marsh Mallow Hibiscus"; from a related species, the ancient Egyptians had made a candy by mixing the sap with nuts and honey.

Among Hosack's odder native medicinals was a carnivorous plant that Caspar had found on Paulus Hook. Called roundleaf sundew (*Drosera rotundifolia*), the plant glistened with a shiny secretion that lured insects. When an insect landed and got stuck, the leaf curled up tightly and slowly digested its victim's body, extracting nutrients in the process. Some Native Americans fashioned a wart remover from sundew, and also a love potion. Hosack's climbing bittersweet (*Celastrus scandens*) was a species sometimes used by native women as an analgesic for labor pains. In the late spring, bittersweet twined its rich green leaves and off-white flowers around the trees and fences; in the fall and winter, its round crimson seeds splashed brilliant color against the gray-brown landscape.

Caspar had found still other native medicinals growing right in New York City—for example, a white vervain (*Verbena urticifolia*), used by some women to reduce menstrual flow, and a sun-seeking elderberry (*Sambucus canadensis*), whose purply-black berries songbirds loved to gorge on; it also made a delicious wine. Physicians used the berries to brew a throat-clearing cough syrup prescribed especially in cases of croup. The bark of the elderberry also produced a very gentle cathartic for a constipated patient; its buds, by contrast, purged the bowels so violently that physicians generally avoided using them.

Hosack and his protégés were peering into the tangled edge of the unknown world, trying to discern the causes of human misery—and the cures. Delile, especially, was trailing Hosack everywhere these days. He had enrolled at Columbia to pursue an advanced medical degree under Hosack's mentorship, allowing him to synthesize his botanical knowledge with expertise in medicine. Delile was conducting research for a thesis on pulmonary consumption (tuberculosis), and he knew that Hosack often cared for consumptives on his rounds at the state prison, the almshouse, and the hospital. He relied heavily on Hosack's clinical experience as he analyzed the symptoms, causes, and treatments of consumption. The first signs a patient might notice, Delile wrote in his thesis, are a "slight chill, a quickened pulse, a burning of the hands and feet, a circumscribed redness of the cheeks." Then, as the patient slowly began to waste away, "the balls of the eyes sink into their sockets" and the "nails . . . curve inwards." Delile acknowledged that the causes of consumption were complex and varied, but he singled out several suspect practices. Sedentary hours spent hunched over a writing desk constricted the chest, for example, as did cobbling shoes all day, or wearing a corset. Better to spend one's days hunting, building ships, or riding horses. Or gardening.

Delile's admiration for Hosack shone through in his thesis, particularly when he talked about the known treatments for consumption. After briefly considering mineral-based medicines such as mercury (which he dismissed as ineffective) and blue vitriol (an emetic he liked), Delile turned to botanical approaches to consumption. Foxglove (*Digitalis purpurea*) was a toxic diuretic growing at Elgin. In the earliest stage of the illness, Delile noted, Hosack had seen foxglove work "like a charm," while in later stages "it manifestly did harm." Other Elgin plants reduced fever in consumptive patients, including horehound (*Marrubium vulgare*), Seneca snakeroot (*Polygala senega*), Virginia snakeroot (*Aristolochia serpentaria*), and fringed gentian (*Gentiana crinita*). Hosack's favorite medicinal plant, boneset (*Eupatorium perfoliatum*), was useful for fever, too.

It was around this time that François André Michaux returned to New York for a long visit. Delile reported in a letter to a friend at the Jardin des Plantes in Paris that he, Hosack, and Michaux frequently gathered to talk in Hosack's "lovely library." Here, as they debated what genus and species a particular plant belonged to, they could leap up at any moment to pull down volumes by Linnaeus, Jussieu, André Michaux, Curtis, Smith, and others. They could also consult Hosack's precious Linnaean specimens. Delile informed his friend in Paris that Hosack was very generous in sharing his books and his herbarium.

Hosack had finally succeeded in re-creating the ambience of the Brompton Botanic Garden, where one could amble in from Curtis's flowerbeds to consult a botanical volume in a peaceful nook that had felt "a thousand miles from London." But a thousand miles from Manhattan—to the west—lay the prairies of North America, teeming with flowers and grains Hosack yearned to see and to study. He eagerly awaited the return of Lewis and Clark.

Chapter 10

"I LONG TO SEE
CAPTAIN LEWIS"

ALONG LOWER BROADWAY, ROWS OF LOMBARDY POPLARS SHOT high into the air like a grand fountain in a royal garden. Lombardy poplars also lined the old park on Broadway just north of St. Paul's, where the new City Hall was still under construction. William Hamilton of The Woodlands had introduced the Lombardy poplar (*Populus dilatata*, today known as *Populus nigra*) to North America in the 1780s, and it had quickly become a favored tree for ornamenting city streets.

In June 1806, New York's Lombardy poplars erupted in ashy-brown caterpillars with forked tails. The caterpillars crawled over the tree trunks and dropped on pedestrians strolling along the sidewalks, as well as on passengers riding in open carriages. The insects spewed a poisonous-looking green oil that inflicted severe pain. Mitchill immediately ran an article in the *Medical Repository* trying to dispel the "general panic" that was racing through the city as New Yorkers informed one another that this "asp" was "one of the most poisonous of all reptiles" and that people were dying from the encounters. (If the reports of fatalities were accurate, the cause may have been an anaphylactic reaction to the poison in the insect's tiny spines.) In Philadelphia, too, the poplar caterpillars were causing terror. Several men conducted an experiment on a young cat; it sniffed at the worm they placed before it, was stung in the nose, and died forty-five minutes later, after suffering "violent internal pains." The men then took some of the caterpillars to Peale for his natural history museum. In Washington, a friend of President Jefferson sent him two specimens he had plucked from a Lombardy poplar, explaining that "as this subject has lately excited some speculation, I supposed it would be gratifying to you to observe the worm particularly."

Hosack had Lombardy poplars at Elgin. Some New Yorkers were felling their poplars because they didn't want caterpillars dropping on their heads, but it is unlikely that Hosack did. He was much too comfortable with the natural world. He was also far too proud of the collection of native and exotic trees slowly maturing around the perimeter of his property. Someday, he would achieve the enclosing effect that Curtis had sought with his poplar "walls" at Brompton.

As Hosack nurtured his Elgin tree collection—what would later be called an *arboretum*—he had the ear of the world's greatest expert on American trees: his friend François André Michaux. Hosack was also benefiting from the expertise of the late André Michaux, because François André had edited and published his father's *History of the American Oaks* in France in 1801. It was probably Hosack who wrote the anonymous review of this book that appeared in the *Medical Repository*, lavishing praise on foreign governments for supporting their botanists and botanical gardens. "It is almost incredible how much good may be done in this way by a government, at a very trifling or insignificant expense. . . . How differently nations act!" The review also charged Americans with botanical illiteracy: "While our native citizens are ignorant or incurious of the leafy tenants of their forests, the more enterprising and industrious sojourners from foreign countries discover, describe and arrange them, and teach us how to know and understand them."

Hosack was as enterprising as any immigrant. Just as he had in every other part of his garden, he was cultivating species of medical interest among his trees. The bark of his English elm could be boiled into a decoction useful in cases of skin ailments such as *lepra ichthyosis*, a condition in which the skin turned scaly as a fish. The Cherokee people chewed the root bark of bristly locust for an emetic effect. Even the dreaded Lombardy poplars were useful. Their buds, gently bruised and mixed with fresh butter, produced a cooling ointment for wounds, and the leaves could be soaked in vinegar and applied to ease the agony of gouty feet.

Two months after the caterpillar scare, Delile wrote to a friend at the Jardin des Plantes saying that Hosack had gathered "the majority of the plants and above all the trees that grow at great distances across the territory of the United States." Among the American tree species at Elgin were red maple, snakebark maple, and sugar maple; white oak, scarlet oak, black oak, chestnut white oak, post white oak, and pin oak; sweetgum, tuliptree, sassafras, quaking aspen, yew, bristly locust, and red cedar. Hosack had planted many exotic trees as well, including English elm, European mountain ash, and an oak from China. He was always on the alert for interesting new species. One day at a friend's house, Hosack saw a writing desk made of an unusual wood. The desk had come from the friend's brother, who lived in Bermuda, and Hosack was soon writing to Bermuda himself. "I have seen in the possession of your brother a writing desk of your native Cedar. Be so good as to direct one for me. As a Botanist I'm desirous of specimens of this sort," he wrote. "I will also trouble you for a few Bermuda Cedars."

***

IN THE SUMMER OF 1806, Charles Willson Peale consulted Hosack's brother-in-law, Caspar Wistar, about a painful condition. Peale was suffering from a hydrocele in his scrotum—had suffered on and off for years, in fact. He had already told Wistar about it in 1804, when Peale had tried to treat it with ice. "Doctors as well as Lawyers should know the whole truth," Peale wrote Wistar cheerfully. "I have the pleasure to inform you that the effect from the application of the Ice to my *Scrotum* Six times yesterday" seemed to be a positive one. Peale decided ejaculation would help ease the pressure, too, but he was now a widower, and he thought that if he slept with a woman out of wedlock he might lose his children's affection and respect. He therefore concluded that he "must submit to the lesser evil!"—masturbation. He wrote

Wistar about the results. "After the discharge of that redundancy . . . the organs were in a better state to be healed." Peale remarried in the fall of 1805, which, according to his theory, should have taken care of the problem for good. But the following June he was again suffering from a hydrocele. This time, Wistar performed the injection method that Hosack had been the first to perform in the United States. Peale was soon reporting to Wistar that his pain was vastly diminished and his temperature had returned to normal.

Not long after this, Hosack pioneered another surgery in the United States: the first documented ligation of a femoral artery. His patient was an oysterman with a visibly pulsating mass in his thigh. Hosack sliced the man's leg open and—as he later explained to Delile—"with my fingers I readily detached the artery [and] passed the needle probing with a double ligature beneath it. I then tied the ligatures leaving about an inch between them and divided the artery between the ligatures." Within two months, the oysterman was back at work.

With his innovative surgeries and his mesmerizing lectures, Hosack was starting to gain a national reputation. As word of Elgin spread across the country, too, more and more young Americans were coming through the garden gate in search of Hosack. He liked to think of his garden as a nursery not only for all the world's flora but also for his own annual crop of engaged citizens. He explained his philosophy in a letter to one of his early botany students, Amos Eaton. "There is no study so well calculated to occupy the young mind, as the study of natural history," he told Eaton. "Since my acquaintance with the principles upon which the subjects of natural history are arranged, I certainly look with very different eyes upon every object that falls under my view, whether it be the production of nature or of art." He believed that studying the natural world saved young men from the "vicious propensities and pursuits" of city life. The rewards of teaching botany, meanwhile, flowed right back to him. He loved spending time with his students, whom he affectionately referred to as "my young gentlemen."

These days, as he walked along the garden paths with his students, Hosack saw an admirable scientific institution. He lectured on the plants that stretched in all directions around them, each species carefully labeled with its Latin and English names. He now had more than fourteen hundred exotic species in the garden and more than two hundred fifty native New Yorkers. At the end of his spring botany course, Hosack instituted an annual tribute to Linnaeus: a strawberry festival. In 1750, Linnaeus had managed to recover from a terrible attack of gout after eating wild strawberries, which he credited for his cure. John Francis later recalled expressing concern to Hosack about the annual expense of the strawberry festival, but he replied that "the disciples of the illustrious Swede must have a foretaste of them, if they cost me a dollar a piece."* Hosack also predicted that strawberries would eventually become "abundant and cheap" in the United States.

Proud as he was of his Elgin collections, Hosack knew that Linnaeus had already named six thousand species in the 1753 *Species Plantarum*; in the half century since then, naturalists around the world had found many more. Hosack wanted them all. His circle of friends and students—Francis, Delile, Michaux, the Eddy boys, and others—kept up a hectic pace of plant collection in the fields and woods around New York, while Hosack continued fetching new shipments down at the harbor. Caspar was now helping his uncle compile a catalogue of the Elgin collections. They sat with plants around them, writing out species names, crossing out those that later seemed wrong, and putting asterisks next to those they had found growing locally.

Some of these specimens survive today in the herbarium of the New York Botanical Garden. One day Caspar collected a wild grass specimen of some sort, for example, and Hosack and Delile labored over its identity. It seemed to be a species of *Festuca*—but what was the correct species name? They settled on *fasciculata*. On September 6, 1806, Hosack wrote the name and the date on a scrap of paper and affixed it to the specimen. The work of botanists was incredibly difficult, full of riddles and blanks. Still, it struck some of Hosack's medical colleagues as frivolous. The activity behind the stone walls at Elgin seemed too conveniently severed from the bloody mess of clinical practice. But Hosack believed that saving lives required more of American doctors

---

* The New-York Historical Society still holds an annual strawberry festival, which some historians have suggested may be a legacy of Hosack's strawberry festivals at Elgin.

Detail from Hosack's Festuca *specimen, collected in 1806*

than sawing away at a diseased limb. It also demanded the rusty reds of dried flowers, the bone-brittle boughs lying on a table, the veined and mottled pages covered in Latin binomials.

Soon Hosack had the Elgin catalogue ready. Now, when he wrote to his botanical contacts around the nation and the world asking for plants, he would be including this catalogue to show what specimens he could share by way of exchange. He had an ever-widening circle of naturalists he could write to, even within the United States. Thanks largely to his own example, more Americans were beginning to realize that botanical gardens were both urgent and possible in the young Republic. Scattered efforts were underway to launch gardens in other towns. In 1805, a group of private gentlemen in Cambridge, Massachusetts, had pooled funds to endow a professorship of natural history and to launch a botanical garden connected to Harvard College. In Philadelphia, Rush was lobbying the state legislature for a garden—although he was having as little luck as Hosack in securing public funds.

Hosack had a strong competitive streak, especially about New York, but his allegiance to the world of botany triumphed over his hometown pride. In September 1806, he sent his new Elgin catalogue to President

Jefferson with a long letter and a bold idea. He told the president he had already managed to grow cotton at Elgin and he hoped to naturalize other Southern and tropical plants to the Northeast before long. If the government were to fund a chain of botanical gardens across the country, American botanists could study the plants of each different region and share their findings nationally, with benefits for the entire nation.

It was a lot to ask of an embattled president. Perhaps Hosack should have stopped there, but he made two more requests of Jefferson. The first had to do with François André Michaux, who was still in New York. Hosack praised Jefferson for having dispatched Lewis and Clark, and then he confidently suggested a new expedition, to the southwestern part of the Louisiana Territory. "In Natural History much is also to be expected from exploring the territory in the course of [the] Red River"—in today's Texas, Oklahoma, Arkansas, and Louisiana. "That latitude is always rich in vegetable productions." Hosack explained to Jefferson who Michaux was and proposed that he be named leader of the new expedition. Evidently no one had ever told Hosack about the botanical debacle that had unfolded in 1793 while he was studying in Britain, when Jefferson helped organize a western expedition for François André Michaux's father that had been aborted by a political scandal involving the French minister Genêt.

Worse still, Hosack hadn't heard that Jefferson had *already* sent an expedition to the Red River five months earlier, with a surveyor named Thomas Freeman at its head. The president had asked sixty-five-year-old William Bartram to serve as chief naturalist, but Bartram had declined. Freeman had next turned to Benjamin Smith Barton, who also said no but suggested his student Peter Custis. In April 1806, the Freeman–Custis Expedition, with thirty-five men in its ranks, had left Fort Adams near the town of Natchez on the Mississippi and navigated to the Red River. As the explorers floated along, they admired the deep thickets of willows lining the riverbanks and beyond that endless groves of cottonwoods, sycamores, cypresses, oaks, and mistletoe-draped pecan trees. In late July, however, the trip was abruptly cut short by the one thousand Spanish soldiers who had been chasing them through the countryside. Guides from the Caddo Nation informed Custis and Freeman that the Spanish commander was "a Cross and bad man" who wanted to capture and kill the members of the American expedition for trespassing on what he considered Spanish territory. After a tense encounter with the Spaniards, Freeman and Custis turned their group around and hightailed it back to Natchitoches (in today's Louisiana). They had arrived there only two weeks before Hosack advised Jefferson to send an expedition to explore the Red River.

Hosack blundered on in his letter to Jefferson with a bold request concerning the Lewis and Clark expedition: "If sir the gentlemen who are at present on their travels to the Missouri, discover any new or useful plants I would be very happy in obtaining a small quantity of the seeds they may procure." Hosack didn't realize it, but he had probed yet another sensitive spot. Jefferson was at that moment agonizing about the long silence from Lewis and Clark. Jefferson's response, written one week later, was all of two sentences long: "Th. Jefferson presents his compliments to Mr. Hosack & his thanks for the catalogue of his plants. Should he have it in his power to be useful to his institution at any time he shall embrace the occasion with that pleasure which attends every aid given to the promotion of science." It wasn't very encouraging. But it was just enough for Hosack to keep nursing his hope that he might one day receive some of the Lewis and Clark specimens for Elgin.

Neither Hosack nor Jefferson knew it, but Lewis and Clark were finally emerging from the wilderness. Five days after Jefferson wrote Hosack, Lewis sat down in a hotel room in St. Louis and composed a letter to the president announcing "the safe arrival of myself and party." In the course of a journey lasting more than two years, the men had battled grizzly bears, run canoe-crushing rapids, crossed the Rockies, and lain helplessly ill beside the trail, wracked by fever and diarrhea. They had traveled—by foot, on horseback, and in their boats—over seven thousand miles. Lewis reported to Jefferson that "we have penitrated the Continent of North America to the Pacific Ocean," but he also broke the news that there was no all-water route to the Pacific coast. It took a full month for this letter to reach the president in Washington. When Jefferson received it, he was filled "with unspeakable joy," as he told Lewis in an immediate reply. Hosack probably heard the exciting news a few days later when the *New-York*

Herald published a letter from a St. Louis resident who had seen Lewis and Clark. "They arrived here about one hour ago . . . three cheers were fired. They really have the appearance of Robinson Crusoes—dressed entirely in buckskins."

When Charles Willson Peale heard that Lewis had returned, he was as desperate as Hosack to see specimens. Already in the fall of 1805, Jefferson had forwarded to Peale skeletons, skins, and live animals from a tantalizing early shipment of plant, animal, and mineral specimens Lewis and Clark had sent from Fort Mandan, the encampment (in today's North Dakota) where they had waited out the winter of 1804 to 1805. Now, however, while Clark went to visit friends in Virginia, Lewis went to Washington to see Jefferson and enjoy a long winter visit packed with festive dinners and excited conversations about the expedition. Peale kept needling Jefferson from Philadelphia, telling him in February 1807, "I long to see Captain Lewis." In April, Lewis finally traveled to Philadelphia, where he sat for a portrait by Peale and showered him with specimens for the museum. Peale rejoiced to a friend, "I have animals brought from the sea coast" that are "totally unknown."

Peale soon began preparing and mounting the latest trove of animal specimens and drawing pictures of them for a volume on the Corps of Discovery's natural history findings that Jefferson and Lewis planned to issue, along with two other volumes that would contain a narrative of the journey and a record of Lewis and Clark's geographical findings. As for the expedition's botanical feats, the number of plant species new to Western science that Lewis and Clark had collected and described during their voyage came to nearly one hundred eighty, among them many plants used as medicines by Native Americans.

Jefferson sent none of these treasures to Hosack. The latter may have annoyed the president with his demands, or perhaps Hosack's close association with the late Hamilton nettled Jefferson. Hosack would be shut out of the most thrilling botanical discovery of his era. After setting aside certain seeds for his own gardens at Monticello, Jefferson divided the bulk of the plants among three Philadelphia naturalists whom he and Lewis knew well. The first was William Hamilton of The Woodlands; the second was Benjamin Smith Barton. The third was Bernard McMahon, whose *Gardener's Calendar* Jefferson loved and with whom he had been exchanging letters and plants for several years. In March 1807, Jefferson sent some of the Lewis and Clark plant specimens to McMahon, who thanked Jefferson and promised to keep him updated on the progress of the plants he was growing from Lewis's seeds. Among them were Mandan tobacco, snowberry, Osage orange, prairie flax, and many more species completely novel to naturalists— at least to those of European descent.

Just as Lewis was arriving in Philadelphia, McMahon sent him a note. "There is at present a young man boarding in my house, who, in my opinion, is better acquainted with plants, in general, than any man I ever conversed with on the subject." McMahon told Lewis he thought this young man, Frederick Pursh, would be the right person to help catalogue and draw illustrations of Lewis and Clark's botanical specimens for the planned natural history volume. Pursh was an immigrant from Saxony who had recently spent several years as head gardener at The Woodlands, but McMahon told Lewis—"*between you and me*"—that William Hamilton hadn't treated Pursh very well. Pursh had decamped in 1805 to work for Benjamin Smith Barton, who needed help collecting and drawing specimens for a planned book on the flora of North America. Barton, the second naturalist to whom Jefferson and Lewis directed plant specimens from the expedition, was also the man Jefferson envisioned writing the text for the volume on the expedition's natural discoveries. Since the talented Pursh was already working for Barton, these two could easily join forces to organize, illustrate, and describe Lewis's plants.

Lewis took McMahon's advice and met with Pursh, and within a month he had hired him to produce drawings and descriptions of the expedition's specimens. Pursh was in botanical heaven. He suddenly had in front of him around one hundred fifty specimens from across the North American continent, only about a dozen of which he had ever before encountered. Among the specimens were plants whose names today honor the explorers who had set out in May 1804—for example, *Lewisia rediviva* (bitterroot), *Linum lewisii* (Lewis flax), and

Clarkia pulchella (ragged robin). Another of the expedition's finds, a plant called antelope bitterbrush, today honors Pursh himself with the name *Purshia tridentata*.

All through 1807 and into 1808, Pursh sat with seeds, sprigs, and roots spread out around him, choosing the best angles for sketching the Lewis and Clark specimens and the right words to describe them. But he kept running into obstacles. People found Pursh warm, brilliant, and cheerful—the naturalist Benjamin Silliman described him some years later as "frank and generous" and "full of fire, point, and energy"—but he also seems to have been an alcoholic. In April 1806, Barton wrote to his brother in the Shenandoah Valley to say that Pursh was on his way to botanize there, "if he dont die drunk on the way."

Barton himself posed another obstacle—he was a cranky perfectionist and also had chronic health problems that slowed most of his projects to a glacial pace. Finally, there was Lewis. In February 1807, Jefferson had nominated Lewis to be governor of the Louisiana Territory—who knew the region better?—but the appointment was to prove a terrible distraction to the work of getting the expedition volumes done and published. After Lewis left Philadelphia to take up his post, Pursh found it impossible to make steady progress on his plant descriptions. He kept running into questions he needed to ask Lewis about the specimens he was supposed to be cataloguing.

McMahon soon saw that Pursh was growing unhappy, and he fired off a volley of anxious letters to Lewis and Jefferson asking when Lewis would be returning to Philadelphia to help with the plants. Pursh, however, began yearning for a more secure position somewhere. He thought he might inquire at Dr. Hosack's garden in New York.

Chapter 22

"STRANGE NOISES, LOW SPIRITS"

Hosack's enthusiasm for New York was infectious. Mary, his Philadelphia-born wife, had come to adore her adopted city. In April 1807 she wrote her foster sister, Catharine Wistar Bache, "I feel so much interested in everything that concerns New York, that I shall not be surprised to hear you ask where I was born." Catharine and her daughters were about to pay a visit to the Hosacks, and Mary told Catharine that she was excited about showing them around. She proposed a sea-bathing excursion, and she also wanted them to see the paintings and sculptures at the Academy of Fine Arts, "as well as a great many other things which I think will be gratifying to you."

Mary belonged to a circle of New York women who were as devoted to improving New York as their husbands were. She was the secretary of the Society for the Relief of Poor Widows with Small Children, which was run by Susan Pendleton and Maria Clinton. Eliza Hamilton had recently helped this society to found the city's first orphanage, which she codirected. (An organizational descendant of this orphanage is still in existence today.) These energetic and compassionate women were helping turn New York into a proper metropolis—not merely the country's chief commercial port, but a hotbed of philanthropy as well.

Observing all this activity, Hosack's friend Samuel Latham Mitchill had decided New York deserved a guidebook, so he had just written one himself. He called it *The Picture of New-York* and published it in April 1807, the same month Mary wrote Catharine about her upcoming visit. It laid out all the city had to offer—its banks, markets, insurance companies, parks, theater, library, schools, colleges, newspapers, bookshops, and more than thirty fraternal and benevolent institutions. It also sketched itineraries that crisscrossed the Manhattan countryside to the north of the city. Mitchill advised visitors to stop at a "picturesque and romantic spot" on the East River where "porpoises are often seen sporting among the foam and eddies," although he noted that the lobster population wasn't what it had once been. He thought this stretch of the East River was best viewed from near Archibald Gracie's mansion, designed in 1799 by Mangin and McComb, the duo who had gone on to design the new City Hall, still under construction in the park.* Mitchill suggested longer tours to scenic locales on Long Island such as Cobble Hill, Jamaica, Rockaway, and Islip, and also to New Jersey. The road to Newark—"one of the most beautiful and thriving villages in the United States"—was flanked by meadows covered in native species "with which the florist and botanist will be delighted." In the late summer and early autumn, Mitchill wrote, "the andromeda and hibiscus on each side of the road are sometimes very frequent and beautiful."

Mitchill, Hosack, and their friends were serious about their work on behalf of New York's national and international reputation, but a young lawyer on Wall Street was watching them with amusement. Washington Irving, together with his brother William and William's brother-in-law, had just launched a literary magazine that made fun of these ambitious New Yorkers and their dreams of greatness. It was in these pages that Irving first applied the name *Gotham* to New York City and the name *Gothamites* to its inhabitants, borrowing from an old English legend about a town named Gotham whose citizens acted like fools on purpose to avoid taxation. Irving and his friends found it

---

* In the early 1940s, Gracie's mansion was turned into the official residence of New York City's mayors; Mayor Fiorello La Guardia was the first to move in.

entertaining that men with no proven literary or artistic talent gathered to pass judgment on cultural works. They lambasted Mitchill's *Picture of New-York* in a takedown also targeting the Academy of Fine Arts, whose statues were in temporary storage in the old City Hall basement, a "poor place for the gods and goddesses—after Olympus." Irving was working on a full-length satire of Mitchill's *Picture of New-York*, which he would publish in 1809 under the pseudonym Diedrich Knickerbocker, thus giving New Yorkers another of their most enduring nicknames.

Irving and his friends could make fun of men like Hosack, but it was thanks largely to the latter and his circle that Europeans were starting to take New York more seriously in science and culture. The decision of the French imperial botanist Alire Raffeneau Delile to pursue his study of plants and medicine in New York was a clear indication of this change. Delile had stayed in town specifically to study with Hosack at Columbia and Elgin. In early May 1807, he completed his medical degree with his thesis on medical botany and the treatment of consumption. He dedicated the work to Hosack, for "devoting the earnings of your profession to the establishment of a Botanic Garden, which being the first that has been instituted in the United States, does you infinite honour." Five days later Delile sent a copy of his thesis to Jefferson, along with a piece on Egyptian plants he had written while sailing home from Egypt to France. Delile informed Jefferson he would be returning to France by order of the emperor to work on the Egyptian natural history volumes. He included his current address for Jefferson—"at Dr. Hosack's, N.Y."—and the president replied within two weeks, telling Delile that "the objects which will employ [him] on his return to Paris will be some indemnification for the short stay he makes with us."

Hosack was very proud to call Napoleon's botanist his student and friend. But it bothered him that Delile was one of only two students who received a Columbia medical degree in 1807. The city needed more doctors, yet the top medical school in New York had never managed to recover its pre-Revolution stature and was now graduating just one or two students most years—and some years none. By Hosack's estimate, around sixty had graduated from the medical school in Philadelphia in 1807. Columbia was a hidebound kind of place, moldering away in its antiquated hall. So Hosack was pleased to learn that year that his Columbia colleague Nicholas Romayne, with whom Hosack himself had studied chemistry in 1790, was launching a new medical school that would surely take a botanical garden more seriously than Columbia had. As Delile was finishing his thesis that spring, Hosack joined the campaign to secure a state charter for the new school. Samuel Bard and Mitchill were both involved, as was Hosack's younger brother Alexander, who was now a doctor, too. Dozens of physicians attended organizational meetings and signed petitions lobbying for a school charter. The groundswell was tremendous, the future of medicine clear: uniform statewide standards and increased clinical specialization.

On March 12, 1807, Governor Morgan Lewis signed a charter creating a new College of Physicians and Surgeons. Gratifyingly for Hosack, the charter pointed out the necessity of a botanical garden for medical students. To help fill the ranks of the faculty while the new college was getting itself organized, he signed up as a professor of botany and *materia medica* and as a lecturer in surgery and midwifery. But he still held his position at Columbia. He needed the money, for his family and to pay his men at the garden. In June, Hosack also had Andrew Gentle place a notice in the *Commercial Advertiser*. "ELGIN BOTANIC GARDEN," it read. "Our citizens are now informed that they can be supplied with Medicinal Herbs and Plants, and a large assortment of Green and Hot House Plants, &c—Application to be made at the Garden."

MITCHILL WAS SENDING SEEDS and plants from Washington to Elgin. He may have been a busy senator, but he still found time for natural history. In the fall of 1807, he wrote his wife to say that the explorer Zebulon Pike had recently brought a pair of young "grisly Bears" from the Rio Grande as a present for the president. It was now "quite the style to visit the Bears after the manner of going to see the Lions in the Tower at London." But they were already starting to outgrow their cage, and Jefferson decided they should move to more suitable quarters. He sent them to Charles Willson Peale, who added them to the little zoo at his museum. The grizzlies quickly drew fascinated crowds, but after one of them ripped off the arm of a monkey and escaped into the family's kitchen, Peale shot it. Not long afterward, the second bear went the way of the first, and he stuffed and mounted their bodies for display in the museum.

Jefferson had more pressing matters than grizzly bears to deal with that autumn. Tensions with the British were on the rise. The previous year, in the spring of 1806, a British cannonball had spilled American blood in New York Harbor. HMS *Leander* had been sailing off Sandy Hook in search of Royal Navy deserters when she encountered an American sloop named the *Richard*. According to the *Richard*'s captain, Jesse Pierce, who swore a deposition before Mayor Clinton the next day, it had been around five in the evening of the twenty-fifth when the *Leander* fired three shots in his direction. The third shot decapitated his brother John, who was at the *Richard*'s helm. John Pierce's corpse was brought ashore and displayed at the Tontine Coffee House. New Yorkers were so rattled by the attack, Clinton observed, that "even the news of the death of Washington did not produce a more solemn effect" than the "electrical shock" of Pierce's murder.

The *Leander* incident had underscored just how vulnerable New York City was to a full-scale invasion—an event many New Yorkers feared was imminent, given the rapidly deteriorating relations between the United States and Great Britain. Suddenly Fort Jay, which had been erected on Governors Island while Hosack was in London, seemed like a flimsy defense against a fleet of British frigates. Men flocked to join the existing US Army regiments, while additional ones were quickly organized. Hosack signed up as a surgeon with the Sixth Regiment. Some of the men who practiced drills on the Parade Ground—located north of the city, on the way to Elgin—had, like Hosack, been born in the colonies and could remember the first great struggle against the British. As for the younger men, they were preparing for battle against a foreign menace that threatened the only nation they had ever known. In the summer of 1807, a British warship chasing Royal Navy deserters fired on the American frigate *Chesapeake* in the waters near Norfolk,

Virginia, killing three of her crew and wounding eighteen. By December Jefferson was asking Congress to pass a drastic measure to protect American ships and goods from the depredations of the British: a ban on the departure of all ships from American ports.

The Embargo Act went into effect on December 22, 1807. It quickly knocked the wind out of New York City. Exports came to a near halt early in 1808, with imports also plummeting. In April, an English visitor was shocked by how desolate the city looked: "Not a box, bale, cask, barrel, or package, was to be seen upon the wharfs." Hosack was lucky. He had amassed the core of his Elgin collections well before the embargo. It would now be nearly impossible to ship specimens to his botanical correspondents and thus also unreasonable to expect to receive shipments in exchange. Yet even as the embargo threatened to sever Elgin from the world, Hosack realized he could turn it to his advantage. For the previous six years, he had labored to create a local source for foreign plant-based medicines and a clearinghouse for known native medicinals. He and his students were also trying to identify new native substitutes for foreign medicines. Now, with imports dried up, Hosack looked prescient. His work at Elgin seemed more urgent than ever, and it wasn't only his medicinal plants that mattered. Hosack had assembled hundreds of agricultural specimens at Elgin, among them species of oats, barley, wheat, rice, sorghum, and sunflowers. He also had dozens of grasses and clovers that were valuable for feeding livestock or restoring exhausted soil.

Hosack saw yet another advantage as he walked the Elgin grounds. "Our embargo," Jefferson wrote Lafayette, "has produced one very happy, and permanent effect. It has set us all on domestic manufacture, and will I verily believe reduce our future demands on England fully one half." Elgin was bursting with plants indispensable to the manufacture of clothes, furniture, and houses. For weaving Hosack had cotton (*Gossypium herbaceum*) from the West Indies and flax (*Linum usitatissimum*) from Europe. For dyeing finished cloth he had native plants such as bastard indigo (*Amorpha fruticosa*). From Europe, he had dyers' chamomile (*Anthemis tinctoria*); from Egypt, safflower (*Carthamus tinctorius*). For making baskets he had European basket willow (*Salix viminalis*). For candles he had candleberry myrtle (*Myrica cerif-*

era), a Southern shrub whose boiled berries produce wax. Among his trees were dozens of species useful in the construction of furniture and houses—and ships, should the nation's shipyards ever come back to life. Hosack thought he ought to be able to convince the State of New York to buy the garden and make it a public establishment. Then it would have been worth it—all the years of toil and expense, the scores of letters written, the horses worn out going up and down the Middle Road. Hadn't he labored all along not for his own glory but for the public good? He would bestow upon the people of New York a scientific institution worthy of any great city in Europe.

In the meantime, however, he needed more money. Hosack now applied for the vacant professorship of surgery and midwifery at the College of Physicians and Surgeons. His qualifications were impeccable, and he would earn more than from his current post as a mere lecturer in surgery and midwifery. But then he heard a rumor that Nicholas Romayne planned to snub him. In February 1808, Hosack drafted an irate letter. "I will not be neglected," he warned Romayne. "If so, I have only to remark, that I abandon the Institution in toto. *I do not choose to be Botanist, or Midwife* to it, if unaccompanied with a more respectable appointment." Romayne appointed Mitchill instead. Hosack resigned from the College of Physicians and Surgeons and went back to Columbia.

He confided in DeWitt Clinton about his woes—he was beginning to fear he would have to abandon the garden. Only a rich man could finance such an ambitious project with his own money for years on end. Clinton knew Elgin intimately from his many conversations with Hosack and from his own walks and picnics there. He now lobbied other powerful New Yorkers on Hosack's behalf, declaring it "mortifying" that Elgin might collapse for lack of public support. In April 1808 Hosack submitted an appeal to the State Assembly asking the government to buy Elgin. A committee was formed, and its members reported back two days later that they believed the garden was "highly honourable" to Hosack and a boon to New York. But the legislative session was almost over. They recommended that Hosack try again the next year.

On June 22 Hosack wrote a generous, wistful letter to John Warren, a doctor in Boston who was working with William Peck to build the new Cambridge Botanic Garden. "I anticipate a great medical school at Boston," Hosack told Warren. "I hope you will immediately proceed to complete your botanical establishment in such manner as will . . . bear honourable testimony of that liberality which has been manifested by the people of Boston upon this as upon all other subjects which call for public spirit." Hosack made a rash pledge to Warren. "I shall gladly part with my collection if the state of New York remains inactive another season."

He dropped the pretense of good cheer completely a few days later in a slump-shouldered letter to a Pennsylvania doctor named William Darlington. Darlington had recently brought a collection of seeds and specimens all the way back from the Calcutta Botanic Garden and had then kindly directed some to Elgin. Now Hosack sent him an appreciative note. He included a copy of the Elgin catalogue, in case Darlington wanted to request any plants in exchange. "I wish every gentleman in similar pursuits with yourself would follow your example," Hosack sighed. "We should in a short time have no occasion to resort to the Botanic Gardens and Museums of Nat. History in foreign countries for our instruction upon these subjects." Hosack confided to Darlington his hopes that the state would buy Elgin from him. "But in that expectation I have been disappointed."

ON JUNE 23, 1808, the day after Hosack wrote to John Warren, Mary gave birth to another healthy daughter. They named her Eliza Bard Hosack, after one of Samuel Bard's daughters. Eliza was their fifth child, not counting the deceased Samuel Bard Hosack. There were also Mary, James, Alexander, and a boy born in December 1806 whom they had named Nathaniel Pendleton Hosack, for Hosack's comrade at the Hamilton–Burr duel.

Reminders of Hamilton were all around Hosack. When he left 65 Broadway to go to Columbia or up to Elgin, he could now see a stately monument looming over the slim headstones in the Trinity church-yard, just steps from his front door. It had been completed in the fall of

(right) A youthful-looking Hosack in 1815, when he was in his mid-forties, by the painter Thomas Sully. Hosack chose to be portrayed with a bust of his most cherished mentor, the Philadelphia physician Benjamin Rush.

(left) In the 1820s, a British botanist, David Douglas, named a genus of wildflowers native to the United States for Hosack, calling it *Hosackia*. Douglas considered Hosack the Sir Joseph Banks of America because of his decades of devotion to botany.

(above) This 1798 view of Manhattan Island looks southwest from the countryside north of New York City. Hosack drove past this bucolic spot each time he went up to the botanical garden. Today it is a busy intersection in SoHo.

(below) An undated view up Broadway from a few blocks north of David and Mary Hosack's townhouse, with the allée of Lombardy poplars stretching into the distance. (Just out of sight beyond them, slightly to the east, is the location pictured above.) The grand new City Hall was completed in 1812 and is still in use today. When Charles Willson Peale visited the city in 1817, he conceded to Hosack that New York had finally eclipsed Philadelphia as the most impressive city in the United States.

(above left) Eliza Hamilton was about thirty years old and had been married to Alexander Hamilton for seven years when she sat for this portrait by Ralph Earl in 1787. *(above right)* This portrait of Alexander Hamilton was done after his death by John Trumbull, who had earlier painted Hamilton from life. Likely commissioned by David Hosack around 1806, it eventually passed into the possession of Hosack's daughter-in-law Sophia Church Hosack.

(below) The Grange as it appears today, seen through the branches of a young sweetgum. Hamilton occasionally stopped at the Elgin Botanic Garden on his way up to the Grange, consulting with his friend Hosack about his horticultural plans.

Rembrandt Peale, one of Charles Willson Peale's many talented children, painted this portrait of Vice President Thomas Jefferson in 1800, shortly before he succeeded John Adams as president.

(above) Charles Willson Peale set this 1822 self-portrait in his celebrated Philadelphia museum of art and natural history. Peale's portraits of famous contemporaries line the gallery, while the giant mastodon skeleton that made Hosack jealous is partially visible beyond the curtain.

(left) William Bartram inherited his father John Bartram's love of botany as well as the family's famous garden and nursery outside Philadelphia. This 1808 portrait painted from life by Charles Willson Peale conveys William Bartram's sweet, shy demeanor.

(above left) This portrait of Hosack's friend Sir Joseph Banks depicts him as a dashing young botanical explorer in the early 1770s, shortly after his return from a long ocean voyage with Captain Cook. *(above right)* James Edward Smith (pictured here in 1793) welcomed Hosack to the Linnean Society of London, of which Smith was the founding president. Hosack spent many hours in 1793 and 1794 studying the botanical collections of Carl Linnaeus at the Linnean Society.

(left) Hosack studied with Dr. Benjamin Rush in Philadelphia in the 1790s and stayed close with him through visits and letters until the latter's death in 1813. Hosack commissioned Thomas Sully to paint a portrait of Rush in 1812 and suggested to Sully that Rush be pictured with the Pennsylvania Hospital in the distance. By the time he sat for Sully, Rush had practiced medicine and taught at the hospital for thirty years.

(above left) Aaron Burr in 1802, when he was vice president of the United States. This portrait by John Vanderlyn was painted eight years after the death of Burr's adored wife, Theodosia, and two years before his duel with Hamilton. (above right) Carl Linnaeus's house in Uppsala, Sweden. Burr made a pilgrimage here in August 1809, visiting the room where Linnaeus had died in 1778.

(below) This temple-like building at the Botanical Garden in Uppsala, known as Linneanum, was dedicated to the study of botany. It opened officially in 1807, the centenary of Linnaeus's birth, and Burr visited just two years later.

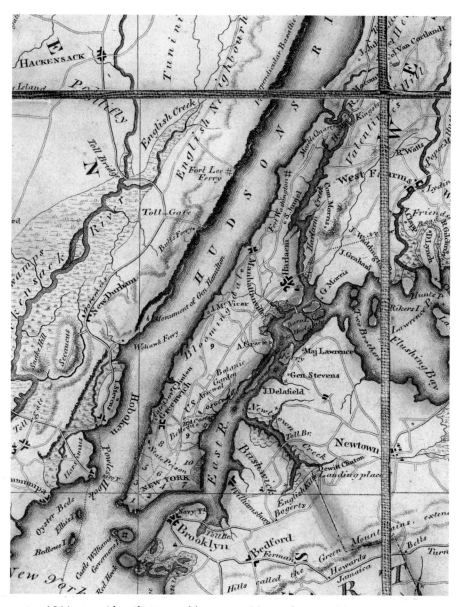

An 1811 map (detail) created by Hosack's nephew and botany assistant
John Eddy. Eddy indicated significant sites in and around New York,
among them the "Botanic Garden" and the spot on the shores
of the Hudson where Burr shot Hamilton in the 1804 duel.

(left) Mary Eddy Hosack with David Hosack Jr. in 1815. Hosack suggested to the painter, Thomas Sully, that three-year-old David be pictured listening to a watch on a chain held by his mother. Hosack thought this pose "brings out all his earnest expression of face."

(below) After Mary's death in 1824, Hosack married the widow Magdalena Coster and helped raise her many children alongside his own. In 1829, with her support, he bought the dramatic *Expulsion from the Garden of Eden* from the young painter Thomas Cole, who had been unable to sell it for a year after he first exhibited it in New York in May 1828.

Photograph © 2018, Museum of Fine Arts, Boston.

1806, more than two years after Hamilton's death. On the front wall of the tomb's rectangular base, an inscription had been chiseled:

To The MEMORY Of
ALEXANDER HAMILTON.
The CORPORATION of TRINITY CHURCH Has
erected this MONUMENT

In Testimony of their Respect

FOR

THE PATRIOT OF INCORRUPTIBLE INTEGRITY.

THE SOLDIER OF APPROVED VALOUR.

THE STATESMAN OF CONSUMMATE WISDOM:

WHOSE TALENTS AND VIRTUES WILL BE ADMIRED

BY

Grateful Posterity

Long after this MARBLE shall have mouldered into

D U S T.

He died July 12.th, 1804. Aged 47.

Atop the rectangular base of the tomb, anchored by four funeral urns, a thick obelisk rose skyward. Beneath all the heavy marble lay Hamilton's body, with its shattered spine.

As for Aaron Burr, by the summer of 1808 he was on a ship bound for Europe. After leaving the vice presidency in 1805, Burr had involved himself in a plot whose goal—depending on whom you asked—was to conquer Spanish-held territory and claim it for the United States or to lead a rebellion against the nation and crown himself emperor of Mexico. After months of deliberation, Jefferson had secured congressional approval to arrest Burr, who was captured in the Mississippi Territory in February 1807. The news was "as afflicting as unexpected," Burr's daughter, Theo, later wrote. When word of the sensational event reached Philadelphia, Peale congratulated himself on never having painted Burr's portrait. "Amongst a collection of about 80 Portraits there is none to disgrace the Gallery as yet," he told a friend.

Burr went on trial for treason in Richmond, Virginia, and Theo and her husband, Joseph Alston, traveled from South Carolina to offer their support. At the end of the summer of 1807, Burr was acquitted, but he was no longer safe in the United States, so he made plans to leave the country. The following May, Theo went to New York to see her father before he sailed. She also wanted to consult with Hosack about her poor health. In the aftermath of the duel, Hosack had defended Hamilton's character by lending support to the theory that the latter had fired his pistol involuntarily. Yet Hosack's bond with the Burrs, forged during his early years as the medical partner of their family physician, Samuel Bard, had survived even Hamilton's death at Burr's hands. It was the most striking possible illustration of Hosack's maxim that doctors should remain above party politics.

When Theo arrived to see Hosack, he fussed over her with affectionate anxiety. She had been ill almost without pause since the spring of 1802, when she had suffered a prolapsed uterus during the birth of her son. In 1806, she had seemed to regain her health, only to fall ill again after the shock of her father's arrest in February 1807. The first

Theodosia Burr as a young woman

doctor Theo consulted proved "to be a blockhead, & would have killed her," but she "got clear of him." By the time she arrived in New York in the spring of 1808, she had not menstruated for more than a year. She was regularly gripped by "various colours & flashes [of] light before the eyes, figures passing round her bed, strange noises, low spirits, & . . . periods of inconceivable irritability & impatience." She thought she was going mad. Her skin was so sensitive that she couldn't stand the touch even of her own hands, and she had "shooting pains through all her joints, particularly those of her hands & feet." Her head ached and she had a constant "taste of blood in the mouth, sometimes a little bleeding at the nose."

Hosack examined Theo in May 1808, shortly before Burr's departure. He persuaded her to give up the all-vegetable diet she had been on for the past four months, and he prescribed a stimulating tincture of aloe, saffron, and myrrh—all of which he was growing at Elgin—as well as a course of bottled water from a popular mineral spring at Ballston Spa, north of Albany. Hosack wrote Theo's husband to say that he hoped these initial steps would restore the "general state of her health." After that, he planned to try to address her specific complaints. Burr seconded Theo's reliance on Hosack, as he wrote her in a note before sailing, and he also approved of the mineral-water treatment. "The spring waters of Ballston or Saratoga are the best. . . . Have faith, and you shall be saved."

As he prepared to leave for Europe, Burr packed a portrait of Theo in his luggage. It was later reported that Hosack lent Burr money for his ocean passage. On June 9, Burr set sail for Nova Scotia and from there crossed the Atlantic to England. He made for London, where he initially booked a room at an inn. Soon, however, the philosopher Jeremy Bentham extended Burr the extraordinary hospitality of inviting him to stay with his family. Burr proudly showed Bentham the portrait of Theo he had brought with him from New York, and Bentham pronounced her a "dear little creature." At some point after Burr had gotten himself settled in London, he met up with one of Hosack's younger brothers, William. William Hosack was tall and swarthy, according to a letter Burr later wrote. The two men began socializing regularly; Burr would come to prize William above all for the latter's willingness to play chess with him as they whiled away many long hours of travel across Europe.

Immediately after her father left the country, Theo went to spend a few weeks at Ballston Spa. She had previously told Hosack that the bottled mineral water agreed with her, and he had therefore urged her to go to the spa so she could bathe in the water as well as drink it. As the summer wore on, the city grew stifling and humid, and Hosack began to worry that yellow fever would return. "Our atmosphere never was in a condition more adapted to the diffusion of poison of this nature," he wrote to a doctor in Philadelphia. Hosack still believed, erroneously, that yellow fever was contagious, but he was correct in thinking that the standing water left by the rains in roadside ditches, on barrelheads, and in the marshy lowlands created prime conditions for the fever to flourish. It was a blessing that Theo was up the Hudson, out of harm's way. From the city, Hosack was still trying to keep in touch with her, sending her letters, medicine, and dietary advice. By August, he was nudging her to "let me have the pleasure of hearing from you occasionally."

Hosack didn't know that Theo had already turned elsewhere for help. She had begun her stay at Ballston Spa by writing a brave, cheerful letter to her father on June 21, her twenty-fifth birthday. But one month later, she wrote a despairing letter to Dr. William Eustis, an old friend of her father from the Revolutionary War, who lived in Boston. When she returned to New York late that summer, Theo was still waiting to hear back from Eustis—and still keeping Hosack in the dark. She was disconsolate in New York. Intitially, she had been welcomed by some of her old friends, but after the first frisson of hosting the daughter of the most infamous of all Americans, they seemed to lose interest in her. One woman, Theo later wrote her father, even had "the cruelty to tell me that I had been so long ill and so long friendless, that I could not feel as keenly as others would." She missed her father terribly.

In London that autumn, Burr fretted about Theo even while he tried to settle into his new life, telling a confidante that her health was a "constant source of distress to me." When he received a letter from Theo saying she had found no relief at the Ballston springs, Burr

decided to consult an eminent London physician named John Coak-
ley Lettsom.* Lettsom urged Burr to summon Theo to England for a
change of climate. Burr wrote to Hosack, entreating him to persuade
her to make the journey immediately. "Show this letter to Theodosia,
and act in concert with her," he directed Hosack. "You are aware that
this is the most interesting concern of my life. Do by me as I should
by you." In the same letter, Burr asked Hosack's advice on a pressing
matter of his own. "Favour me also with your opinion as to the policy
of my returning to the United States, and the most suitable time."

It was an uneasy autumn and winter of 1808 for Burr. In addition to
his anxiety about Theo, he was finding the doors of many fashionable
Londoners closed to him for his political sins. But in a detailed journal
he was keeping with Theo in mind, as well as in his many letters to her,
he focused on more cheerful topics. He saw the Tower of London and
London Bridge. He bought a pair of boots and then took them back.
He bought a wig because he didn't have time to have his hair dressed
properly while traveling, and he lost an umbrella, which was returned
to him by a coachman. When Theo learned Burr was keeping the journal for her, she was touched. "What a feast it shall be for me. How sen-
sible I am to your goodness and attention in writing it."

Confident that Theo would soon be boarding a ship in New York,
Burr asked her to pack some botanical specimens in her luggage. "My
amiable friend [Jeremy] Bentham is a botanist," he told her, requesting
that she bring "one very large and handsome ear of Indian corn," and
"a handful of the Pecan or Illinois nut; of the butternut, the black walnut, the hickory nut." Then he thought better of his instructions. "A
handful, indeed! why, your little paw could not possibly hold more than
two black walnuts or butternuts. Bring, then, forty or fifty of each."
Burr also told Theo it might interest Bentham to see some English walnuts that had grown in American soil. The encounter between the New
World and the Old World was a theme he often sounded in his letters

---

* As it happened, Hosack had studied with Lettsom during his year in London and kept up
a warm correspondence with him; when Lettsom died in 1815, there was an oil portrait of
Hosack among his possessions.

and journal. In a letter to a Swedish friend, Burr jokingly described William Hosack and another companion, Thomas Robinson, as "the two American savages," adding that "they are so far tamed as not to bite, unless greatly angered by some strong passion, as love or anger." When he went to Edinburgh in January 1809 for a visit, he attended a dinner party filled with doctors, lawyers, and their wives; he described himself afterward in a letter to Bentham as the only "Moheigungk"— Mohegan—at the table.

While Burr was enjoying Edinburgh, Theo was enduring a miserable New York winter. Neither she nor Hosack knew of Burr's desire that she travel to England, because they hadn't yet received the letters he had written on this subject the previous November. Theo was now so ill that Hosack had taken to visiting her every day. For all her optimism, Eustis's medical advice, when it finally arrived, had proved fruitless, and she had reluctantly agreed to take Hosack's mercury. She soon rebelled, however, declaring in a letter to her father, "*I will not* take any more mercury. It . . . ruins my teeth, and will destroy my constitution." Burr was of the same mind, writing to her once he found out, "I condemn utterly the use of mercury." In spite of the mercury treatments—perhaps Hosack's other measures were offsetting them—Theo seemed gradually to be regaining her health. By late January 1809, when Hosack and Theo finally received Burr's request that she come to England, she was feeling "so much improved" that Hosack's first words were, as she later reported to Burr, "I presume you will not go." Theo wrote to Burr on February 1 saying she would not travel to England and reminding him that Hosack was still "ignorant of my application to Eustis (so let him remain, I entreat)." Her strength back, Theo finally felt well enough to turn her attention to Burr's botanical request for his friend Bentham. She wrote her father a few weeks later to say that although she would not be bringing any specimens in person, she thought she could get some to Bentham by an acquaintance who would soon depart for Europe. Although Theo was still in New York, her married years in South Carolina seem to have put Southern plants foremost in her mind. She proposed sending Bentham okra seeds and also some seeds of benne, a grain from Africa that was widely cultivated in kitchen gardens in the South.

When Burr eventually received Theo's letter, he was displeased with her choice of plant specimens. "Pray show a little more sensibility to the attention of Jeremy Bentham. Of seeds, search for the beautiful or curious rather than useful."

IN ADDITION TO TRYING to care for Theo, Hosack spent that winter of 1808 to 1809 collecting—but not just plant specimens. He was going around the city gathering pledges of support for Elgin. His colleagues at Columbia, at the College of Physicians and Surgeons, at the New-York Hospital, at the new Medical Society of the City and County of New York—they were all willing to help. The testimonial that Hosack secured from some of his former colleagues at the College of Physicians and Surgeons was especially useful. It alluded to the upsetting situation in which the state found itself, thanks to Jefferson's ruinous trade policies. "Many medicinal articles, belonging to the vegetable class, which are of indispensable use in diseases, and which might be readily produced in this country, have become so scarce, and so exceedingly enhanced in price, as to create apprehensions with respect to the sufficiency of future supplies."

Hosack submitted his sheaf of written testimonials and a new petition to the legislature. It was "with pain" that he informed them he was on the brink of abandoning the garden. He felt it was unfair to his family to keep spending money on it. The medical students who were Elgin's most direct beneficiaries could hardly be asked to pay for its upkeep, and the colleges and hospital were themselves strapped for cash. The only real alternative organizational model—a subscription society of the sort that funded the Academy of Fine Arts—was a hard sell in a city where botany enthusiasts were still thin on the ground. The garden's future was now handed over to a select committee of five assemblymen, who took just one day to make up their minds. They reported back to their colleagues that the Elgin Botanic Garden was "the first establishment of the kind ever attempted in the United States." Given that most European governments supported botanical gardens, the State of New York should certainly buy the garden from

Hosack. As for the economic wisdom of such a purchase, they noted, "the land will probably increase in value."

Now the committee members introduced a bill they called "An act for promoting medical science in the state of New-York." It was an inspired title. Even people who didn't care about botany cared about living and dying. When the State Assembly voted, the bill passed, sixty to twenty. Three legislators were appointed to spearhead the appraisal process. One of them was Hosack's dearest friend in the world, DeWitt Clinton. Elgin's future was being placed in the hands of the man he trusted most.

The next day, though, everything began to go wrong. One assemblyman was apparently upset by the idea of spending public money on a garden, and he successfully moved to reopen debate on the bill. Politicians who knew nothing about botany began opining on the monetary value of Hosack's plant collections, charging the estimates were wildly inflated. Hosack lamented afterward that "these erroneous impressions had the effect, of changing the sentiments of a few." The bill was rejected on March 14. He had lost by only six votes. Nicholas Romayne later said that it wasn't surprising that lawmakers should have balked at funding a garden. These were men who traveled to Albany from their hometowns across the state through landscapes of wild meadows, cultivated crops, and deep woods. "In a country where every farm and forest affords a variety of plants sufficient to illustrate the principles of Botany, public animosity may be aroused," Romayne sniped.

Few people noticed, and even fewer cared. By March 4, 1809, while the assemblymen were arguing over Elgin, Congress had lifted the embargo. A ban on trade with Britain and France was put right in the embargo's place, but New York's merchants could anticipate at least a partial recovery. Another distraction was the event that took place that same day in Washington, where crowds gathered to watch the new president, James Madison, take the oath of office. Jefferson was watching, too—and relishing the prospect of walking the grounds of Monticello within days. Hosack, meanwhile, returned to his own garden. His beloved New York had disappointed him, but he refused to give up.

Chapter 12

"SUCH A PIECE OF
DOWNRIGHT IMPOSTURE"

IN APRIL 1809, A FEW WEEKS AFTER HOSACK'S LATEST DISAP-
pointment in Albany, things began to brighten. Samuel Latham
Mitchill ran a piece in the *Medical Repository* praising Elgin as "that
ornament of Manhattan isle." That same month, Frederick Pursh,
fresh from working on the Lewis and Clark specimens in Philadelphia,
arrived at Elgin.

No one could blame Pursh that the natural history volume from
the Lewis and Clark expedition hadn't yet appeared. Pursh had stayed
in Philadelphia all through 1808, working as far as he could on the
drawings and descriptions while he waited for Lewis to return from
St. Louis with his crucial firsthand knowledge of the specimens. Pursh
was scrimping by boarding with Bernard McMahon, but he wasn't
being paid as long as Lewis was away. In January 1809 McMahon had
dispatched an anxious letter to Jefferson asking for news of Lewis, but
by April the latter still had not returned. McMahon felt he couldn't
keep Pursh in Philadelphia any longer, and he sent him to New York
with a letter of introduction for Hosack. Although Andrew Gentle
was a fine plantsman, Hosack felt he couldn't pass up the chance to put
Pursh's expertise on American plants to work. Elgin now had a new
head gardener.

Six months later Meriwether Lewis committed suicide in a cabin in Tennessee. When Charles Willson Peale heard the news, he wrote to his son Rembrandt, reporting that Lewis had been "on his way to Washington, when at a tavern he shot himself by 2 shots. This not being effectual he . . . compleated the rash work with a Razor." Peale speculated that Lewis had been devastated by the refusal of the federal government to honor debts Lewis had incurred on behalf of the expedition after he and his men had returned to St. Louis. "This mortification compleated his despair," Peale told Rembrandt.

McMahon wrote to Jefferson about the shocking event. "I am extremely sorry for the death of that worthy and valuable man Govr Lewis, and the more so, for the manner of it." McMahon assured Jefferson that he had kept all of the expedition's specimens safe in Philadelphia, because Lewis himself had hinted darkly that some grasping naturalist might steal them. "I never yet parted with [on]e of the plants raised from his seeds, nor with a single seed . . . ," McMahon told Jefferson, "for fear they should make their way into the hands of any Botanist, either in America, or Europe, who might rob Mr Lewis of the right he had to first describe and name his own discoveries, in his intended publication."

Pursh, meanwhile, settled in at Elgin. He managed Hosack's men, tended the plants, and went on expeditions in New York and New Jersey to bring back more native species for the garden. Hosack was delighted to see the daily labor at Elgin in the hands of such a talented plantsman, in part because he now found more time to pursue his own medical and botanical research. He began writing dozens of letters to his contacts in the United States and the West Indies, trying to collect systematic data on whether yellow fever behaved differently in different climates. Hosack asked his correspondents for weather patterns, symptoms, mortality rates—anything they could think to share regarding epidemics they had witnessed or simply heard about. He debated his theories with everyone he could buttonhole in person or by post, among them Mitchill, Rush, and Noah Webster. His cherished mentor Rush was the most genial of these interlocutors, and Hosack predicted cheerfully that his own yellow-fever research would soon "bring us closer together in our views of this subject."

As Hosack badgered his friends and acquaintances for medical data, Elgin was always in his thoughts. He closed a typical letter on yellow fever to a doctor in Jamaica with a request for plant specimens. "If you can aid my Botanical Establishment by sending me some seeds from the Bot. Garden of Jamaica, or by procuring me the interest of its present superintendent, you will also do me a favor. I hope you will not fear a troublesome correspondent in me." Hosack's thinking on yellow fever was informed by his botanical experiments at Elgin. To a physician in Georgia, he recommended decreasing the fever risk from stagnant water by planting the rice fields with "grasses, trees, and shrubs at the same time intersecting [them] by drains." This approach, Hosack advised, "will be the most effectual means of restoring the purity of the atmosphere as well as the health of your inhabitants."

On September 4, 1809, Hosack set aside his research and went to the old City Hall on Wall Street. Two hundred years earlier, Henry Hudson had sailed into the harbor, and the New-York Historical Society was celebrating the day with speeches and a dinner. Governor Daniel Tompkins, Mayor Marinus Willett, and State Senator Clinton participated in the events, along with many other distinguished and ordinary New Yorkers. After the speeches, at the City Hotel on Broadway, members of the New-York Historical Society dined on good local fare—wild pigeon, fish, and succotash. Mitchill was the toastmaster. When Hosack got the floor, he raised his glass to the settlers who had founded New Amsterdam. "May the virtuous habits and simple manners of our Dutch ancestors be not lost in the luxuries and refinements of the present times." Then Nathaniel Pendleton toasted *him*. "May the same virtues and the same industry continue in our land which have converted an Indian cornfield into a Botanic Garden."

Hosack was feeling optimistic again. All over the city, his friends and colleagues were mobilizing on behalf of Elgin. His most recent failure to secure state aid had "created strong sensations of regret" among his friends and colleagues. Mitchill was urging him to lobby all the medical societies across the state to write to the legislature, and Hosack followed his advice. It soon bore fruit. In October 1809 James Tillary, president of the Medical Society of the City and County of New-York, gave a speech calculated to appeal to his members' competitive streak. "Unless a botanic garden of dimensions befitting national views, be established near this city, and supported at the public expense, Pennsylvania will inevitably become, as it relates to the states, what she deserves to be, if we cannot rival her, the Edinburgh of America, the seat of science and chief nursery of the arts." Tillary's constituents promptly authorized him to sign and publish a statement of support. A month later Samuel Bard, president of the Medical Society of Dutchess County—the home county of his Hyde Park estate—gave a similarly helpful speech. "If we suffer this garden of Dr. Hosack's to sink as sink it must, if left in the hands of an individual, we give a decided advantage to every medical school in the United States, as well as in almost every other country, over our own." Eighty-four medical students signed a petition to save Elgin. Even the Common Council of New York City issued a unanimous statement of support. DeWitt Clinton, mayor of New York once again, signed it on behalf of the city.

Hosack, floating on the wave of goodwill, permitted himself to daydream about a new project. Like Curtis before him with his *Flora Londinensis*, and James Edward Smith with his *English Botany*, he would bring native plants, glorious and humble alike, into the farthest-flung households. He thought he might pay tribute to both his teachers at once by calling it *American Botany, or a Flora of the United States*. Like theirs, it would be a multivolume work. The drawings and descriptions would be based on the American specimens in the Elgin collections. But in order to bankroll the volumes, he would need the proceeds from the sale of the garden to the state. "Till then," he joked to a friend, "I must go on in my ordinary occupation of pulse feeling."

BY THE LATE SPRING OF 1809, Theo had returned to South Carolina in better health. She was perusing a newspaper one day when she came across a bulletin announcing that her father had been forced to leave England. It was reported that the pressure had come from the Spanish ambassador in London. Theo wrote to Burr that "for some minutes I remained stupified, as if stunned by the blow."

In early May, Burr and William Hosack sailed for Sweden. They spent most of the summer in Stockholm, but in August Burr indulged his love of botany with a trip to Uppsala, the town where Linnaeus had lived, taught, written, and gardened until his death in 1778. Burr took William Hosack with him, and they played chess on the river barge as a "perfectly wild and picturesque" landscape floated past. Burr found the Swedish countryside strangely familiar, as he wrote Theo—"the beauty of their roads being everywhere like that from New-York to Harlem." In Uppsala, Burr met two of Linnaeus's former students, now famous botanists themselves. One was Adam Afzelius, a friend of James Edward Smith and Sir Joseph Banks who had lived and botanized for years in Sierra Leone, eventually returning to Uppsala to become a demonstrator at the Botanical Garden there. The other was Carl Peter Thunberg, a professor of medicine and botany at the University of Uppsala who had once traveled through Ceylon, Africa, and Japan in search of plants. In 1807, the centennial of Linnaeus's birth, Thunberg had presided over the inauguration of a neoclassical building at the Botanical Garden that was devoted to the study of natural history and christened Linneanum. Burr was pleased to meet this "man whose works I had read with pleasure," and he planned to pepper Thunberg with questions about Japan, "a country which has always excited my curiosity."

Thunberg took Burr on a tour of the Botanical Garden, and Burr was astounded by the collections. He spent three joyful hours "examining the ten thousand things which are here," as he wrote in his journal. He could hardly tear himself away to keep an evening engagement. "It requires a month to examine these collections with satisfaction," he marveled, dropping his usual sardonic tone. "My head is still giddy with the number and variety. It would take more time and paper than I can now afford to enumerate the most striking." Burr did make a note about one specimen that intrigued him: a black walnut tree from America that had been "grown from a nut planted here; the only tree of the kind which I have seen on this side of the Atlantic."

A few days later, on August 16, 1809, Burr made a botany-lover's most significant pilgrimage—to the house and garden that had once belonged to Linnaeus himself. Here he saw the large hall in the greenhouse where Linnaeus had given his lectures and the chamber where he had drawn his last breath.

These garden visits seem to have reminded Burr how much he loved reading on natural history. He addressed Theo in his journal about a fascinating French book on tree-grafting he had just discovered. "Where have I laid that book? Will find it to-morrow and give you the author's name. It is a new discovery by which you give to any tree the sap and nourishment of another or of some branch of another, and by this means you may *change* and improve the *colour*, size, and flavour of any fruit. The results are curious and useful; pray try it. You see, Madame, I have not been idle; now allow me to attempt sleeping." Then, in early September, Burr procured a French translation of a new work by Alexander von Humboldt called *Ansichten der Natur* (in English, *Views of Nature*). The book changed Burr's travel plans.

In July 1804, when Humboldt had left Philadelphia—skipping the chance to go to New York to meet Burr, Hamilton, and Hosack—it was largely to work on *Views of Nature*. This was Humboldt's tribute to the natural world as he had experienced and studied it on his travels in South America. Burr, as it happened, had hoped to meet with Humboldt in the summer of 1804 to learn about Mexico. Now, on the afternoon of September 8, 1809, after arriving home from William Hosack's lodgings, Burr "threw off my coat and sat down near the window to read '*Tableaux de la Nature*'"—a French translation of *Views of Nature*. He quickly "got much engaged with my book" as he voyaged with Humboldt across oceans, up volcanoes, into rain forests, and through deserts. Two days later, Burr informed a friend that he had decided to go to Berlin to meet Humboldt in person. After that, he would visit Paris, "to see the people and the things of which all the world talk so much."

William Hosack and Thomas Robinson accompanied Burr as he traveled south into Germany that fall of 1809. At some point, Burr seems to have learned that Humboldt was now living in Paris, where he had settled at the end of 1807 after a brief interlude in Berlin. For political as well as intellectual reasons, Burr turned west toward the French border. At Weimar, he socialized with the poet Johann Wolfgang von Goethe and with Humboldt's older brother, Wilhelm von

Humboldt. Burr arrived in Paris on February 16, 1810, in horrible weather. "I suffer and freeze," he complained, but even the chilly drizzle couldn't keep him from taking in the Paris sights.

Two months of business affairs and social calls elapsed before Burr finally noted in his journal that he had seen Alexander von Humboldt at a dinner party. Humboldt was the toast of Paris as he made the rounds among the city's great scientists—among them Hosack's friend Alire Raffeneau Delile, who wrote Caspar Wistar in Philadelphia to say that he saw Humboldt as well as François André Michaux from time to time. Humboldt, Delile reported, was as talkative and entertaining as ever. The dinner party at which Burr encountered Humboldt was at the home of Helen Maria Williams, one of his translators. Burr didn't editorialize on their interaction when he mentioned Humboldt in his journal, observing only that he had passed "a very pleasant day." Whatever Burr's personal impressions of the man, he remained captivated by Humboldt's books, going to great lengths to procure copies of them in Paris.

THE YEAR 1810 wheeled into New York City on a cloud bank of mild, wet days. In late January, however, the air suddenly turned frigid, as Hosack recorded in his weather journal. For the next five days the mercury refused to budge above zero, and "a constant gale" buffeted people and animals, houses and ships. Hosack couldn't remember its being so cold for so long since his student days. Mercifully, the cruel weather lifted before the end of the month, when Mary gave birth to their sixth child, a daughter whom they named Emily.

A few days later, Hosack put Mary and the baby in the care of his medical protégé John Francis and left for Albany on Elgin business. He probably took the steamboat, an innovation debuted in 1807 by one of his patients, Robert Fulton. The steamboat left from a wharf on the Hudson and churned up the river at a speedy five miles an hour. Hosack considered it a "perfect success," although his nephew John Eddy thought passengers paid for the increased speed with a teeth-rattling ride. When John went by steamboat to Albany a few months later, he complained in his journal of the "never ceasing jar of the engine." His deafness spared him the noise, but he still suffered for the entire trip. "I cannot by any means recommend it to my friends who intend to travel for pleasure."

In 1810, the beauty of the Hudson River Valley worked its magic on travelers just as it had on the Dutch settlers two centuries earlier. When Hosack traveled up the Hudson that February day, he saw long stretches of forested shores slipping past, punctuated here and there by pretty hamlets and thriving towns. There was the old Dutch village of Sleepy Hollow, and farther north they would pass the town of Hudson, where oceangoing ships like the *Mohawk* were built and then floated south to the Atlantic. The riverbanks between New York City and Albany were dotted with country estates, including Samuel Bard's Hyde Park estate, with its white mansion poised high on a bluff above the river. Hosack knew the place well; he and his family frequently visited the Bards there, and his daughter Mary went for an annual summer stay. Across the river in the distance lay the Catskill Mountains, looking smooth and serene from the water but in fact filled with wild, secret places. Artists and writers were clambering through the Catskills with ever-greater enthusiasm these days as they tried to capture on canvas and paper the rocky crags, tumbling waterfalls, and twilight skies shot through with glowing streaks of orange. The most gifted of the artists were the British-born Thomas Cole and a New Jersey painter named Asher Brown Durand. Washington Irving was the prince among the writers.

About one hundred fifty miles north of Manhattan, the steamboat approached Albany. When Linnaeus's student Pehr Kalm had visited this town in 1749, he had found it almost entirely populated by Dutch-speaking people living in Dutch-gabled houses. He criticized them for being fussy about their immaculate floors when their streets were "very dirty, because the people leave their cattle in them, during the summer nights."* By 1810, when Hosack arrived in Albany, there

---

* Kalm loved botanizing in the countryside around Albany, which was full of apple trees and red maples. He also kept track of the fauna he encountered on his travels, noting, "The porpesses seldom go higher up the river Hudson than the salt water goes; after that, the

were still Dutch names everywhere—Schuyler, for example—but the Dutch-gabled houses had mostly given way to the straight rooflines of English-style townhouses. The steamboat pulled alongside the sloops at the river landing and sent merchants, lawmakers, and sightseers streaming up the hill into the city on foot or in carriages. The handsome new state Capitol, just completed in 1809, stood high above the Hudson at the top end of State Street, a steep avenue lined with shops and offices.

On February 7, a few days after Hosack had left New York, John Francis sent him a happy update. "Mrs. Hosack is recovering her former state of health as rapidly as usual, and as to the little one, nothing can be more promising. The other members of your amiable family are in their usual good health and I am requested to present their most affectionate regards to you." But by the time Hosack received Francis's letter, he was in a nail-biting wait to learn the fate of another of his children—Elgin.

He had worked feverishly up to the very last minute preparing his case. One day before the State Senate was scheduled to consider his appeal, he had managed to wrangle a statement of support from the Medical Society of the State of New York, the parent organization of all the county medical societies across the state. Its president was Hosack's nemesis, Nicholas Romayne, who had earlier blocked his appointment to the surgery chair at the College of Physicians and Surgeons. Romayne later claimed he was coerced into signing the Medical Society's statement. His sour attitude toward Hosack may also have been responsible for the damaging statement on Elgin that the board of the College of Physicians and Surgeons had recently drafted. Hosack had asked for a strong vote of confidence, and instead they had damned him and his flowers with faint praise. The college was "impressed with the advantages to be derived from the Botanical Garden," but Hosack had built Elgin too far away from New York City to be of much use to their medical students. At any rate, the utility of a botanical garden was "far inferior to that derived from chemical apparatus [i.e., labora-

---

sturgeons fill their place. It has sometimes happened, that porpesses have gone quite up to Albany. There is a report, that a whale once came up the river quite to this town."

tory equipment], an anatomical museum, [and] a medical library"—all of which together would likely cost the state less than purchasing the garden. If the College of Physicians and Surgeons could choose anything to lobby for, they said, it would not be the Elgin Botanic Garden.

These barbed remarks amounted to an about-face from the college's earlier statements of support, and they were undeniably awkward for Hosack. But his *pièce de résistance* was so impressive that it didn't really matter. He had come prepared with a petition to save Elgin signed by sixty-eight prominent New Yorkers—Federalists and Democratic-Republicans alike. His principle of political aloofness had served him well. Among the signatories were men who had been governors, mayors, and city council members—men whose names would live on in the city's streets, landmarks, and memories for centuries to come: Bayard, Beekman, Colden, Cutting, Fish, Gracie, Livingston, Pendleton, Remsen, Rikers, Rutgers, Schuyler, Van Rensselaer, Varick, and many more.

It was probably on February 8, 1810, that Hosack climbed the fifteen steps of the Capitol and passed between the fluted columns of the towering front portico, for that was the day that the Senate was to consider his latest appeal. In the cavernous hall of the Capitol, legislators in dark suits strode across a diamond-checkered floor of gray-and-white Italian marble—some turning right to enter the Assembly chamber, others moving left toward the smaller Senate chamber. Visitors like Hosack had to climb a lobby staircase to watch the proceedings from the second-floor galleries, where they looked down at the backs of legislators' heads but faced the speaker on his dais at the front of the room. The rooms were elegantly decorated with symbols of the nation, including the American eagle.

The senators listened as the Elgin testimonials were read aloud. Then they appointed a bipartisan select committee to deliberate on Hosack's appeal and report back. DeWitt Clinton and Edward P. Livingston were the most prominent among the Democratic-Republicans on the committee, while Jonas Platt, a former United States congressman, led the Federalists. On February 12, Livingston took the floor to speak for all his committee colleagues. "Your committee state[s] with pleasure, that this is the first establishment of this kind that has ever been attempted in the United States, and that this praise-worthy example has already been followed by several of our sister states, particularly, Massachusetts, Maryland, and South-Carolina." To raise the funds to purchase the garden, the committee recommended organizing a lottery, a procedure often used at the time to fund roads, bridges, schools, and colleges. The next day, Livingston introduced a bill bearing the same title as the one that had almost passed the Assembly the previous year: "An act for promoting medical science in the state of New-York."

As debate on the legislation opened, grumbling and sniggers began circulating through the Capitol and out into the city. One legislator later said it should be called "An act for the relief of Doctor Hosack," ignoring the fact that his work at the garden had been motivated by a concern for his fellow citizens. Other men conceded that botany had its scientific merits, but with a British war looking increasingly likely, the legislature simply had to focus on more urgent issues—for example, building fortifications against a possible invasion of New York and arming the state's militia. When the citizens of Albany got wind of the Elgin business, a local paper published a scathing letter received by an Albany resident. The letter had been written by a man who had gone to see the Elgin Botanic Garden for himself. Opinion pieces of the day were often published anonymously or under pseudonyms, and this author identified himself only as "Mr. B." He dismissed the entire pursuit of botany as "the least useful of all the branches of learning, a mere *science of hard* words," with its "*monogynia* and *pentagynia* and *pistilla* and *stamina*. . . ." But Mr. B. saved his most lacerating language for the garden itself. "I have been as anxious as yourself," he wrote his friend,

> to know something of *this great object of national concern* . . .
> you may readily imagine that I expected to find something, if
> not rivalling, at least not inferior to, what you and I have wit-
> nessed in Europe. I was prepared to see a garden possessing all
> the various exotics of the celebrated *Jardin national des plantes*,
> and outstripping in the splendor of its disposition the *Thuiller-
> ies*, the *Champs Elisees*, the Bois-de-Boulogne, of France, and

Hyde Park and *Kensington*, of England. My fancy pictured to me something very magnificent. I imagined an entrance of massive gates, crowned with crouching lions; winding woods whose recesses were adorned with winged Mercuries, Cupids, Naiads and timid Fauns. I fancied grottos, and knolls, and mossy caverns, and irriguous fountains, and dolphins vomiting forth huge cascades, and griffons, and chateaus. . . . Thus I was musing, as we passed along what is called the middle or New Boston road, when Mr. W. suddenly roused me with "Here's Elgin." I looked around me, but saw no Elgin. . . . It is impossible for me, my dear friend, to describe to you my sensations, when assured that what I saw, was the *Botanic Garden*. . . . My sensations were indescribable, tumbled as I was in a moment from the very acme of ardent expectation, into the Trophonian abyss of disappointment. I did not know whether to vent my execrations, or my laughter. There never was in the world, such a piece of downright imposture as this Botanic Garden, or as it is dignifiedly called Elgin. Take away from it, the "Orangerie" or Greenhouse, which stands at the remote end of it, and it looks more like one of those large pasture-grounds near Albany, in which the western drovers refresh their cattle, after a sweaty march, than a Botanic Garden.

Now an Albany resident calling himself "Mr. D." rushed to Hosack's defense, attacking Mr. B. as an ignorant fool. "If *Linnaeus* were to rise from his grave, with what pity would he view your diatribe against botany, and your unjust attack upon his nomenclature! . . . If you had lived and written in the time of *Kalm*, you would have been represented by him in his correspondence with *Linnaeus* as a *monstrum Americanum*, the upper part man and the lower part beast, combining the most cultivated mind, with the most Gothic hostility to science." Mr. D. knew Mr. B. personally, he said, and he was mortified to realize the true depths of Mr. B.'s ignorance. "Jet d'eaus, artificial cascades, purling streams, mossy caverns, porticos, knolls, grottos, griffons and dolphins vomiting forth water, are foreign from the nature of a Botanic establishment; and however pleasant they may be at a gentleman's country seat, or in a pleasure garden, yet surely nothing is more ridiculous than to require them in a scientific institution."

Mr. D. also informed Mr. B. that, as it happened, he was old enough to have had the privilege of botanizing in his youth with Kalm during the latter's Albany visit sixty years earlier. Mr. D. could still recall "the animated pleasures which glistened in the eyes of that great man, when he discovered a new plant," and he also remembered how Kalm had predicted that "America contained more vegetable riches than Europe" and "New York alone has at least 2000." Hosack had acted on that botanical promise with greater vigor and vision than almost anyone before him. What he had accomplished at the Elgin Botanic Garden was astounding. When Mr. D. had visited there he found "the greatest collection of valuable vegetables [that is, plants] which I ever witnessed; and whether there were knolls or grottos, I did not indeed take the trouble to inquire; for which sin of omission I must most humbly crave your indulgence." He shuddered at the thought that men as uninformed as Mr. B. held any sway over the future of American science.

At the Capitol in Albany, the Senate debated the merits of the Elgin Botanic Garden on and off for a week. Then, on February 20, DeWitt Clinton took the floor. A physically imposing man his whole adult life, he was now in his prime. He carried himself with such hauteur that his

The Elgin Botanic Garden as it appeared around 1810

opponents had nicknamed him the Great Apollo. In some circles, men had started to consider him a promising candidate for president.

Clinton began his remarks in the Senate that day by lecturing his fellow senators on the history of European botany, and then he lectured them on the history of North American botany. He mentioned the most famous botanists who had traveled and collected in America, and some of the more obscure, too—"Colden, Cutler, Belknap, the Bartrams, Muhlenberg, the Michauxs, Barton, and others." He talked about how critical the study of plants was to American progress in medicine, agriculture, and commerce, and finally he closed with a passionate plea for Elgin. The state should purchase the garden. The lawmakers would not only improve the state's medical schools, they would also burnish the intellectual reputation of the entire nation. Hosack confessed to a friend afterward that although he didn't much agree with Clinton's politics, he was deeply moved by his friend's devotion to the cause of natural history.

Two days after Clinton's speech, a clerk walked next door to the Assembly chamber to deliver the news. The Elgin act had passed the Senate. In the Assembly, a lawmaker by the name of Pratt now tried to block consideration of the bill until after the plant collections and the land could be appraised. But Hosack had loyal allies here, too. Among those leading the charge for Elgin in the Assembly was his friend Mitchill, whose political career had him alternating between federal and state offices.

On March 9, 1810, the assemblymen took a vote on the bill: fifty-seven to forty-two. They sent a clerk back next door to tell the senators. Elgin would survive. Three days later, the act became law. Five managers—one of them Mitchill—were appointed to take charge of organizing a public lottery to raise money for the purchase of the garden from Hosack. The news of Hosack's victory raced south from Albany to New York City.

Chapter 13

"YOU KNOW, BETTER
THAN ANY MAN"

ROSES BLOOMED ALL AROUND HOSACK AS HE STROLLED
through Elgin in the summer of 1810. There were roses from the Carolinas, Spain, Provence, China, and the farthest reaches of Russia.
Their delicate pinks and deep reds accented the flower beds and their
velvety petals fell on the hothouse floors. The garden looked especially
beautiful to Hosack that summer. He observed with satisfaction that
his property was clearer of weeds than the neighboring farms, thanks
to a British technique that involved sowing twenty pounds of clover
seed per acre of grass. A huge old oak stood in the middle of Elgin's
smooth lawns. Here and there, winding paths led through thickets of
flowering shrubbery, and as he walked along them he saw an ever-
shifting composition of textures and colors.

Hosack was in high spirits. The state appraisers had just submitted their estimate for the garden, suggesting $103,137 (about $2 million today) as a reasonable payment to him. He calculated that he had
spent more than that on the land, labor, gardening supplies, plant collecting, buildings, and interest payments on his mortgage—but not
so much more that he found the appraisal upsetting. When he sub-
tracted the appraisers' estimate for the buildings, fences, and other

improvements, and then divided the remainder by his twenty acres, he found the appraisers' offer per acre was a robust $3,700 (about $75,000 in today's dollars). After a decade of uncertainty, the next steps unfolded before him with perfect clarity. The state would take ownership of the garden soon, and the lottery would be drawn, after which they would pay Hosack the agreed-upon amount. In the meantime he would continue to oversee the garden, exchange specimens with his botanical contacts, and conduct research with his students. Then, when the state had paid him, Hosack would hand the reins to a talented plantsman chosen on his recommendation. The garden would be a fully public institution, its scientific research programs assured by an enlightened government for the benefit of his own and future generations.

Hosack would be handing over a stunning botanical collection. He had gathered hundreds of additional specimens since publishing his first Elgin catalogue in 1806. There were a dozen new species of rose alone, but they were just the beginning. When he walked through the conservatory now, he passed a dark flowering catchfly (*Silene ornata*) from the Cape of Good Hope and a honey locust tree (*Gleditsia sinensis*) from China. He was growing rare plants found by the Bartrams and André Michaux in the American South, including yellow-flowered anise (*Illicium parviflorum*) from Florida and the red side-saddle flower (*Sarracenia psittacina*) from the Carolinas, also called pitcherplant.

Perhaps Alexander von Humboldt had made good on his earlier offer to send specimens, for Hosack had a strange plant with spotted, heart-shaped leaves that Humboldt and his friend Aimé Bonpland had found in Mexico and named *Smilax cordifolia*. In the hothouses were two more species that Humboldt and Bonpland had named: a sarsaparilla they called *Smilax syphilitica* and another plant they called *Inga microphylla*, both from South America. Hosack had four different species of passionflower (*Passiflora*) and a bird-of-paradise (*Strelitzia reginae*). There was a date palm (*Phoenix dactylifera*) from Syria, a banana tree (*Musa sapientum*) from the West Indies, and a breadfruit tree (*Artocarpus incisa*) from the South Sea islands. Hosack also had a West Indian plant called *Tephrosia toxicaria* that had the power to paralyze fish.

Outside the conservatory, Hosack's arboretum was flourishing along the perimeter of the property. He had planted new species of oaks, firs, spruces, pines, alders, birches, and elms; among these trees were many whose names the botanical world owed to his friend François André Michaux, including two from the *Juglans* (walnut) genus. Meanwhile, in Elgin's agricultural fields, Hosack had added many dozens of species of grasses and grains, such as buckwheat, oats, wheat, fescue, and sedge. In a pond on the property he was cultivating water lilies, marsh grasses, and water eryngo (*Eryngium aquaticum*), a medicinal plant that the Cherokee and other native peoples used as an antidote to poisonous snakebites.

Hosack had dreamed of making Elgin a source for plant-based remedies that American doctors were otherwise forced to import. He now had licorice root (*Glycyrrhiza glabra*), a staple medicine he had bought at an import shop on Maiden Lane when he set up his own medical practice in the 1790s. He had his own source for camphor, as well—his *Laurus camphora* tree, which had come to him from Japan. He had a gum arabic tree (*Acacia nilotica*) from Egypt, four new species of aloe from the Cape of Good Hope, turmeric (*Curcuma longa*) from the East Indies, and a balsam tree (*Copaifera officinalis*) from South America. He had also amassed more medicinal species native to North America. They grew in his outdoor beds, in his greenhouse and hothouses, and in his arboretum. He had a prickly ash (*Zanthoxylum fraxineum*), sometimes called toothache tree, whose bark when chewed was thought to alleviate toothaches. He had three species of clubmoss (*Lycopodium*), whose water-repellent spores were used by physicians and pharmacists for coating pills. Hosack and his students conducted chemical experiments on the Elgin collections—boiling down roots and leaves, subjecting the resulting liquids to different chemicals, and noting their reactions in an effort to compare medicinal properties. They hadn't made any dramatic discoveries thus far, but they were making progress in mapping the chemical similarities and differences among certain species. Even going down dead ends contributed to the sum of scientific knowledge. Across the United States and Europe people were increasingly rewarding Hosack with praise and respect. Naturalists consulted him, doctors followed his research, and European visitors made sure to include a tour of Elgin on their itineraries. When Hosack stood on his hilltop at Elgin now, he was surrounded by the garden of his dreams.

⸻ ⚭ ⸻

ON JULY 20, 1810, Mitchill sent Hosack a plant specimen and asked for his help identifying it. Hosack wrote back the next day to say it was Canada thistle (*Cnicus arvensis*), adding that he had learned from Curtis that farmers called it "cursed thistle" because of its nasty habit of invading their fields. Two days later, Mitchill and his wife arranged to meet Hosack and Mary at Elgin so they could show some visiting British friends around the garden. The artists James and Ellen Sharples were well known in the United States. On a visit in the 1790s, they had done portraits of Washington, Adams, Jefferson, Hamilton, and Burr, among others. Now they had returned to secure new commissions, visit old friends, and see how the country had changed in their absence. Mitchill felt it was important that Elgin be on their itinerary. Although James Sharples was on a trip upstate, Ellen Sharples and her daughter rode to Elgin with the Mitchills in their carriage. As they strolled through the garden, Ellen was especially intrigued by the rare plants and beautiful blooms Hosack showed them.

The Mitchills also took the Sharpleses on a ferryboat across the Hudson to the Palisades, where they climbed the rocks and picnicked. At the summit they were rewarded with a striking view across the river toward Elgin and the surrounding countryside—although when Ellen looked straight down, she shivered "at the idea of being precipitated," as she noted in her diary. They then wound their way back down to the river, where they hired a boat to take them to another spot on the shore. They had climbed partway up the hillside, Ellen later noted in her journal, when they found what they were looking for: "The neat elegant monument of white marble is placed on the spot where the

General fell, at the foot of stupendous rocks." A Latin inscription had been etched into the marble.

Incorrupta Fides, nudaque Veritas,
Quando ullam invenient parem?
Multis ille quidem flebilis occidit.

These were the very lines from Horace that Hosack had quoted in his account of Hamilton's death. The monument to Hamilton, in the form of a fourteen-foot obelisk, had been erected by the New York chapter of the St. Andrew's Society, a fraternal group founded in the 1750s for the benefit of New Yorkers of Scottish descent. Hosack was a member and later a vice president. Given Hosack's intimate connection with the duel, the Horace inscription, and his St. Andrew's membership, he may well have been instrumental in having the monument placed to honor Hamilton.*

Burr, meanwhile, was still trying to stay in touch with Hosack from Europe, writing in March 1810 to a mutual friend, "Tell Dr. Hosack that I wish to hear from him." As Burr waited for news of Theo and word that he might safely return to the United States, he made the most of his time in Paris.† He went to the Paris Opera and the Comédie-Française. He rambled along the quais of the Seine and browsed through the stalls looking for books to send Theo. He went for walks in the great pleasure gardens of the city—the Luxembourg Gardens, the Palais-Royal, and the Tuileries. At the Palace of Versailles outside Paris, he found the royal gardens laid out "in a style of magnificence surpassing anything I have seen."

---

* The Horace inscription on the Weehawken monument recorded by Ellen Sharples in her diary contains an error (*quidem* for *bonis*) that Hosack also made when quoting these lines in his August 1804 account to William Coleman about the duel.

† William Hosack was still in Germany, where he accidentally shot a man while on a hunting trip. Burr made light of the episode in his journal, writing that William "went a shooting, and, in shooting at a hare, shot a man; not dead, but wounded him badly, which cost him money and gave him much trouble." William fled the vicinity after this incident, but Burr didn't know where he had gone.

*The Jardin des Plantes in Paris as it appeared
not long before Burr's visit*

Burr toured the scientific collections at the Jardin des Plantes, seeking out François André Michaux to talk about botany. Michaux was completing his *North American Sylva* at the moment, and Burr noted in his journal that the work would show that "we (not the whole continent, but the United States alone) have three times the number of useful trees that Europe can boast." On one occasion Burr went to Michaux's house, hoping "to ascertain the identity of a plant and a tree, both vaunted in medicine." Burr didn't record the names of these species, but he did report that "Mr. M. gave me the most perfect satisfaction."

In December 1810, Burr wrote to Hosack from Paris to ask a botanical favor. A French friend had requested Burr's help in procuring a long list of plants from America, and the latter had naturally thought of Hosack. Burr told Hosack, "You know, better than any man" how to track down the specimens. The request had come from Étienne Calmelet, a close friend of Empress Josephine. Burr explained to Hosack he realized that "most probably not one of the articles will be found with any 'marchand de greens' in America." Fulfilling Calmelet's entire order, Burr conjectured, "would require to open a correspondence with perhaps fifty persons in different parts of the continent, and to pursue the thing for years." Burr didn't expect Hosack to undertake all this work, but he did express hope that some of the plants might already be growing at Elgin.

By way of thanks for Hosack's trouble, Burr said he was requesting that a botanist at the Jardin des Plantes package up some specimens, including one of Burr's own choosing: "that precious herb estragore, which you will have the honor of introducing into our country." This herb (*Artemisia dracunculus*) was actually called estragon by the French, as Burr indicated elsewhere in his journal; in English, it is called tarragon. Burr was wrong that Hosack would be introducing it. Already during the French Revolution, a French merchant who had emigrated to New York was selling estragon at his shop on William Street, and Hosack himself was growing it at Elgin by 1806. At any rate, Hosack seems to have assisted Burr with his request, because soon a collection of three hundred seeds arrived at Elgin from André Thouin, the director of the Jardin des Plantes. Hosack was ecstatic over Thouin's shipment, as his student John Francis could still recall vividly decades later.

When Hosack received these treasures from Paris, he was frantically busy trying to get his second Elgin catalogue into print. His goal was to facilitate specimen exchange with other botanists, so it would include the Latin and common names of the garden's two thousand species, with countries of origin provided but no pictures. In a bold announcement of his botanical aspirations for the United States, Hosack entitled his catalogue *Hortus Elginensis* (Elgin Garden). He told friends that the *Hortus Elginensis* was modeled on the *Hortus Kewensis*, a recent catalogue from the Royal Botanic Gardens at Kew. When it was fresh from the printer, Hosack sent copies of the *Hortus Elginensis* to all his botanical and medical friends and correspondents, among them Jussieu, Thouin, Delile, and Michaux in Paris, and James Edward Smith in London. He directed another copy to Bernard McMahon, who was building his own garden outside Philadelphia, named the Upsal Botanic Garden in honor of Linnaeus. Hosack also sent a copy of the *Hortus Elginensis* to Burr's friend Carl Peter Thunberg at Linnaeus's original Uppsala garden.

Hosack was on firm ground at last. He could now launch his most ambitious project of all: his flora of North America, modeled on

William Curtis's *Flora Londinensis* and James Edward Smith's *English Botany*. Hosack envisioned a series of illustrated volumes that would eventually encompass every plant on the continent. With his customary self-confidence, he was undeterred by the fact that it would undoubtedly be decades before all the species of North America had been collected and identified. He brushed aside any memory of Curtis's frustrations with the *Flora Londinensis* and imagined the day he would present the very first copy of the first volume of his *Flora of North America*, as he took to calling it, to the legislature in Albany. The state's payment for the garden would be critical to the project, when it eventually came in. He couldn't hasten that hour, but in the meantime he assembled a team of botanists and artists to begin work.

Leading the project would be Frederick Pursh, with his expertise in American plants, gained through long months of work on the Lewis and Clark specimens. John Francis would join Pursh, as would Hosack's nephews Caspar and John Eddy. Amos Eaton, another former student of Hosack's, had moved back to the Catskills, but Hosack didn't mind because Eaton had just founded a school there that was devoted entirely to botany. "To your pupils and their teacher, as first on the field, much praise is due," Hosack wrote to Eaton, pledging his unconditional support. Pursh, Francis, and the Eddy boys would be joined on the *Flora* by John Eatton Le Conte, a Columbia graduate in his midtwenties, as well as by Isaac Roosevelt, a twenty-year-old medical student from the village of Hyde Park, near Samuel Bard's estate up the Hudson.* Hosack also hired James Inderwick, "a young gentleman of great genius and taste," whose job would be to draw and hand-color the engravings of the plant specimens that would go with the written descriptions.

Hosack excitedly began planning for a second publication, a magazine modeled on Curtis's *Botanical Magazine*. Pursh agreed to edit this, too. Applause for these new projects soon reached Hosack from Britain, where the editors of the *London Medical and Physical Journal*

---

* Isaac Roosevelt would not live long enough to meet his grandson Franklin Delano Roosevelt, born two decades after Isaac's death. The young FDR often visited his uncle John A. Roosevelt at Rosedale, the Italianate villa at Hyde Park his grandfather Isaac had built in 1832.

praised his "ardent perseverance." Hosack's selflessness contrasted sharply with "the cold, calculating, trading spirit of the public body." They hoped he would not lose heart but would instead "make the whole of the American continent his GARDEN."

⸎

HOSACK URGENTLY NEEDED his nephew John Eddy for the *Flora*, but he would have to be patient. In late June 1810 John had gone upstate with DeWitt Clinton, Gouverneur Morris, and a few other prominent New Yorkers, including his own father, Thomas Eddy. The men were members of a state commission appointed to explore the course of the Mohawk River, in search of the best route for a possible canal linking the Hudson with the Great Lakes. Hosack had packaged up medicines for the men to take with them, including emetics and Peruvian bark. Clinton had extra time on his hands for the trip, having recently been removed from the post of mayor in a political shuffle. Morris had invited the artist James Sharples to go along—thus making him miss the Elgin visit—and Ellen Sharples joked in her journal about the fact that Morris and her husband would be traveling by carriage to meet the other men upstate; Morris had put his "French cook and other servants" in a second carriage.

John Eddy was going on the trip mainly in order to botanize along the Mohawk River. He traveled to Albany by steamboat, departing on a lovely summer afternoon with his father and Clinton. By the Fourth of July, they were all settling into a flat-bottomed boat outfitted with awnings, curtains, and benches. They christened the boat the *Eddy*, in honor of John's father, calling the baggage boat the *Morris*. They had invented these names as a joke, but then the boatmen had painted them on the vessels. Some of the men in the party had brought along books to while away the languid summer hours as they floated along the Mohawk. They agreed to consolidate these in a single trunk, giving John the key and appointing him librarian. Clinton soon decided that the most interesting volume in their little lending library was a treatise disproving the existence of magic and magicians. Like Clinton, John was keeping a travel diary, and he noted in it that he was "appointed Segar Keeper general, and received 2 boxes containing 2000 Segars."

In his own diary, Clinton mentioned that John was deaf, and that he and the other men could communicate with John only by signing or by writing things down. Clinton soon decided John was extremely well read and had an impressive memory, but he also found his "temper bad." If John was feeling out of sorts, however, he didn't reveal it in his diary. The views from the *Eddy* completely mesmerized him. "The richly cultivated country" and the "high Hills cloathed with wood among which we had winded this afternoon, had presented us with many rural and romantic views." Near the town of German Flatts, he admired the fertile bottomland that stretched into the distance on both sides of the river. The local Dutch farmers cultivated their farms so beautifully that they looked to John more like gardens than fields of crops.

Clinton got one thing right about John—he loved books. They suffused the quiet world around him with layers of meaning. The Dutch settlers in Schenectady who were sweeping the streets for a Fourth of July parade, for example, made John think of an anecdote from Washington Irving's *Knickerbocker Tales*. A waterfall on the Mohawk reminded him of passages in Jefferson's *Notes on the State of Virginia* about the Shenandoah and the Potomac. John was also a gifted writer, possessing an almost alchemical power to turn images into words. One evening, he witnessed a scene so magnificently American he thought it could never have "entered into the imagination of an european painter." On a glassy lake, a Native American man stood in a canoe poised to spear a fish, his figure illuminated by the red glow of a campfire on the shore. In the distance beyond him, the towering forest faded into the misty darkness.

Whenever the barge pulled to the shore to allow the men to pick up supplies or investigate a new town, John jumped off to go plant hunting. He marveled at the riot of species "growing wild with a luxuriance that can scarcely be imagined by those who have not seen the banks of the Mohawk." On July 9 he was excited to find a specimen of what he noted as *Hypericum pentagyna* (a species of St. John's wort) in bright-yellow bloom. He looked for seeds to take back to Elgin but had to content himself with gathering some of the other plants growing in great masses on the riverbank. He was upset with himself for not bringing any botany books on the trip—he had decided against it for fear they might prove too heavy to lug around. He did his best to jot down accurate accounts of where he had found the various plants growing and what they looked like in the wild. When he got back to Elgin, he could check these descriptions against the volumes in Hosack's library and try to identify his new specimens.

Clinton, too, was tramping along the shoreline in search of plants. He found wild hemp, mandrakes, and a "beautiful wild flower, whose botanical name is *Ocsis*." He spotted Hosack's favorite plant, boneset, growing wild and noted in his diary that it was "useful in medicine." Clinton also kept track of the fish, birds, and other fauna he saw as the team floated along the Mohawk and walked its banks. Inside the dried stalks of a mullein plant, he found a nest of "young bees in a chrysalis state, deposited there by the old ones." Clinton lamented the overfishing of the Mohawk River, which had once teemed with trout but now seemed to have none. Still, birds of all kinds chattered in the trees around him, and he was very pleased to see "great numbers of bitterns, blackbirds, robins, and bank swallows, which perforate the banks of the river. Also, some wood-ducks, gulls, sheldrakes, bob-linklins, king-birds, crows, kildares, small snipe, woodpeckers, woodcock, wrens, yellow birds, phebes, blue jays, high-holes, pigeons, thrushes, and larks." But it was geology that truly made Clinton's heart pound. He was electrified by the drama inscribed in the rocks near Little Falls, where all around him he saw piles of giant boulders scattered and sculpted by the "violence of the waves."

Clinton and his companions returned to New York City later that summer and reported to the public on the main question assigned to them by the state. Yes, it looked feasible to run a canal from Albany to Lake Erie. With enough political will, New York could have the glory of linking the Atlantic Ocean to the very heart of the continent.

Hosack was impressed with Clinton's scientific research on this trip. When Hosack learned of his own election to the American Philosophical Society not long afterward, his first reaction was to nominate his friend. "I wish you would make DeWitt Clinton a member of your society," he wrote the secretary, adding wishfully that science "knows not party politics."

⟡

STARTING IN JULY 1810, while Clinton and his travel companions were still upstate, it began raining almost every day. Along the river, cellars began to fill up and water lapped at the tops of the docks. This deluge was followed by a stretch of steamy weather. Rotting animals and vegetables littered the streets. Hosack wrote to a friend in Philadelphia that he couldn't remember the city ever having smelled quite so awful, but he also felt perversely cheerful about the stench for medical reasons. Those of his misguided colleagues who thought yellow fever was of domestic origin generally argued that it emanated from putrefying organic matter. Yet as of September—prime yellow-fever season—no cases of the disease had been reported, to his knowledge.

The rain and heat may not have brought fever, but Hosack was in for a miserable autumn nonetheless. The first dark hint was a June 1810 letter that had appeared in the Albany paper, assailing the state's plan to pay $100,000 for the "frippery of a botanic garden!" The anonymous author raged that it was enough money to build roads all over Manhattan, or found a university, or arm ten thousand militiamen, or finish fortifying the harbor against a British attack. The next bad news came in August, when Hosack heard a rumor that his perennial foe Nicholas Romayne was making a "vile attempt"—Hosack's words— to intervene in the Elgin appraisal process.

Hosack thought he knew why. Romayne was worried that if the state purchased the garden at such great expense, it would consider its duties to medical education fulfilled, and Romayne's fledgling College of Physicians and Surgeons, a rival to Columbia's medical faculty, would have no prospect of seeing state funds anytime soon. Hosack told Bard privately that Romayne was the sort of man who enjoyed "blowing [on] coals." But he was shocked to learn that Romayne had managed to wrangle some sort of supportive letter out of Samuel

Latham Mitchill. Hosack told a friend he thought Romayne must have duped Mitchill. "Otherwise, I shall hereafter entertain a very despicable opinion of human nature." Mitchill immediately reassured Hosack that he had known nothing about Romayne's true intentions and was "much disgusted." Before long, however, Hosack had to conclude that Mitchill was being disingenuous, as he wrote to another friend. "When I may see you, I have some precious things to lay before you relative to Romayne and Mitchill."

The mounting opposition to the appraisal alarmed Hosack. He had to make sure the deal didn't fall apart at this delicate stage. He was sure the estimate was fair, because both he and the team of appraisers had solicited estimates from landowners all around Elgin and on comparable tracts elsewhere on the island. The Kip brothers, who lived near the East River on their Kip's Bay estate, had told Hosack they thought the Elgin land was worth $3,000 per acre, while a man named Theodorus Bailey said he had recently bought land a mile south of Elgin for $5,000 per acre. The state's Elgin estimate, at $3,700 an acre, was perfectly reasonable. In any case, the land would only increase in value over the coming years, as Hosack confidently predicted to a friend: "N. York cannot under any circumstances fail to become the metropolis of this country."

All that autumn of 1810 Hosack juggled his Columbia lectures, clinical rounds, and work on the *Flora of North America* with meetings about the land deal. On his side, he had District Attorney Cadwallader D. Colden, Governor Daniel Tompkins, and DeWitt Clinton. Hosack also had steadfast allies on the Common Council, men who didn't hesitate to praise his "patriotic contributions of time, talents, and labour" and who insisted on the "immense importance of a botanic garden." They also pointed out that if the city and the state didn't save Hosack's garden, it would be years before anyone else would try to found a New York botanical garden.

At this precarious moment, Frederick Pursh set down his gardening tools and left for the West Indies. He had been ill with some sort of stubborn fever and had finally decided that a change of climate was his best hope for recovery. He planned to return to New York when he was well again, but in the meantime, Elgin had no head gardener. Hosack filled the position as soon as he could. Acting on a tip from the proprietors of a well-known nursery near the Brompton Botanic Garden, he hired a plantsman named Michael Dennison, who had recently settled in New York with plans to open a seed store. Still, no one in the world could stand in for Pursh on Hosack's planned *Flora of North America*.

In early September 1810, only a week after he hired Dennison, Hosack received terrible news from the state land office. They had decided to slash the appraisers' original estimate by nearly $29,000 (around half a million in today's dollars). If Hosack accepted the deal, he would lose every penny he had spent on the conservatory buildings and on all the other improvements he had made to the land. He would also be unable to pay the botanists and artists working on the *Flora of North America*, and the project would collapse. Yet his friends had worked so generously to drum up support for Elgin in the state legislature and the Common Council. Hosack felt cornered. In October, he accepted the lower appraisal.

That winter, while he waited for the state to draw the lottery, Hosack kept himself as busy as ever. He spent time with Mary and the children, taught his Columbia courses, visited his patients, wrote his medical essays—and tried to keep the Elgin plants alive. In January 1811, the state officially took possession of the garden, and Hosack assumed he would now be consulted about its future. But in March, to his shock, the state suddenly handed the garden to the College of Physicians and Surgeons. It was beginnning to dawn on the legislators that "to imitate the garden of plants in Paris or another botanical institution of Europe would require too great an annual expenditure."

Hosack moved quickly. When Nicholas Romayne, his *bête noire* at the College of Physicians and Surgeons, resigned in April from its presidency over an unrelated matter, Hosack seized the opportunity to persuade his mentor Samuel Bard to take Romayne's place. Then he took a bold measure—he joined the faculty of the College of Physicians and Surgeons himself, while continuing to teach at Columbia. The income from the two positions would help fund the *Flora*, and his foothold at the College of Physicians and Surgeons would let him stand guard over Elgin. Although Hosack's motives were honorable, his Columbia colleagues were incensed. In early May he had a disagreeable conversation with James Stringham, one of five professors who had just sent a vituperative letter to the trustees demanding that Hosack resign from Columbia. Hosack's decision to teach at the city's rival medical school was "totally incompatible with the duties he owes to Columbia College." Worse still—from his Columbia colleagues' point of view—Hosack had thrown his support behind a new state initiative to merge the College of Physicians and Surgeons with Columbia's medical faculty.

Years later, Hosack would give a speech to a group of medical students in which he described his most disputatious fellow physicians as being "like tigers concealed in their jungles, [who] lie crouching for their prey." Hosack's conversation with Stringham left him feeling similarly ambushed and aggrieved. His Columbia colleagues had offered him tepid moral support and no financial support throughout the years he had labored on Elgin. Now they were willfully misconstruing his motives for supporting a merger between the two medical faculties. The idea had come from Albany, not from Hosack; he had simply volunteered to present it to his Columbia colleagues. As he insisted to Stringham, his sole desire in doing so had been to improve medical education in New York. "If that be an offense, I have indeed offended, and shall continue to offend."

But Hosack's situation was untenable. In June 1811, he resigned from Columbia and moved to the College of Physicians and Surgeons, where he was appointed to teach courses on the theory and practice of medicine—and on botany, using the garden. With Romayne gone and the supportive Bard in charge, the College of Physicians and Surgeons soon proved a hospitable home for Elgin. Hosack's new colleagues quickly appropriated funds to repair the many cracked glass panes on the hothouses. At a faculty meeting in May 1811, they had officially praised Hosack's "zeal for the propagation of botanical knowledge" and pledged to take care of the garden.

It was a bittersweet moment. For the first time since he had founded Elgin in 1801, Hosack stood shoulder to shoulder with a group of his peers in its ownership and management. On June 3, he invited his colleagues to his house to discuss the garden's immediate future. His drawing room was a pleasant place for a faculty meeting, much more comfortable than the classroom at the college's temporary quarters on Pearl Street. They had a problem that urgently needed solving. The state had agreed only to *buy* the garden from Hosack—not to pay for its upkeep. The college certainly couldn't afford to pay, so the professors had devised a plan to raise the required funds by leasing the property to a competent, biddable gardener. It appeared, however, that there was little appetite for the job among the city's few nurserymen. The professors had just two applications on the table. The first was from a man named William Tough, who proposed leasing the garden for five years and promised to "keep at all times healthy at least three of each species of plants" and to preserve the grounds in as "clean and hand-some a state" as Hosack had left them. In return, Tough asked that he be permitted to cultivate plants at the garden and sell them for his own profit. The other application was from Elgin's current head gardener, Michael Dennison, who proposed almost identical terms. The professors settled on Dennison, but Hosack probably should have noticed that Dennison's application didn't mention anything about keeping the plants alive and well.

The college would now need a complete inventory of the speci-mens Dennison was to inherit with his lease. Hosack took charge of this task, going up to Elgin a few weeks later. June was turning out to be ferociously hot and arid. Not only in the garden but all over the island, plants were suffering from the drought and in some places were "altogether destroyed." He went walking through the greenhouse and hothouses with a little notebook in hand, jotting down the names of the dozens of species he had acquired even in the few months since the *Hortus Elginensis* had gone to press. Elgin now contained more than two thousand species, but when he turned in the inventory he asked the college that just one rare and beautiful flower be set aside for him. He was leaving four *Camellia japonica* plants at the garden, but only temporarily. They were very precious, and absolutely "not to be sold" by Dennison. At the end of June, the searing heat gave way to a strange cold snap, and New Yorkers began rummaging in their ward-robes and trunks for winter clothes. Hosack found he needed blankets at night. If Dennison didn't see to the cracked panes on the hothouses immediately, the tropical plants might perish.

Soon after Dennison took over the garden lease, Pursh returned from the West Indies. He discovered that the city was roiling with anxiety over an anticipated British attack, a situation he quickly concluded was "very unfavorable to the publication of scientific works." He packed up his worldly possessions—including his Lewis and Clark specimens, drawings, and descriptions—and sailed for London. War or no war, the stalwart old Linnean Society would keep the banners of botany flying. In the United States, however, shoring up forts and training soldiers took precedence over funding the arts and sciences. Writing from Monticello in the spring of 1812, Jefferson made this point bluntly to his friend John Bradbury, a British naturalist. Bradbury had written to Jefferson from Washington to say he had learned that the federal government was on the verge of founding a public botanical garden, but Jefferson replied that "it is an idea without the least foundation." He conceded that it would probably be within President Madison's power to allot public land for a botanical garden, but its organizers would have to overcome the "suspicion that it would be converted into a mere kitchen garden, for the supply of the town & market."

Anyway, observed Jefferson, the national mood hardly favored such projects. Botanical gardens might be "desired by every friend of science"—but not by a nation preparing for enemy attack.

Chapter 14

"INSTEAD OF CREEPING
ALONG THE EARTH"

URR DESPERATELY CRAVED A CIGAR. HE HAD LEFT SIX BEHIND in London, "all which I would gladly smoke this evening." But Captain Potter of the *Aurora* refused to go ashore for such trifles. They had only just set sail for America.

Burr had decided he could finally show his face in New York again. For now, he was taking the precaution of traveling under an assumed name, his true identity known only to Captain Potter. It was early April 1812, and they were bound for Boston, where Burr planned to board a smaller vessel for New York. The voyage began on rough seas in a freezing rain that was soon followed by hail and snow. The ship was heaving so violently, Burr noted in his journal, "that we don't attempt to put anything on the table, but eat off the floor." As he struggled to write legibly, the ink blotched and spread on the damp paper. He spent his first week at sea sleeping, reading, and reorganizing his book-filled trunks (in one of which he discovered seven cigars). On April 9, "though half sick all day," he managed to enjoy a volume by Humboldt so much that he raced through and finished it the next day—when the ship was at "Lat. 49°29," he noted in a Humboldtian flourish.

As the *Aurora* entered the open Atlantic, Captain Potter kept watch for British ships. He confided to Burr that he was worried President

Madison might declare war while they were at sea, putting the *Aurora* at risk of capture as a war prize. Burr scoffed at these fears; he thought that "J. Madison & Co." were far too cowardly to take on the British Empire. "I treat their war-prattle as I should that of a bevy of boarding-school misses who should talk of making war; show them a bayonet or a sword, and they run and hide." When the *Aurora* reached Boston safely in early May, Burr found that he was right. Madison had not declared war.

Burr was forced to wait several weeks for passage from Boston to New York, but he finally embarked on the sloop *Rose*. He sailed along the Connecticut coast through Long Island Sound, arriving on June 8 at Rikers Island, at the northern end of the East River. Although Burr was hoping to travel down the river that night and slip unnoticed into New York City, the captain of the *Rose* decided to wait until morning to navigate the tricky strait known as Hell Gate. Annoyed at the delay, Burr hailed a passing sailboat and asked its occupants—two farmers from Long Island—to drop him in Manhattan. They agreed, but when their boat was blown off course toward Long Island, Burr jumped ship again, this time paying "two vagabonds in a skiff" a dollar to take him to shore. He went to Water Street to look for a friend.

Had Burr arrived during the day instead of sneaking across the harbor in the dark, he could not have missed the colossal evidence that New York officials did not share his dismissive attitude toward the possibility of war. Several new forts had sprung up on the southern shores of Manhattan, and another had risen on the Governors Island site of the old Fort Jay, which had fallen into ruins. According to Diedrich Knickerbocker, Washington Irving's alter ego, Governors Island had once been a "smiling garden" but now looked like "a fierce little warrior in a big cocked hat, breathing gunpowder and defiance to the world!" The new fort on the island had been completed in 1810, and by the time Burr came home, officials had also erected a broad, squat stone tower called Castle Williams. When the castle was finished, American frigates fired practice rounds of cannonballs at its walls, after which military inspectors concluded that "no apprehension need be entertained of their being battered down."

They were just in time. Ten days after Burr arrived home, the United

States declared war on Great Britain. Proponents of war had found fuel in the ongoing practice of naval impressment as well as in reports of Native Americans emboldened by the British to attack US citizens in the frontier territories. New Yorkers went into a frenzy of fear and recrimination. On June 20, 1812, two days after the war vote in Congress, a newspaper published the names of prowar delegates from New York who had decided "to plunge this State into a tremendous scene of suffering, devastation and carnage." The most outspoken antiwar New Yorker was DeWitt Clinton, who had been consolidating his control in the city and state since regaining the mayoralty in February 1811. Now he was lieutenant governor. Clinton had amassed so much power through so many different offices that a sarcastic open letter published in a New York paper in May 1812 addressed him as "Lieutenant Governor of the state, (and every once and a-while) Mayor of the city of New York, Judge of the Court of Sessions and of the Mayor's Court, Commissioner of Lock and Canal navigation from Lake Erie to the Hudson, Commissioner of fortifications in and near the city of New York, exoff[i]cio director of the bank, &c." In the months running up to the declaration of war in June 1812, Clinton had been cultivating his position at the head of a coalition of Federalists and anti-Madison Democratic-Republicans. The Great Apollo had decided to run for president. He would lose to Madison that fall.

While New York girded itself for a British attack, Burr set about reestablishing as much of a law practice as he could muster and hoping for news from Theo in South Carolina. He had written her on May 9, announcing his return to the United States and asking her to send his grandson to New York so he could oversee his studies. But on July 12—the eighth anniversary of Hamilton's death—Theo wrote to Burr with heartrending news. "There is no more joy for me; the world is a blank. I have lost my boy. My child is gone forever. He expired on the 30th of June." Burr tried to console Theo as she sank into a bewildering fog of pain. Longing to see her father, she made plans to travel to New York as soon as her husband's affairs allowed her to steal away. Finally, on the last day of 1812, Theo set sail from South Carolina on the ship *Patriot*. She never arrived in New York. The *Patriot* was lost at sea.

Burr had kept up a constant conversation with his darling girl in his mind and on paper throughout his European travels. Now he would never pull a chair close to hers and laugh with her about the fascinating, maddening people he had met in Europe. He would never dazzle her with stories of the palaces, the landscapes, and the gardens he had seen. He would never get to take her little hand in his again, or that of his only grandchild. Theo's disappearance from the world set Burr permanently adrift. He would spend the rest of his life honoring his matchless love for her by surrounding himself with young people.

———— ❦ ————

THE ERA OF POST-REVOLUTIONARY PEACE—however tenuous it had sometimes seemed—was over. Support for Elgin now waned further. In 1812, the year following Hosack's promising move to the College of Physicians and Surgeons, a handful of his colleagues issued a statement arguing that the garden was a financial burden to the college. With gratuitous cruelty, they noted that the most faithful garden visitors seemed to be the two dozen or so cattle Dennison was pasturing on the property. "These animals, to the number of 20 or 30 attend the Botanical Garden, and excite the ridicule of travellers passing the [Middle] Road."

There were at least a few human visitors to the garden that autumn, however. During a spell of mild days at the heart of October, a small team of men appeared. The one in charge was a dark-haired young man who carried a leather notebook in his hand and a new world in his head. He and his men surveyed the Elgin property with their instruments. Later, they came back and placed marble slabs at regular intervals along the Middle Road. Each slab bore a different number on its south-facing side: 47, 48, 49, 50, 51. On their west-facing sides, they all bore one number: 5.

The young man was John Randel Jr., civil servant and sorcerer. He had been hired by the city to map a grid of new streets onto the island. In his wake, boulders shattered and streams went underground. Where he and his team planted their rows of marble, flat ribbons of pavement slowly began to unfurl over hard-packed dirt lanes and across farmers' fields. Speculators snapped up land and sheared it into narrow strips. Villages began to melt together, the memories of old borders fading. The Middle Road was to be renamed: the Fifth Avenue.

⸎

IN MAY 1813, THOMAS JEFFERSON wrote to John Adams with sad tidings. The two men had fallen out after Adams had left the presidency, and it had taken Benjamin Rush's ministrations to heal their wounded friendship. Now Rush had died in Philadelphia of a pulmonary illness. "Another of our friends of 76. is gone, my dear Sir, " Jefferson wrote Adams. "A better man, than Rush, could not have left us." Rush's son James, knowing Hosack's profound attachment to Rush, wrote to him three days later, but Hosack had already heard the news. "To your family, it is an afflicting bereavement and on this occasion I feel myself one of its members," he wrote to James.

Rush's paternal warmth toward Hosack had persisted through two decades of intense medical debates, particularly over the causes of yellow fever. While other physicians who joined in the fray left blood on the floor, Hosack and Rush had stayed extremely close. The previous summer, Rush had sent Hosack a typically generous letter affirming their friendship in the face of professional pressure to fall out. "Let us show the world that a difference of opinion upon medical subjects is not incompatible with medical friendships; and in so doing, let us throw the whole odium of the hostility of physicians to each other upon their competition for business and money." When Hosack received this letter, he was, in fact, arranging his own tribute to their friendship—he had asked the artist Thomas Sully to paint a portrait of Rush. Although Hosack already owned a sketch of Rush by James Sharples, he told Sully that it didn't do Rush justice. Hosack wanted a painting that would perfectly capture "the mind which animates the face of my friend." He also ventured an idea to Sully about the composition: "Would it improve the picture by throwing into the background a distant view of your city Hospital or University to which Dr. Rush's labours have been so much devoted?" When Sully completed his portrait the following year, it showed Rush sitting before an open window with the Pennsylvania Hospital visible in the distance.

For the first few weeks after Rush died, Hosack found it too painful to write to his widow, Julia. He finally sent her a heartfelt letter expressing how stricken he and Mary were; every time the mail coach had left New York, Hosack confessed to Julia, he had felt awful about not having written her yet. Later, when he felt more equal to the task, Hosack wrote a long eulogy for Rush, praising the qualities he had admired in his mentor for decades: the devotion to medicine, the energetic self-discipline, the eloquence, and above all the generosity. Hosack noted that Rush's dying words to his son were, "Be indulgent to the poor." Hosack sent a copy of the eulogy to Rush's family, telling them that Washington Irving had requested it for publication in a new periodical. The eulogy soon appeared there along with an engraving of Sully's portrait of Rush. When Sully painted Hosack's own portrait the following year, he depicted Hosack seated near a bust of Rush, so that mentor and student looked as though they were continuing their decades-long conversation.

THE WAR WAS RUINING all Hosack's plans for Elgin. First Pursh had left for London, and now James Inderwick, the artist Hosack needed for the *Flora*, had enlisted in the US Navy. Inderwick had signed up as a surgeon and boarded a brig in New York Harbor in May 1813. Many other medical students and recent graduates had left for the service after the declaration of war, including Hosack's nephew and chief plant collector Caspar Wistar Eddy, who had graduated from Columbia in 1811. The classrooms at both the Columbia medical faculty and the College of Physicians and Surgeons were emptying so fast that the professors had begun to worry they might have to close down until the war was over. John Francis, now a lecturer at the College of Physicians and Surgeons, observed drily that all the ill-equipped young medical students who were being rushed into posts as military physicians would soon be able to "do as much execution in a nautical way as most of the legalized murderers with which our city abounds."

Hosack was feeling ever more anxious that spring. There was the war, of course, and Rush's death, but something else was gnawing at him. Only part of the lottery had been drawn, which meant he hadn't been paid in full by the state for the garden. He was in debt to ten different friends, among them Brockholst Livingston, Nathaniel Pendleton, and a wealthy merchant named Henry Coster. All of these men had lent him money while he waited for the state to complete its end of the Elgin deal. He was hemorrhaging interest payments, and now Coster was pressuring him for repayment of his loan. Although he had had no choice, Hosack was wracked with regret about his decision to relinquish control of Elgin.

One bright spot did flicker. His student Andrew Anderson completed a dissertation on boneset (*Eupatorium perfoliatum*). Anderson had conducted the research at Elgin, and he dedicated his work to Hosack and his assistant John Francis. In twenty-six experiments, Anderson had subjected decoctions and infusions of the leaves, flowers, and roots to ammonia, sulphuric acid, silver nitrate, and dozens of other chemical preparations. He concluded that the medicinal properties of boneset closely matched those of Peruvian bark, the imported emetic and sudorific that was especially prized for fighting yellow fever. Peruvian bark was in far shorter supply than the native boneset, and this discovery was quickly picked up and disseminated by other doctors, with appreciative mentions of Hosack himself. The garden was beginning to make its mark on American medicine, out in the world beyond his circle of young men.

Hosack sorely needed the validation. His colleagues had just taken the upsetting step of replacing him with Mitchill on the committee in charge of the garden. Perhaps Hosack had been too controlling, in the time-honored tradition of founders. His relations with Mitchill were already strained because of Mitchill's behavior during the Elgin appraisal process. One of Hosack's students later recalled Hosack's shock the day Mitchill barged in during a lecture and began rifling around in the classroom cabinets for a specimen he needed. Hosack sputtered, "Dr. Mitchill, I am engaged," the student later recalled. Mitchill reportedly bowed with cool impudence and replied, "So am I, sir." Hosack was so offended he dismissed class on the spot.

Now sidelined from the Elgin committee entirely, Hosack listened with mounting concern at the faculty meetings to updates on the garden. In July 1813, Michael Dennison, Elgin's current tenant, informed the college that he could no longer afford to honor his lease unless the college helped him with the debt of $2,000 he had incurred while trying to maintain the conservatory and the collections. Dennison complained that because of the garden he was "sinking the funds of my friends and exhausting my youth instead of saving for hereafter." He was spending all his time on the Elgin property, and still it wasn't enough. Dennison pleaded with the professors for help. "As it is with a man drowning, I am ready to catch at a straw." Hosack knew better than anyone else what Dennison was going through, but he was more worried about the irreplaceable plant collections. During a rainy week in late August, Hosack went by carriage to the garden. He took along two colleagues as witnesses.

The moment the carriage rolled in through the south gate, Hosack saw something had gone terribly wrong on Dennison's watch. Here and there, the fences and walls around the edge of the property were falling down. Hosack had always insisted that his men keep the garden pathways rolled out and smooth, but now he felt the carriage jarring and bumping over a bed of rocks. The path running from the gate up the hill to the conservatory looked awful. He saw that grass from the once-neat lawn to either side was rooting promiscuously in the dirt. Another path that traced a winding route alongside his forest trees was almost invisible in a welter of weeds.

Hosack was a man who esteemed all plants, no matter how humble. He had cultivated and celebrated the most obscure American flora— the secret mosses, the dullest sedges—but this was unbridled chaos, imparting delight to neither the eye nor the mind. As he reached the conservatory, the shrubbery plantings in front of the buildings came into view. He saw they were splattered with yellow sunflowers. His *Helianthus* had escaped. He had acquired six species that he classified in this genus over the years, two of them all the way from New Holland (today's Australia). Looking around the garden, he realized that sunflowers had established colonies in other places as well.

Bedraggled potted plants sat on the ground in front of the greenhouse. Dennison had been right to bring them outside for fresh air and sunshine, but Hosack could tell from their state that it had been done only recently, when any competent gardener knew it should have been done months ago, in May. Now he entered one of the hothouses and saw that many of his most precious specimens were gone. He had gathered them from around the globe, from the most intrepid scientific explorers, the most learned botanists, the most admired friends—Thouin, Curtis, Smith, Banks, Michaux, Delile, Jefferson, and so many others. Now these treasures had been heaved onto some farm cart and taken away to be sold like common produce. Hosack reined in his outrage, went back down to the city, and composed a terse statement to his colleagues at the College of Physicians and Surgeons describing what he had seen at Elgin that day.

The trustees rallied to Hosack's side. They put him back on the garden committee and drew up a list of charges against Dennison, who responded a few days later with a defensive letter. Because the hothouse flues were broken, he protested, the missing exotic plants wouldn't have survived anyway. He had dramatically improved the "barren" grounds "by dint of hardship and manure!"—to be specific, two hundred forty wagonloads of manure, purchased at his own expense. He was doing his level best to maintain the garden, but his debts kept dragging him under; his own brother had lured him into a bad loan "under the mask of friendship." Dennison closed his rambling defense with an anxious postscript: "I would thank the College to inform me if they authorized Dr. Hosack to enter a prosecution against me or not." But Hosack was focused on one thing. He had to find a way to get Elgin back.

Not long after Hosack delivered his report to his colleagues about the state of the garden, Dennison sent John Francis—who was still boarding with the Hosacks—a brazen note. Would Francis kindly inform the faculty of the College of Physicians and Surgeons that they could find specimens of the rare *Camellia japonica* in full bloom and for sale at his store on Chatham Street? It was the one species Hosack had requested not be sold.

———⚬⚬⚬———

ON DECEMBER 6, 1813, Thomas Jefferson wrote a moody letter to Alexander von Humboldt. "You will find it inconceivable that Lewis's journey to the Pacific should not yet have appeared, nor is it in my power to tell you the reason." Seven years had passed since the expedition had returned, four years since Lewis's suicide. With Jefferson's blessing, two volumes containing a narrative of the expedition were finally in preparation in Philadelphia. He promised to ship them to Humboldt, along with some tobacco seed the latter had requested— "if it be possible for them to escape the thousand ships of our enemies spread over the ocean."

The third volume, however—the one containing Lewis and Clark's natural history findings—was hopelessly stalled. Benjamin Smith Barton was chronically ill; Pursh had gone to Britain. Jefferson fretted to Humboldt that Lewis and Clark's discoveries would "become known to the world thro other channels." Two weeks later, in London, Pursh presented the Linnean Society with an advance copy of his major new work: *Flora Americae Septentrionalis*—the plants of North America. Pursh noted in its preface that his book contained almost double the number of species that André Michaux had been able to include in his 1803 *Flora Boreali-Americana*. He acknowledged his profitable association with Lewis and Clark, Hosack, Barton, the Bartrams, McMahon, William Hamilton of The Woodlands, and another Philadelphia-based naturalist, Henry Muhlenberg. But Pursh didn't mention that he was scooping Barton by publishing many of the Lewis and Clark plants.

When Hosack learned of Pursh's *Flora*, he seemed not to mind that his former head gardener had published a book on plant species gathered in part while Pursh was employed at Elgin. He didn't even mind that Pursh had beaten him to publishing a North American *Flora*. The "almost untrodden field" of American botany about which Hosack had written excitedly to Rush in 1794 was becoming crowded. Yet since then, Hosack had firmly established his own botanical reputation with Elgin, and he had also grown attuned to how much collaborative research was still needed to collect and identify all the flora of North America. Each *Flora* that appeared, Hosack realized, was destined to be superseded by a new one. Just a year after Pursh published his *Flora Americae Septentrionalis*, Hosack confided in James Edward Smith about his undimmed dreams for his own *Flora of North America*. "I only want from the state my remuneration for my garden viz. 75.000 Dolls to enable me to employ the best artists to execute it upon the scale I wish." For now, Hosack wrote Smith, Pursh's *Flora* was "unquestionably the best work of its kind that has yet appeared." It was a generous and fair assessment of a work that would end up preserving Lewis and Clark's botanical discoveries for future generations of scientists. As was so often the case during his lifetime, Hosack's first love—before vanity, pride, or politics—was the work.

HOSACK'S GUESTS WERE WAITING for him at his house one winter's night early in 1814. He was late to his own party, having been detained by a meeting. In spite of a difficult wartime winter, Hosack was throwing his energy into a new project—or perhaps it was *because* of the war, a defiant show of optimism despite a constant fear of British attack. Two decades after Hosack had sailed home from London with his head full of bold ideas, New York would finally have its own version of the Royal Society. He was one of the chief instigators, as usual, along with DeWitt Clinton and John Pintard. They were calling it the Literary and Philosophical Society of New-York, and Clinton, who was still mayor, agreed to be the new society's first president. Hosack would be one of three vice presidents.

When he rushed in from the cold and joined his guests, Hosack was apologetic for his delay but pleased about the cause. John Francis was watching as Hosack greeted the venerable old Gouverneur Morris. "I have been detained with some friends, who together this evening have founded a Philosophical Society," Hosack told Morris, as Francis later recalled the scene. Morris replied, "Indeed!" Hosack took the bait, saying, "Yes, sir, we have indeed this evening founded a Philosophical Society." Morris then asked Hosack innocently, "But pray, Doctor, where are the philosophers?" It was obvious to Francis that Hosack felt embarrassed in front of his guests. But Morris was playing dumb for the sheer pleasure of needling Hosack. He immediately joined the new society—as did Bard, Mitchill, Post, Francis, Cadwallader D. Colden, Brockholst Livingston, Robert Fulton, Hosack's nephew John Eddy, and dozens more New York lawyers, doctors, professors, and statesmen. Even the eternally skeptical Washington Irving joined.

Two months later, Clinton delivered the inaugural presidential address to the assembled members of the Literary and Philosophical Society. Hosack and Clinton were as close these days as they had ever been. They breakfasted and dined together frequently—sometimes alone, other times with their families. They talked about plants, politics, and the future of New York. Hosack wrote to a mutual friend about Clinton, "You will see him our next Governor or I will be much deceived and as much disappointed." In the meantime, Hosack joked, "our friend Mr. Clinton will be appointed at the next meeting our Baron Humboldt to explore the state and to make a statistical report."

In his inaugural address to the society in May 1814, Clinton unleashed an attack on the old "degeneracy" myth, reminding his listeners that inaccurate theories about American inferiority in every regard still had a foothold among European men of letters. Clinton complained that when Americans took steps to improve themselves, then "the master spirits who preside over transatlantic literature view us with a sneer of supercilious contempt." As hotly as he defended the United States, however, Clinton conceded in his speech that European critics were correct with regard to Americans' intellectual accomplishments. Americans had shown so much commercial initiative that their "enterprising spirit" was now admired the world over. Clinton lamented that if this famous spirit "had soared to the heavens in pursuit of knowledge, instead of creeping along the earth in the chase of riches," the United States would be as respected for its cultural achievements as were the greatest European nations—which it certainly wasn't.

The Literary and Philosophical Society was New York's optimistic answer to this failure. Even as it became clear that the earliest chapter of the nation's history was drawing to a close, the next generation of its leaders looked toward the future. There was so much still to be done in every field: history, literature, art, medicine, archaeology, mineralogy, zoology—here Clinton approvingly mentioned Charles Willson Peale's mastodon—geology, geography, animal husbandry, and botany. And now, in front of New York's most powerful citizens, the mayor paused to praise Hosack for his tireless work at Elgin, which "contains seven hundred and thirty-three genera, and two thousand four hundred species of plants." Clinton also noted that young Americans were being trained at Elgin for the critical task of identifying native plants and their medicinal, commercial, and agricultural uses before the continent was completely overrun by invasive species. "It has already become difficult to discriminate between our native and naturalized plants," Clinton warned. "With the progress of time the difficulty will increase."

BY THE SPRING OF 1814, Hosack was the patriarch of a large young family at 65 Broadway. He enjoyed domestic life but never slowed his breakneck professional pace. "He was indefatigable," one of his sons would later recall. "He always spent hours in his study after the labors of the day, and seldom retired to rest until after midnight, either devoting himself to medical study, reading over the lecture he was to deliver the following morning, or answering letters to his numerous correspondents." He sometimes went back out into the city late at night to check on a patient whose condition was worrying him. One of his favorite sayings, John Francis later recalled, was "the more a man has to do the better he does it."

Distracted as Hosack was by his medical and civic commitments, he loved his children and in turn inspired affection and reverence in them. He and Mary now had eight children. In 1812, after a frightening pregnancy that had confined her to bed for weeks, she was safely delivered of a healthy boy whom they named after his father, referring to him for clarity's sake as "little David." He was their fourth surviving boy, given the 1801 death of first-born Samuel Bard Hosack. The next baby, their fifth boy, had arrived in November 1813, and they had named him Thomas Eddy Hosack, in honor of John Eddy's father. But they had realized immediately that something was not right with the baby. He often started suddenly, as if frightened by something no one else could see. His breathing was labored, and he seemed to be in pain each time he urinated. He was constantly ravenous, even after he had eaten as much as the older children. Hosack was completely mystified by Thomas's condition, as was his medical colleague Dr. Wright Post, who examined the baby at Hosack's request.

Thomas died when he was six months old, in May 1814. Hosack wrote to Samuel Bard with the news, in a letter mingling paternal anguish with clinical candor. "The day after you left town our dear little babe continued ill . . . the whole nervous system became affected showing itself in a fixed state of the eyes, irregular contractions of the

Hosack's favorite medicinal plant, boneset (Eupatorium perfoliatum)

muscles of the face, and occasional strabismus." Hosack faulted himself for not being able to save the baby, and he made the difficult decision of asking Post to conduct an autopsy. In an age when autopsies were the surest way to decipher the physical sources of human misery, an autopsy on one's own child was an act of supreme altruism. Hosack bravely remained in the room as Post prepared the body for dissection. "Upon opening the abdomen," Hosack wrote to Bard, "we were instantly surprised at the great size of the stomach, which was as large as that of most adults—and what was remarkable—the spleen grew exclusively upon the stomach, having no connexion whatever beside." The autopsy results brought Hosack and Mary a sliver of solace. Hosack told Bard that, had Thomas survived, he would "have been a constant sufferer."

THE NEWS WAS RELENTLESSLY grim outside 65 Broadway, too. The war had wrecked New York's commerce and shipping; food was now scarce and exorbitantly priced. The charitable institutions founded by Hosack and his friends teemed with out-of-work laborers and their families. The almshouse behind the new City Hall reported that it had twelve hundred residents, including five hundred fifty children. So many people had sought refuge there over the past few years that the city was now on the verge of building a larger almshouse out near the Bellevue quarantine hospital on the East River.

It was a terrifying spring and summer. On April 8, 1814, British forces made landfall at Essex, a shipbuilding town on the Connecticut River, where they burned twenty-eight ships. Soon after this news reached New York, a seventy-four-gun British frigate was sighted off New Jersey. Then, in late August, the British burned Washington. The New York papers sounded the alarm: "*CITIZENS! TO ARMS!!*" Militiamen drilled as volunteers worked feverishly to erect new fortifications. Members of the New-York Historical Society hastily boxed up their library, so that they would be ready—if the British torched New York—to spirit the city's past to safety.

But when enemy forces invaded in September 1814, they chose

to do it upstate, near Plattsburgh. The Americans repelled the British at the Battle of Lake Champlain. The Madison administration, however, still faced the specter of a foe reinvigorated by the defeat of Napoleon in the spring of 1814. On Christmas Eve, American representatives in Belgium signed a treaty with the British that gained little for the United States besides officially ending the hostilities. When news of the Treaty of Ghent reached New York in February 1815, the city celebrated with parades, parties, fireworks, and the energetic resumption of commerce. Laborers put down their muskets and took up their hammers, paintbrushes, wheelbarrows, and pails. Captains began inspecting their dry-docked ships for seaworthiness. Gradually, cargo piled up on the docks again and shelves filled with goods.

The men in Hosack's circle greeted the peace with as much renewed vigor as anyone in New York, and not only because so many of them had commercial interests. Perhaps now the public would finally look more favorably on expenditures for the arts and sciences. In May 1815, however, Hosack, Pintard, and the other members of the New-York Historical Society were evicted from their rooms in the Government House, a building they shared with the Academy of Fine Arts, which was also ordered to pack up its paintings and sculptures. The building had just been sold at auction. Pintard wrote to John Francis to ask for his help with the move and suggested he temporarily stash the books at the College of Physicians and Surgeons. "Let me intreat you to get our Historical Soc[iety] Library out of the way. There will be the very devil to pay after tomorrow. & let the gods and goddesses take care of themselves."

Pintard, meanwhile, was lobbying the Common Council for use of the old almshouse, now standing empty after its residents had been moved to the new site on the East River. But others had the same idea—more than half a dozen institutions were clamoring for space in the almshouse. Hosack was deeply involved with three of them: the Historical Society, the Academy of Fine Arts, and the Literary and Philosophical Society. In June 1815, a Common Council committee recommended that these three, along with several others, be gathered in the almshouse under one roof.

Samuel Latham Mitchill, who was now in the Common Council, later bragged to Peale that he had been the councilman most instrumental in supporting the almshouse petitions. Mitchill himself was founding a society called the Lyceum of Natural History, and he hoped to get space for it in the almshouse. Whoever did write the committee's report, it sounded two of Hosack's favorite themes—civic competition and the lack of public funding for the arts and sciences. "The Citizens of New York have too long been stigmatized as phlegmatic, money making & plodding—Our Sister Cities deny we possess any taste for the sciences." The report pointed out that even the Royal Society of London had begun as a casual circle of friends and had depended on government support to rise "to that splendid zenith at which a Halley was Secretary & a Newton President."

The Common Council hadn't been able to resist the idea of a New York Newton. They handed over the almshouse keys, and the New-York Institution was born—a consortium of societies and museums fittingly located just north of what Pintard called "our magnificent new City Hall, the proudest Edifice in the U[nite]d States." Hosack soon brought his herbarium down from Elgin, including the original Linnaean specimens that Smith had given him in 1794, to put in the New-York Historical Society's new rooms, which they had leased for the token sum of "one peppercorn, if lawfully demanded."

THROUGHOUT THE SPRING and summer of 1815, men flooded back to the United States from the war. Peale's son Linnaeus returned to Philadelphia after participating in a naval battle off New Jersey. In New York, fathers, sons, and brothers showed up suddenly at their own doorsteps, injured and exhausted but happy to be home. James Inderwick, Hosack's artist for the planned *Flora of North America*, was not among them. About six months after the peace treaty was signed, his ship was lost at sea in the Straits of Gibraltar. Hosack had managed to publish just one plant engraving by Inderwick before the war took him away. It was a picture of the Canada thistle specimen Mitchill had given Hosack to identify in 1810.

James Inderwick's engraving of Canada thistle, done for Hosack

Some men were coming home from the war; others seized on the return of peace to plan European tours. In May Hosack wrote to James Edward Smith that "my friend Mr. Washington Irving of this city proposes to pass some time in London. . . . He is among the finest belles lettres writers of our country." Three months later Hosack wrote to Smith again, this time saying that John Francis would soon be in London: "His desire for an opportunity of seeing & knowing the great men of the Earth leads him across the Atlantic." By now Smith was officially a great man. The previous summer, the Prince Regent had laid a sword on his shoulder and dubbed him Sir James Edward. "I really rejoice in the accession to your honors that of knighthood," Hosack wrote graciously to Smith. While Smith had received a knighthood for his services to natural history, Hosack's great botanical achievement for his own nation was being ridiculed as a cow pasture. He hadn't even been fully paid for his land yet.

JOHN FRANCIS REACHED LONDON that fall, and Hosack's old circle at the Linnean Society welcomed him warmly. He soon wrote to
Hosack to say that Smith was Hosack's "most ardent" friend and had
asked after his namesake, James Edward Smith Hosack. Francis spent
that winter studying medicine and natural history in London, pausing
long enough in his work to travel to Edinburgh, where he met some
of Hosack's acquaintances on the medical faculty. Soon after this,
a friend of Francis from the College of Physicians and Surgeons in
New York, Peter Townsend, wrote to ask whether the medical professors in London and Edinburgh truly merited their international fame.
Townsend told Francis he thought that no professor "during the last
winter in any part of the globe" could have received more acclaim than
Hosack. More than a hundred students were now enrolled at the College of Physicians and Surgeons, and "I can assure you (privately) that
Dr. H. is the great magnet of their attraction."

In May 1816, Francis wrote to Hosack to report a delightful remark
Sir Joseph Banks had made to him over breakfast. "The reputations
of the U.S. as a whole are more enlightened than that of any other
country; they possess more enterprise than any other people under the
canopy of Heaven; they are inferior to no people in regard to physical
and intellectual capacity." Banks had recently finished reading the two
published volumes documenting the Lewis and Clark expedition. He
expressed to Francis his astonishment at what the Corps of Discovery
had endured and achieved. There would be no sneering about American efforts at self-improvement from this quarter. Banks felt that even
his own accomplishments as a young explorer paled in comparison.
He pronounced the expedition "a great performance," a feat unparalleled in the annals of Western exploration. "Sir," he told Francis, "the
fatigues of a single day would have killed almost any European."

As far as Francis was concerned, however, the richest London gossip concerned Frederick Pursh. He was "his own worst enemy: drunk
morning, noon & night." He had been received warmly at first, but
once rumors of his bad botanical citizenship concerning the publication of the Lewis and Clark specimens had reached the Linnean Soci-

ety, Smith refused to be in the same room with him. Francis thought Pursh had treated Hosack himself very unfairly by taking credit for Hosack's idea of publishing a *Flora*. Pursh had also failed to thank some of the younger botanists who had assisted him while he was collecting for Elgin, including Caspar and John Eddy.

Pursh did make one generous and honorable gesture toward Hosack. He tried to name a plant for him—*Hosackia Louisiana*. But Francis told Hosack that Smith, one of the world's chief arbiters in such matters, was rejecting the classification as "somehow or other a doubtful species." Smith didn't think the plant was worthy of Hosack anyway, he assured Francis, who reported the conversation to Hosack. "The plant is a very unimportant one," Smith told Francis. "We must give our fine Dr. Hosack a garden Genus. We'll find one out for him. He deserves a conspicuous plant."

In fact, Smith had been working for months to secure for Hosack another very conspicuous tribute. Francis wrote Hosack excitedly about the plans afoot in London. "They in this country think FRS the highest honour that can be conferred on any individual." On May 23, 1816, with Banks presiding, Hosack was elected a Fellow of the Royal Society—not a Foreign Member, but a full-fledged Fellow. It was an extraordinary honor for an American. When Hosack received the news of his election later that summer, his friends showered him in congratulations. He reveled in the attention, confiding to Nathaniel Pendleton that "in some measure I must candidly own my pleasure is increased at the effect it has produced upon my professional brethren. Our friend Mitchill is teased at every corner that he has been omitted." Sometime that summer, Hosack made a small alteration to the preprinted tickets that admitted his medical students into their courses. Below his name, he wrote three initials in florid script: *F.R.S.*

A few weeks after Hosack had achieved his triumph, Bernard McMahon died at his Upsal Botanic Garden outside Philadelphia. William Hamilton of The Woodlands had died in 1813, around the same time as Benjamin Rush. Benjamin Smith Barton had died in December 1815 at the age of forty-nine, despite the best efforts of Hosack and other physicians to prolong his life. Barton had trained several very promising students, among them Jacob Bigelow, a young medical botanist who lived in Boston and had recently published a guide to the plants of the Boston area. But aside from William Bartram, there was now no living American who had done more than Hosack to teach the United States about the world's flora. From Paris, François André Michaux wrote Hosack, "Every learned man in Europe knows that you are one of the most zealous promoters of Sciences & arts in North America."

IN LATE JULY 1816, Hosack received a package of seeds from Monticello. It had been a decade since Jefferson had ignored his request for specimens from the Lewis and Clark expedition. In the intervening years, however, Hosack's work at Elgin had earned Jefferson's respect. Now, when André Thouin sent Jefferson a shipment of seeds from the Jardin des Plantes, Hosack was the first person he thought of. Jefferson was as enthusiastic about botany as ever. "Botany I rank with the most valuable sciences," he had written to a friend in 1814, "whether we consider it's subjects as furnishing the principal subsistence of life to man & beast, delicious varieties for our tables, refreshments from our orchards, the adornments of our flower-borders, shade and perfume of our groves, materials for our buildings, or medicaments for our bodies."

In a cover note to Hosack, Jefferson explained that he hadn't even opened the package to inspect the seeds, "knowing I could not pack them so well again." Hosack thanked Jefferson promptly, saying that it was too late in the season to sow the seeds but "with the aid of manganese most of them will probably grow the next year." Certain that Jefferson would appreciate the praise, he also copied out Banks's glowing remarks about the Lewis and Clark expedition. But when Hosack turned to the subject of the garden, he grew morose. "It gives me pain to state to you that altho New York has done herself great credit by the *purchase* of the Botanic Garden she has made no provision whatever for the support or the improvement of it."

Hosack's misery about Elgin was deeper even than he confessed to

Jefferson. In 1814, the state had wrenched Elgin away from the College of Physicians and Surgeons and handed it to Columbia in lieu of a badly needed government loan. This decision made little sense, because the state had recently transferred Columbia's medical faculty to the College of Physicians and Surgeons. Hosack had strongly supported this merger. But it meant that, now, he and all the other professors with medical or botanical expertise were based at the new College of Physicians and Surgeons building on Barclay Street. At Columbia, there was no one left who was likely to fight for the garden with anything like Hosack's energy and expertise. The state had made matters worse by burdening the grant of the Elgin property to Columbia with a contractual condition completely at odds with the preservation of the garden itself. Columbia was ordered to turn the property into its new campus within twelve years.

Columbia's trustees were as annoyed as Hosack by the state's decision. They considered the garden a terrible substitute for a loan and complained to the legislature that, rather than bringing the college any profit, it would be "a Source of Expense." They dragged their feet on taking possession of the land, while Hosack, from the College of Physicians and Surgeons, desperately tried to keep watch over the plant collections. Elgin was still in Michael Dennison's clutches and now in even worse shape than the awful day in 1813 when Hosack had toured the garden with two colleagues as witnesses. Dennison himself was in such dire straits that he had taken a job at a glass and window store downtown. On a visit to the garden in the summer of 1815, Hosack found the pathways so "neglected and filled with stone . . . that a Carriage is in danger of being upset." He also discovered that Dennison had ruined the tidy botanical ordering in which he had maintained his specimens—and of what pedagogical use was the garden without it? The plants that Dennison hadn't already removed were dying, and the conservatory buildings were falling apart. Broken glass lay everywhere.

Hosack longed to get his garden back. He told Jefferson that he was working on plans to do so, and that when he did he would relaunch work on the *Flora of North America*. Suddenly, however, the Columbia trustees decided to take possession of the land after all. When Dennison learned of the power transfer, he wrote to a member of the college's new botanical garden committee, a classics and divinity scholar named Clement Clarke Moore.* Dennison alerted Moore to the repairs direly needed on the hothouses, warning that "there has been last night as severe a frost as to injure Cucumbers, [and] the tender plants of the tropical climates will not be any the better of it." He also informed Moore that, although he wanted to renew his lease for the garden under Columbia's ownership, he would have to have a friend sign it on his behalf, owing to "my Insolvency occasioned by the harsh treatments of an unnatural Brother."

To Elgin's new owners, Dennison was an unappetizing tenant, and Hosack feared that he would adopt a scorched-earth policy toward the garden on his way out. It was likely at Hosack's prodding—he was close to Moore—that Columbia's garden committee chose Andrew Gentle, Elgin's former head gardener, to take over the lease from Dennison. In October 1816, Hosack sent Gentle to scout out the situation. Gentle made an appointment to meet Dennison at the garden on a Saturday morning, but Dennison wasn't there when he arrived. After a long wait, Gentle gave up. He then plotted a surprise attack, returning unannounced a few days later. This time he found Dennison at the garden, along with a wagon and some carts. It was obvious to Gentle that Dennison was planning to steal the remaining plants, and he commanded him to leave the property immediately. Dennison refused, saying he didn't have the authority to turn the garden over to Columbia because he was employed by Matthew Hawkins, the proprietor of the downtown glass and window store where he now worked. This was an outrageous claim. Hawkins had nothing to do with the garden lease.

Gentle left the premises. He immediately composed a letter to Moore about the situation and sent it to Hosack for his prior approval, along with a brief, elegiac message.

---

* Moore would later be credited with the posthumously published poem "'Twas the Night before Christmas," although the Livingstons insisted it had been written by a member of *their* family.

Dr. Hosack

Sir if you think proper you may hand the contents
*   on the other side to C. C. Moore*
the Impression
on my mind is
that he means to
strip the place as far
as he can of any thing that is
rare & valuable etc.
I am sir yr obt

Andrew Gentle

Hosack forwarded Gentle's letter to Moore at Chelsea, the family estate north of New York City, near the village of Greenwich. In a cover note, Hosack assured Moore that he stood ready to "unite with you in the preservation of the plants at the Botanic Garden." Hosack wrote to a friend in Philadelphia about the situation at Elgin, declaring with grim determination, "I shall again get charge of it."

The Columbia trustees deputized the garden committee to oust Dennison, but he refused to leave the garden and wouldn't let anyone else touch the plants. The garden committee made no headway in evicting him that winter. Then, in March 1817, Dennison's employer, Matthew Hawkins, placed an advertisement in the *National Advocate* for his glass and window store at the corner of Murray and Broadway.

FOR SALE

AN ELEGANT ASSORTMENT OF

GARDEN *and* FLOWER SEEDS,

WARRANTED GOOD AND TRUE OF THEIR SORTS;
ALSO PEAR, PEACH, ENGLISH CHERRY, . . .
WALNUT OR MADEIRA NUTS, &C.

WITH A HANDSOME COLLECTION OF

Green-House Plants, &c.

Chapter 15

"YOUR FORTUNATE CITY"

EVEN AS HOSACK'S BELOVED ELGIN WAS BEING RIPPED UP AT the roots, New York City, with his help, was blossoming. His one-time rival Charles Willson Peale arrived in town, looked around, and praised Hosack for what he saw.

It was on a sunny day in May 1817 that Peale boarded a north-bound steamboat in Philadelphia with his wife, Hannah. He had some museum business to conduct in New York. Peale had never managed to secure state or federal funding, and he was hoping to find a possible solution to his quandary in Hosack's New York circle. Peale and Hannah took their seats on a bench under a broad awning, and as the steamboat picked up speed he admired the way it slipped through the water "like magick." He gave mental thanks for Robert Fulton.

In New Jersey, Peale stopped to visit an old friend, Elias Boudinot, who advised him to sell his museum to the City of New York. Peale was annoyed at this suggestion, as he wrote in his diary. "I told him that I thought the City of Philad[elphi]a ought to keep it as being the first City in the U. States in the advancement [of] arts & Science." Boudinot insisted that with the recent founding of the New-York Insti-tution, New York City had finally pulled ahead of Philadelphia. Peale was incredulous.

Peale and Hannah took a stagecoach to Paulus Hook, where they

caught the ferry to lower Manhattan. Peale was feeling sick to his
stomach after a breakfast of oysters and toast, but the next day he felt
well enough to investigate the New-York Institution. On the ground
floor, he found a little museum of natural history run by a man named
John Scudder. Peale was impressed with the preparation of the stuffed
specimens; he especially admired a bald eagle poised to seize a snake
with its beak. He went upstairs to tour the Academy of Fine Arts. Its
members had knocked out the top story of the old almshouse building to create a soaring, light-filled space in which to hang their paintings and arrange their sculptures. Peale studied the paintings by John
Trumbull and Benjamin West with an appreciative eye, but he decided
that the best picture in the room was a portrait of Hosack's mother-in-
law by Samuel Lovett Waldo, a Connecticut-born artist.

Over the next few weeks Peale and Hosack saw each other frequently, sometimes breakfasting at Hosack's house and sometimes
meeting at the New-York Institution, which Peale was now visiting
almost daily. Their old quarrel over the New York mastodon was long
since forgotten, perhaps eased by the fact that Hosack's pioneering
hydrocele surgery, performed on Peale in 1806 by Hosack's brother-
in-law, had rescued Peale from chronic pain. Hosack gave Peale a copy
of a treatise he had written about the history of Elgin as well as piles of
reports published by all the different New York societies. He encouraged Peale to write something in the Philadelphia press about "the
improvement of N. York." He also urged Peale to bring his museum
to New York and put it in the New-York Institution. Peale demurred.
He thought the whole building was too small for the great museum of
natural history New York would surely one day have, and the same
was true of the art museum.

The more Peale saw of New York, the more the city impressed
him—with its natural beauty as well as with its civic progress. One day
he went to the edge of the Hudson and sketched a picture of the New
Jersey Palisades. He went to the East River and admired the views
of Long Island, which lay covered with thick green groves that were
punctuated by white farmhouses. Another day he positioned himself at
the southern end of City Hall Park and made a quick pencil sketch that
showed the grand allée of Lombardy poplars along Broadway. When

he climbed the stairs to the cupola atop City Hall, he looked out over "a great City with its many Churches & Elegant buildings." He began to think Boudinot had been right about New York.

On June 2 Peale toured the Lyceum of Natural History, recently founded by Mitchill for the purpose of collecting and studying natural history books and specimens. Mitchill was extremely pleased to see Peale and insisted that he attend a Lyceum meeting, at which he praised Peale as "the father of Natural History in America." The Lyceum's most precious specimen was a partial set of mastodon bones from the same area near the Shawangunk Mountains where Peale had acquired his own set years earlier. Knowing what an arduous task the Lyceum members would face in completing the skeleton, he donated $5.

Peale next went to see the rooms of the New-York Historical Society. Here he met John Pintard and walked through the society's four rooms with him. Two of them were lined with glass cases filled with natural history specimens from each state in the Union. Peale was amazed at the luxury of the society's rooms. He chalked it up to the fact that the members were by and large rich men. In the main meeting room, he was shown a chair that had belonged to Marie Antoinette and had been brought by Gouverneur Morris from France along with other "Rich furniture that had belonged to the unfortunate Lewis 16th." Depending on how Peale looked at it, this was either a patriotic display—the spoils of a toppled monarchy repurposed for a democracy—or just pretentious. He kept his opinions to himself.

The following week Peale was invited to attend a meeting of the New-York Historical Society. Hosack, its vice president, was presiding that day—presumably in Marie Antoinette's chair at the front of the room—and he insisted that Peale sit next to him. Partway through the meeting, Hosack leaned over and whispered to Peale that he wanted to nominate him for membership. Peale protested that he would be useless to them. He was nearly deaf and had trouble following meetings, and he lived so far from New York anyway. But Hosack prevailed. Peale was escorted out of the room while the members elected him. He then offered some brief remarks, thanking them for the honor. Afterward, Hosack and Pintard took Peale out for coffee.

The next day, June 11, the new president of the United States arrived in New York. After James Monroe's inauguration the previous March, James and Dolley Madison had returned to Montpelier, their country estate in Virginia, where they lived "like Adam and eve in Paradise"—as a friend of Dolley put it—although they did so only with the help of slave labor. President Monroe had embarked on a tour to inspect the nation's military defenses and to learn firsthand about the concerns of its citizens. He arrived at Manhattan on a steamboat from Staten Island. The cannon thundered on Governors Island and the city's flags fluttered in a strong wind. Thousands of citizens jammed Broadway, with many others leaning out the windows as they strained to catch a glimpse of the president.

Peale went back to the New-York Institution the day after the celebrations, having heard that President Monroe was scheduled to take a tour. Discovering that the event had been postponed, he decided to go ahead with his planned return to Philadelphia the next morning. He bought six mackerel at the fish market to preserve for his museum, and then he left town with Hannah. At Trenton, the Peales ran into a widow named Rebecca Blodget, who was well known as one of Burr's lovers. She made clear to them that she was on her way to New York to see Burr, joking to Peale that he ought to put her on display in his museum "as a curiosity" because of her complete disregard for public opinion. In his diary, he observed that if he did put her in the museum, it would not be for "her extraordinary virtue."

That same afternoon, President Monroe arrived at the New-York Institution for his tour. Accompanied by Clinton—the incoming governor of New York—and other dignitaries, Monroe strolled through the Academy of Fine Arts, the Literary and Philosophical Society, the New-York Historical Society, and the Lyceum of Natural History. These last three organizations had voted to make Monroe an honorary member, and the New-York Historical Society had entrusted Hosack with the honor of presenting the president with his certificate. Before Monroe left town, he also visited the prison, the hospital, and Eliza Hamilton's orphanage. As he toured all these institutions, the president repeatedly praised New York's intense commitment to science, the arts, and charity. From Philadelphia, Peale wrote Hosack a gracious letter expressing similar sentiments. Thanks to the pile of pamphlets Hosack had given him and his own visits to the New-York Institution, Peale had to admit that "New York is already more advanced in learning, arts & science than Philad[elphi]a." Hosack and his friends were bringing "honor and fame to your fortunate city." Between Peale's visit and President Monroe's, Hosack had passed a highly gratifying month. For twenty years he had been laboring with his friends and rivals to make New York the greatest city in the nation. Now, in the space of a few weeks, their work had been praised by the foremost living American naturalist and the president of the United States.

Neither Peale nor Monroe, however, seems to have visited Elgin. It was just as well. Hosack's masterpiece was now a wreck, as the French naturalist Jacques Milbert discovered when he went to see the garden around this time. Milbert noted in his travel journal that "the conservatory alone bears witness to its original location."

IN EARLY APRIL 1817, two months before Monroe's visit, Clinton had written Hosack a brief note from Albany. "In politics all is calm," he reported. "I shall be detained here some days by the new Canal bill which I think will pass." He was right—the bill passed on April 15, and Clinton won the race for governor a few weeks later. On Independence Day, a groundbreaking ceremony for the Erie Canal was held in the town of Rome, about a hundred miles west of Albany.

Two weeks later, on July 18, Clinton wrote Hosack a long letter about an interesting species of wheat he had seen in Rome, where it grew wild in the swamps and was now being cultivated by a local farmer who had told Clinton about it. Clinton enclosed some specimens that had been gathered only the day before, saying he thought this might be a native species—one hardier than the wheat New York's farmers usually planted, which was often damaged by frosts. He complained to Hosack that too many indigenous American plants were being misidentified as exotic. Hoping that this wild wheat did not yet have an official botanical name, Clinton proposed a patriotic binomial: *Triticum americanum*. "The opinion on this subject of such an eminent

Botanist as yourself will be very acceptable," he told Hosack. In four
and a half pages, Clinton never once mentioned the canal.

At Elgin, Hosack was growing many species of *Triticum*. But
whatever he told Clinton about his specimen—Hosack's reply has
not survived—his immediate response was to write a generous letter
about Clinton to Sir James Edward Smith. Hosack informed Smith of
Clinton's election as governor and his progress on the canal, and he
predicted that Clinton would be the next president. Hosack thought
the nation would be in excellent hands. Few American men were as
learned as Clinton, who upon Hosack's nomination had been elected
to the American Philosophical Society. Hosack was now hoping to
secure a higher honor still for Clinton. "It would be highly gratifying
to me to see that gentleman elected a Fellow of the Royal Society and
of the Linnean Society," he told Smith, pointing to Clinton's expertise in geology and adding that "Mr. Clinton's botanical collections are
also extensive."

Hosack also reported that with the blessing of the other members
of the New-York Historical Society, he had just placed a bust of Smith
himself atop the herbarium cabinets holding the Linnaean specimens
Smith had given him in London in 1794. The bust of Smith was a con-
crete expression of the transatlantic botanical exchange Hosack had
worked so hard to keep open over the previous quarter century. Smith,
taking up Hosack's idea, soon wrote Banks about the idea of electing Clinton to the Royal Society, but the response wasn't positive.
"With Every degree of Respect for our American Friends I have many
doubts about Electing Govr de Witte Clinton into the R.S.," Banks
replied. Banks thought Hosack's "Talents and his Progress in Science"
appeared to have justified his election, but he doubted Clinton was "as
much a man of Science."

Hosack had earned this scientific renown above all through his
work at Elgin. These days, although the garden had slipped from his
grasp, some of his most important scientific contributions were alive
and well and walking around New York. One was Jacob Dyckman,
who had come to the city from his family's farm in northern Manhattan
in 1810 to study medicine and botany with Hosack. In 1818, Dyckman
published the first American edition of the *Edinburgh New Dispensa-*

tory. He preserved the original European entries but interspersed them with many plant-based remedies native to North America, which he had learned about at Hosack's side. It was a milestone in American medicine, and Dyckman dedicated the book to David Hosack and his assistant John Francis.

Of all Hosack's students, it was John Torrey who was destined to have the greatest impact on American science. Hosack referred to him as "the young man with an old head." Torrey's father was the governor of the prison overlooking the Hudson at Greenwich, where Hosack had long been an attending physician. As a boy Torrey had loved exploring the countryside between Greenwich and New York City with his brother William, who later recalled that they would scour "the hills on each side of Broadway for garnets, which we found in considerable numbers." Like so many other young men Hosack trained, Torrey was studying medicine and botany simultaneously. In March 1818 he received his medical degree from the College of Physicians and Surgeons, joking to the botanist Amos Eaton, "I have now got my Sheepskin & have full powers granted to kill & destroy in any part of the earth." Torrey could be just as biting about botanists as about doctors, describing Frederick Pursh to Eaton as "such a notorious liar & plagiarist that we can put no confidence in his assertions." He also heaped

John Torrey, one of Hosack's botanical protégés

scorn on an émigré French botanist named Constantine Rafinesque, thinking him too quick to take credit for the discovery of new plants. "I expect that he will soon issue proposals for publishing the botany of the moon with figures of all the new species!"

Now liberated from the classroom, Torrey spent the spring and summer of 1818 collecting plants all over Manhattan, along the Hudson, on Staten Island, and on Long Island. He went out almost daily, bringing home piles of native and naturalized species and taking careful notes on when and where he had found them. He botanized everywhere—including in the streets of New York, where he cut specimens from Lombardy poplars. In Lispenard's Meadow, he harvested rock cress (*Arabis reptans*). He went up the new Eighth Avenue to a field three miles north of the city and found a medicinal plant called elecampane (*Inula helenium*). He went to a swamp near the old Elgin land and collected nodding lizard's tail (*Saururus cernuus*) and arrow arum (*Arum virginicum*). He wandered along a brook that wound through Bloomingdale, a swath of lovely countryside whose name came from an old Dutch name meaning "valley of flowers." Here he gathered sweet white violets (*Viola blanda*), white milkweed (*Asclepias variegata*), tall thimbleweed (*Anemone virginiana*), and many more species. Today this is the Upper West Side.

Hosack admired Torrey's obsession with plants and soon invited him to take charge of the old Elgin herbarium, now housed at the New-York Historical Society. But Torrey confessed to Eaton that he found Hosack overbearing. Eaton wrote back sternly, "You ought by all means to listen to Hosack. . . . do not slight his attentions." Torrey relented and agreed to organize the herbarium. Soon Hosack was sending a package of specimens prepared by Torrey to Sir James Edward Smith, who wrote Torrey a very kind letter in reply and helped him with some plant queries. Hosack loved forging these links between great European naturalists and the rising generation of Americans, and he invited the talented Torrey to serve as his assistant in a new course of botany lectures. Torrey told Eaton privately that he felt competitive with Hosack, because he dreamed of writing a *Flora of North America* one day—although Hosack, stymied by the war, his many duties, his lack of funds, and the crushing loss of Elgin, had never managed to

publish his own *Flora*. Eaton unfairly blamed this failure on procras-
tination, which he thought was "an essential characteristic of N. York
writers. They seem to think that if a book is but announced, it will
grow up itself without any labor." He told Torrey, "If you ever begin
the work, you must calculate to finish it."

If anyone was procrastinating, it was the State of New York. It took
until Valentine's Day of 1818—seven years—for state officials to settle their debt to Hosack for Elgin. The years of work on the garden
and the doomed *Flora* had bled him so dry that he had signed over
the state's funds in advance to his creditors, one of whom warned him
around this time that "it will be greatly and cruelly injurious if I do not
get the 8000 dollars tomorrow. Your credit will also suffer."

Receiving the final payment from the state did nothing to quell
Hosack's desire to rebuild the garden. But the Columbia trustees,
meanwhile, had been devising a plan to earn income from the property.
First they would need to wriggle out of the state's order that they move
the college up to the Elgin property. Then they would be able to lease
out the land in small parcels to many different tenants, thus drawing
far more income than they could ever hope to earn from a single tenant
like the gardener Andrew Gentle. In February 1818, the trustees submitted a petition to the legislature, cagily warning of the "Evil Consequences" that would arise if the students were forced to live so far away
from their parents in town. At the same time, the trustees argued that
the garden had come to them in a "state of Dilapidation and Decay"
and would cost thousands of dollars a year to maintain as a research
and teaching facility. They acknowledged that the state had meant to
do them a favor by giving them the land, but they insisted that "the
real Value of the Property was probably over-rated." For a full year,
the trustees lobbied the state. In February 1819, the legislature finally
removed the conditions on the Elgin land grant. Columbia would not
have to turn the property into its new campus. The trustees could lease
the land without concern for the remains of Hosack's garden. Gentle,
Hosack's ally, was to be evicted from the grounds.

When Hosack learned that Columbia would be leasing the Elgin
property to a new bidder, he made his move. Together with the other
members of yet another new society he had helped organize—the

New-York Agricultural Society—he submitted an application in June 1819. The society would be needing a large plot of land on which to plant experimental crops, test new farm machinery, and pasture live-stock. Columbia rejected the application. They preferred to hold out for a more reliable tenant than a society of gentlemen who planned to dabble in farming. By now, the trustees were allowing the greenhouse to be stripped of its plants. They were also laying plans to uproot "such *ornamental trees* and shrubs as might be removed without injury to the place."

Hosack's trees and shrubs were to be carted up the Bloomingdale Road to a new site about four miles north of Elgin. There, on a beauti-ful spot high above the Hudson, the officials of the New-York Hospi-tal were building a new kind of institution: an insane asylum. Hosack was involved in the project, but it was the Columbia trustees who had decided to pillage Elgin's collections to beautify the asylum grounds. Soon the architect in charge of the asylum project applied to Columbia for permission to strip one of the Elgin hothouses of its remaining glass and tiles. His request was granted.

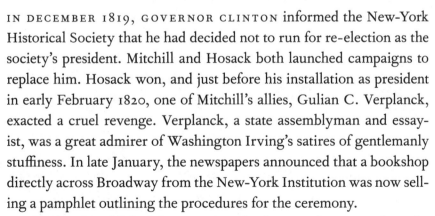

IN DECEMBER 1819, GOVERNOR CLINTON informed the New-York Historical Society that he had decided not to run for re-election as the society's president. Mitchill and Hosack both launched campaigns to replace him. Hosack won, and just before his installation as president in early February 1820, one of Mitchill's allies, Gulian C. Verplanck, exacted a cruel revenge. Verplanck, a state assemblyman and essay-ist, was a great admirer of Washington Irving's satires of gentlemanly stuffiness. In late January, the newspapers announced that a bookshop directly across Broadway from the New-York Institution was now sell-ing a pamphlet outlining the procedures for the ceremony.

Verplanck published it anonymously, but word got out about its authorship. The insults to Hosack began right on the first page: "Cer-emony of the Installation of David Hosack, M.D. L.L.D. F.R.S. London, Edinburgh, Hayti, and Pekin; First Vice-President of the New-York Society of Conchology and Indian Earthen-Ware." After a few preliminaries, the pamphlet explained, there would be speeches in honor of Clinton and Hosack in Greek, Italian, French, Dutch, Swedish, Irish, Chinese, and Hebrew. (The Italian speech was to be delivered by Lorenzo Da Ponte, a former librettist for Mozart who was now a language professor at Columbia.) After the speeches, a chariot borrowed from the theater just across City Hall Park would roll into the room. It was actually a prop from a play depicting "the triumphal entry of Alexander the Great into Babylon." But now, according to Verplanck, it would be filled with Hosack's certificates of membership in the Royal Society, the Linnean Society, the American Philosophical Society, and all the other societies to which he belonged. Next, following some chants in pig Latin—including an insult about Hosack's intellectual inferiority to Alexander von Humboldt—Hosack was to be draped in robes by officials representing the fields of medicine, law, and divinity, taking care that each robe left a little of the previous layer visible. Verplanck went on to describe another series of chants, ending with:

> *Hail to great Hosack—triple Doctor thou!*
> *Of Law and Physic erst—of Sacred Letters now!*
> *Thrice hail to Clinton—greater Doctor still!*
> *Who wields the State—as Hosack wields the pill!*

This section of the ceremony was "to conclude with a grand CRASH," and now two generals would come forward—one of them Attorney General Cadwallader D. Colden—to attach epaulettes to Hosack's robed shoulders and put the "hat worn by Napoleon at the battle of Waterloo" on his head. Hosack would be knighted, a band would play martial tunes, "and punch shall be handed to the ladies." William P. Van Ness, accompanied by six other New Yorkers from old Dutch families whose names began with Van, would come up and present the scepter of the New-York Historical Society—the "gold-headed cane of Rip Van Dam" (a real-life eighteenth-century governor of New York). Finally, Hosack would give an inaugural address on "the comparative merits of the two patron Saints of the Institution," Santa Claus and DeWitt Clinton. (The New-York Historical Society had for years celebrated St. Nicholas Day on December 6 in honor of the city's Dutch origins.)

Hosack saw Verplanck's satire in January along with everyone else, but John Pintard soon wrote Clinton in Albany to report that Hosack was shrugging off the attack. Clinton replied, "I am glad to find that our friend the Dr. feels himself above the little satire of literary buffoons. I am sure that they render nobody ridiculous but themselves." On February 8, in the lavish meeting room of the New-York Historical Society, Hosack was installed as its fourth president. He spent most of his prepared remarks that day exhorting the assembled members to keep gathering every possible artifact relating to the history of the state and the nation. Hosack pointed out that in the course of human history whole civilizations had come and gone, their existence known to later peoples only through their material effects, especially their coins and medals. Coins and medals didn't disintegrate like paper—or, for that matter, like plant specimens. "The march of nations is before us," Hosack warned, "and the gloomy night of darkness and ignorance that obscures their paths, to the eyes of posterity, may one day envelope our own, however bright the sun of civilization may now beam upon us." Even great republics like the United States were not invincible.

A few months later, Hosack proved he was a good sport about the satirical pamphlet by presiding over the election of Alexander von Humboldt to honorary membership in the New-York Historical Society.

SIR JOSEPH BANKS, the only possible rival to Humboldt's scientific fame, died the following month in London. Until shortly before his death, Banks was still sending Hosack interesting new books from London, and the two men were exchanging letters about mutual friends and botany. In 1817, Hosack had been elected one of the first foreign members of the Horticultural Society of London, of which Banks had been a founder.* The following year, Hosack had shipped

---

* Today this organization is called the Royal Horticultural Society.

the Horticultural Society of London eighteen trees of an American pear species whose fruit he explained was "admitted by all, to be one of the most exquisitely and highly flavoured we possess." Hosack's specimens were distributed to nurserymen around London, and in 1819 the Horticultural Society awarded Hosack a gold medal for introducing the Seckel pear to Britain.

Banks died just as the guard was starting to change in Hosack's New York circle as well. In the summer of 1821, Samuel Bard's wife of fifty years passed away, and Bard, who was already very ill, didn't last a full day without her in the world. Not long before this Bard had confided to Mitchill that he knew "he had become a weak and crazy vessel, and could not hold together much longer." Of Hosack's old mentors—Bard, Rush, Curtis, Banks, and Sir James Edward Smith—only Smith was still alive.

Hosack's world was shifting on its foundations. Even the island around him was disappearing as the grid of avenues and cross streets rolled northward from the former edge of the city. In the summer of 1818, laborers had begun laying a road—the Ninth Avenue—through Clement Clarke Moore's Chelsea estate. Moore was livid, and not only about the loss of his fruit orchard. He and many other New Yorkers were watching in helpless fury as the island's lovely hills were sliced away and its green valleys filled up with dirt—all to make the city more convenient for commercial traffic. Moore published a pamphlet blasting the city officials in charge of the grid as men "who would have cut down the seven hills of Rome." Some people's houses were suddenly submerged in earth, while others now dangled in the air above a low new street bed. Along the edges of burial grounds, exposed coffins stuck out like rows of rotting teeth. Moore worried that as the earthworks disrupted the natural courses of the island's streams, the risk of fever epidemics would rise, and he reproduced in his pamphlet a letter from Hosack praising the underground sewer system of Philadelphia for its public health benefits.

All the uncertainty about what Manhattan's landscape would look like from one day to the next began to creep into Columbia's attitudes toward the Elgin property. The trustees had been thinking about selling the land and using the proceeds to pay off some of the college's debt, but in the chaos unleashed by the new streets and avenues, they suspected that a sale wouldn't "bring more than six or seven thousand dollars." They decided to hold on to their property a bit longer. Hosack, meanwhile, was still waiting for his chance. He continued to collect seeds and specimens in anticipation of the moment he could plant them at the garden. From Monticello, Jefferson continued to forward shipments he received from the Jardin des Plantes. As the rain came pelting through the shattered roof of the one remaining hothouse, Hosack kept hammering home the importance of plants to human health and happiness whenever and wherever he could. In the fall of 1820, to a captive audience of medical students, he gave a lecture arguing that the city's polluted air and water could be purified through a citywide campaign to plant forest trees. He particularly recommended plane trees, horse chestnuts, elms, lindens, black walnuts, and catalpas.* Up the new Fifth Avenue, his magnificent arboretum was being torn out and transferred to the Bloomingdale insane asylum.

Yet as Hosack brooded over the destruction of the garden, he was discovering more and more kindred spirits in the growing city. In 1818, a group of local seedsmen and gardeners—Andrew Gentle among them—had begun gathering informally to share ideas about horticulture over drinks at a hotel on Broadway. They were soon joined by some of their wealthy employers—gentlemen with country estates who loved rearranging meadows and forests whenever they could flee their affairs in the city. Hosack had no country house, but when he officially joined the New-York Horticultural Society in 1822, he brought a more comprehensive knowledge of the world's flora than any other member. He also brought his genius for organizing and his access to a global network of plantsmen. In return, he found a congenial group of men who liked nothing better than clustering around a tavern table to marvel at such horticultural curiosities as "three heads of Celery measured 20 inches in length" or the seeds of "chocklate corn" sent in by Cadwallader D. Colden.

As word of these gatherings spread, the membership rolls grew

---

* Hosack thus predated by two centuries the city's "Million Trees NYC" project, completed in 2015.

longer. It was an unusual kind of society for New York, where gardeners and shopkeepers could hobnob with men like John Jacob Astor's
son William—whom Hosack himself put up for membership. Some
of the humbler members later acknowledged their gratitude to Hosack
for inviting them into his house to meet with his distinguished scientific
friends and for letting them browse his peerless botanical and horticultural library whenever they wanted. One of the society's most active
nurserymen, Michael Floy, expressed his appreciation by naming a
camellia for Hosack: *Camellia hosackia*. With Hosack involved, the
New-York Horticultural Society quickly grew the customary organizational carapace—charter, constitution, bylaws—and began meeting
at the New-York Institution. Before long, the society started a journal
called *The New-York Farmer, and Horticultural Repository*, which they
packed with reviews of botany and horticulture books and answers to
such practical questions as "What is the simplest, cheapest, and most
sightly mode of guarding single trees planted in parks or lawns from
the depredations of deer or other animals?"

Hosack connected the New-York Horticultural Society with the
Horticultural Society of London. The latter enthusiastically embraced
the idea of a transatlantic exchange by sending over a botanist—a Scot
named David Douglas. As he sailed into New York Harbor in early
August 1823 after a difficult Atlantic crossing, Douglas was enchanted
by the natural beauty of the city's setting. "This morning can never be
effaced," he recorded in his journal. He saw "the fine orchards of Long
Island on the one side, and the variety of soil and vegetation of Staten
[Island] on the other. I once more thought myself happy." That same
day, Douglas visited Hosack and his student Torrey and found both
men very welcoming. One week later, Douglas went up the Middle
Road to see the famous Elgin Botanic Garden and was dismayed to
find it "in ruins." He noted that Hosack had chosen a wonderful site for
the garden, with its hills, bottomlands, and soil variety, but he found
the only remarkable specimens left were two Southern magnolia species (*Magnolia cordata* and *Magnolia macrophylla*). One hothouse was
completely torn down. The other was a mere frame, all its glass panes
gone. The grand central greenhouse was barely hanging together and
contained no plants.

In addition to visiting Elgin, Douglas made several other botanical pilgrimages during his stay in the United States. He went to Philadelphia to see the late Bernard McMahon's Upsal Botanic Garden and the Bartrams' legendary garden, where he missed William Bartram by one month—the latter had died on July 22. Douglas next took a steamboat up the Hudson to botanize in the footsteps of Linnaeus's student Pehr Kalm, but when he arrived in Albany on October 8, he found the place "all in an uproar." Crowds packed the streets, church bells pealed, and cannon boomed from a hill outside town. Douglas had accidentally arrived on the day the first boat ever was scheduled to pass into the Erie Canal, a portion of which had recently been completed. A huge deputation from New York City had come up to celebrate. Douglas soon ran into Hosack, whose dear friend Clinton, though out of public office now, was the undisputed man of the hour. Even the inaugural canal barge was named the *DeWitt Clinton*.

Hosack was proud of Clinton, but it may well have nettled him that Mitchill had been awarded the second most prominent role in the festivities. Once the *DeWitt Clinton* had borne the canal commissioners past the crowds, Mitchill poured two bottles of water into the canal — one from the Atlantic Ocean and one from the Indian Ocean. Then he gave a speech in which he purported to speak for Neptune himself. The "Sovereign of the Deep" had sent "these samples of his saline element from the latitude of 36 degrees South, and from 40 degrees North." He informed the crowd that a mingling of waters was the way the Doge of Venice annually celebrated the marriage of the city to the Adriatic Sea. Then the chemistry professor in Mitchill won out over the poet, and he concluded by noting that a report on the chemical composition of the water would be filed along with the official documentation of the day's proceedings.

Back in New York City after the canal celebrations, Douglas saw Hosack for breakfast almost daily. By the time Douglas left the United States that fall, he had decided that Hosack was his favorite American, the one who treated him with the most unflagging generosity and warmth. He was also impressed with Hosack's work at the New-York Horticultural Society, which Douglas thought was in "a state of perfection." By now, many men close to Hosack had joined, including his son James Edward Smith Hosack and his former student John Francis. Even the ubiquitous Mitchill was involved; around this time he sent Hosack a long letter on the question of whether Americans would ever be able to make good wine. (Mitchill was optimistic.) Absent from the membership rolls was the plant-loving Aaron Burr, who had settled permanently in New York but remained unwelcome in many organizations and drawing rooms, thanks both to the duel and to his subsequent political and romantic intrigues.

THE FOLLOWING SPRING, Hosack took out the family Bible and wrote a mournful inscription. "Mary Hosack died on the 19th day of April, 1824, after a long and painful illness which she sustained with pious resignation and that equanimity of mind which characterized her through life and endeared her to every member of her family and an extensive circle of friends." She had left seven children behind— three girls and four boys. The oldest, Mary, was now a young woman of twenty-four, while the youngest, David, was just twelve. Jefferson wrote to Hosack from Monticello with his condolences. "On the subject of your recent affliction, experience in every bereavement which can rend the human heart has enabled me to sympathise sincerely with those who are suffering as I have suffered . . . time, silence & occupation are the only remedies. Of the last you have so much that much may be hoped from it's salutary operation."

Jefferson had been following Hosack's career from Monticello. The previous October, when Jefferson's son-in-law, Thomas Mann Randolph Jr., was planning a visit to New York City, Jefferson wrote to Hosack that Randolph "naturally wishes to be made known to some of it's prominent characters, and to no one with more desire than to yourself." In August 1824, three months after Jefferson's condolence letter on Mary's death, he wrote again from Monticello and noted in his letter that Hosack is "so justly esteemed there and here." Still another letter had Jefferson assuring Hosack of "my sense of your eminence in useful science."

Jefferson was right about Hosack's response to grief—after Mary's death, Hosack threw himself into his work again, especially at the historical society and the horticultural society. Late in the summer of 1824, Hosack was awarded a great honor on behalf of the entire New-York Institution. The Marquis de Lafayette, hero of both the American and the French Revolutions, sailed into New York Harbor in August to begin a tour of the United States with his son, George Washington Lafayette. Hosack was chosen to lead the two guests and a deputation of municipal officials from City Hall to the New-York Institution. When they entered the building, the New-York Historical Society rooms were filled with waiting men and women. They rose to their feet and bowed to Lafayette and his son as the two men walked with Hosack to the front of the society's meeting room. Here, the newspapers reported, Lafayette was shown to a seat at Hosack's side. Perhaps on this occasion Hosack ceded Marie Antoinette's chair to his guest, a man who had known the queen personally and had helped restore order in Paris in the bloody days immediately following the storming of the Bastille.

Hosack spoke first, addressing his remarks to Lafayette. "I have the gratification to announce to you your election, as an honorary member of this institution." Hosack reminded those present that the point of the society was to preserve the history of New York and the nation, especially the history of the great struggles that had given birth to the United States. "Every heart in this assembly throbs with inexpressible emotions at the sight of the hero who this day enters their hall." Lafayette now rose to his feet and spoke to Hosack in accented English. "With the most lively gratitude I receive the honour which the Historical Society of New-York have conferred by electing me one of their members," he said, going on to praise the citizens of the United States for showing the world how to create a stable democracy. Hosack bestowed a certificate of membership on George Washington Lafayette, and then he took the two guests on a tour of the Academy of Fine Arts and the Lyceum of Natural History. Afterward, Hosack climbed into a carriage with Lafayette and his son and rolled a short distance down Broadway to the City Hotel. Lafayette seems to have enjoyed meeting Hosack; in a letter he later wrote to Hosack from France, he called himself "your affectionate friend."

After Lafayette and his son left town on their tour of the United States, Hosack refocused his attention on the New-York Horticultural Society. On August 24, 1824, one week after the Lafayettes' visit, he nominated Adams, Jefferson, and Madison as honorary members of the horticultural society, and it was very likely Hosack who was also behind Sir James Edward Smith's election as an honorary member. Then, on August 31, at the horticultural society's annual dinner, Hosack was installed as its new president. It was his fifty-fifth birthday. The ceremony was held at William Sykes's new coffeehouse on William Street, where about a hundred members and their guests sat down to tables decorated with bouquets. Garlands of flowers festooned the walls all around them.

At three o'clock, Hosack stood up and looked out over the sea of faces. He had sailed into New York Harbor thirty years and five days earlier, with his head full of botany and with Linnaeus's specimens in his luggage. Ever since, he had labored to share his passion for plants with his fellow Americans. He had built Elgin and introduced New Yorkers to species from every part of the globe. Drawing on his British education, he had taught pioneering remedies and procedures to a new generation of American doctors. He had done more than any other citizen of the United States to call into being a generation of professional botanists where there had been almost no one. All the while, he had been helping turn his city into a national treasure in the arts and sciences.

Hosack began speaking. Horticulture encompassed three great endeavours, he told the audience: the cultivation of edible plants, the cultivation of ornamental plants, and the art of landscaping. These were among the noblest and oldest of human pastimes. Had he more time, he would begin the story from the very beginning. "But even the charms that Milton has attached to the blissful abode of the first happy pair" would not prevent him, Hosack told his listeners, from getting to his main point. "It is obvious that a garden should be established in the vicinity of this city."

Hosack never mentioned Elgin by name, but everyone knew what he was driving at—and not just Mitchill, Gentle, Clement Clarke Moore, and all the other members who had had direct dealings with the garden. Elgin was so famous among New Yorkers that this advertisement had been running that very summer in the *New-York Evening Post*:

> Sea-bathing and Public Entertainment. Coney Island House, lately erected . . . is this day opened for the reception of company. . . . The merchant, the man of business, and all who are exposed to the debilitating effects of a hot summer's day in New York, will find a tonic in sea bathing, and even in the sea air, which they may look for in vain among the drugs and medicines of the shops, the medical herbs and plants of the celebrated Elgin Garden, or the Materia Medica of the schools.

Now, before this huge audience of plant-loving New Yorkers, including many of the most powerful men in town, the indefatigable Hosack laid out his vision once again. The New-York Horticultural Society, he said, needed a parcel of land ample enough to grow fruit trees, vegetables, medicinal plants, and plants useful in manufacturing. For the purpose of "exciting the attention of our youth of both sexes to botanical inquiries, and of contributing to the beauty and elegance of the establishment," Hosack said, the horticultural society would need a vast collection of native and exotic plants. They would also need a conservatory, a library, a lecture hall, an herbarium, and a professor of drawing to teach plant illustration. New Yorkers enjoyed unparalleled access to the nation's naval and commercial networks, Hosack pointed out, so in short order they could easily obtain plants from all over the world.

Astonishingly, Hosack's dream for Elgin remained undimmed. It burned even brighter, in fact, when he thought about the Erie Canal, which was opening the way to thrilling new American landscapes. "The secrets of nature are yet to be unfolded," he told the audience. "Her hidden treasures, her countless varieties, and her unnumbered beauties are yet to be presented."

Hosack sat down, and Sykes's staff bustled around the room, serving a lavish dinner. Later, after the plates had been cleared, guests began proposing toasts—to the late Sir Joseph Banks, to Sir James

Edward Smith, to André Thouin at the Jardin des Plantes, to the three living ex-presidents of the United States. On and on they went. John Francis gave a toast to "Adam, who watered the first plant, and Eve, who plucked the first fruit." James Edward Smith Hosack raised a glass to "Liberty—a plant indigenous to no soil—it flourishes wherever assiduously cultivated." All the while, Hosack sat at the head table beneath an arbor made of flowers, looking like the president of the republic of botany.

He plunged into his new plans for the horticultural society. At the next meeting, he nominated Lafayette, Delile, Michaux, and Thouin for honorary membership; a few weeks later, he wrote to Madison and Jefferson about their recent election as honorary members. Both former presidents wrote kind replies to Hosack, although Jefferson's struck a wistful note: "I love the act, but age has taken from me the power of proving it by any services." With all these nominations of eminent Americans and Europeans, Hosack was cannily linking the New-York Horticultural Society to the reputations of the greatest political and scientific figures of the day. Now he would try to get Elgin back.

Chapter 16

"EXPULSION FROM THE
GARDEN OF EDEN"

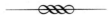

O N FEBRUARY 26, 1825, HOSACK WALKED INTO GRACE CHURCH, at the corner of Broadway and Rector Street, leaving the noisy chaos of horsecarts and carriages behind him. He was about to marry again.

His bride was Magdalena Coster, the widow of Henry Coster, one of the ten men who had lent Hosack money while he waited for payment from the Elgin lottery. Hosack had lost two wives and three children over the previous three decades. He now had four boys and three girls who were motherless. Magdalena Coster had seven children who were fatherless, and she had welcomed the marriage proposal from the sociable doctor who ministered to so many New Yorkers. As the congratulations poured in—Washington Irving even wrote from Paris—Hosack moved into the Coster townhouse on Chambers Street, along with most of his children. (His second-oldest son, Alexander, was studying medicine in Britain.) Hosack and Magdalena were soon presiding together over large family dinners. Their children sat in an alternating pattern around the table: Hosack, Coster, Hosack, Coster.

Hosack was now fifty-five years old, but he still struck his friends as youthful and energetic. An American writer named Anne Royall met Hosack the winter he and Magdalena were married and found him attractive: "His complexion is dark, his hair and eyes of the deepest black; his face is oval, with a high retreating forehead, of the finest polish." Hosack, she noted, was "manly and dignified" but also "affable and engaging." Royall admired his devotion to natural history. "He ranks amongst the first of great men" and was "one of the greatest botanists of the age; to his labor and indefatigable industry may be ascribed the success of that study in New-York." Royall also met Samuel Latham Mitchill that winter, and she thought that by comparison Hosack seemed "quite a young man," although Mitchill was only five years older.

Soon after the wedding, Princeton College commissioned a portrait of Hosack from Rembrandt Peale. The college had awarded Hosack an honorary degree in 1818 and wished to display his picture. Hosack was so busy these days, however, that Peale had trouble corralling him for sittings. He joked to a Princeton professor that Hosack "seemed disposed at first to think that I should create his Portrait chiefly by an effort of genius, rather than submit to the imprisonment of my room." Peale eventually prevailed. In the finished portrait, Hosack looks confident and alert, his face framed by graying sideburns but crowned with the familiar black waves. Hosack was pleased enough with Peale's work to hire the best gilder in New York to make a frame at his own expense.

He could now indulge in such luxuries. Henry Coster had died a wealthy man, and Magdalena had brought his land and assets to her new marriage. A few weeks after the wedding, the newspapers reported with astonishment that a parcel on William Street belonging to the Coster estate had just sold for $93,000—the equivalent of more than $2 million today. Flush with funds for the first time in his life, Hosack began hosting glittering parties with Magdalena that quickly became the most fashionable social events in New York. After an evening at their townhouse, guests went home and wrote breathless letters and journal entries describing the packed library, salon, game room, and dining room. Silk curtains hung at the windows, Brussels carpets covered the floors, and glass chandeliers hung from the ceilings. There were paintings and sculptures everywhere, including a bust of Hamilton. Hosack was the consummate host. He circulated happily among his guests, who included Clinton, Francis, Mitchill, Pintard, the paint-

ers Asher Brown Durand and Thomas Cole, the writer James Feni-
more Cooper, and many other old and new friends.

He seemed to know all New York. An acquaintance observed that as
he stood talking with Hosack on Wall Street one morning an "incredible" number of people greeted him—nearly everyone who passed.
Hosack now tapped this vast network for the Chambers Street soirées.
When the American writer William Cullen Bryant attended one night,
he found "a crowd of literary men—citizens & strangers—in fine
apartments splendidly furnished—hung with pictures." Another distinguished guest described a party he attended at the Chambers Street
townhouse as "one of the most brilliant assemblies I have ever witnessed." Hosack and Magdalena served a sumptuous late-night supper
that "embraced every available delicacy." The "wines were of the best
quality, [the] conversation animated."

At the Hosacks' parties, European aristocrats mingled with American merchants, judges, doctors, writers, and artists. Around the time
of Hosack's marriage, he had befriended Prince Charles Bonaparte, a
nephew of Napoleon who was in the United States studying ornithol-
ogy. Hosack praised the young prince to Sir James Edward Smith as
"zealously devoted to science"—their favorite kind of man. Another
French nobleman who enjoyed Hosack's hospitality was Alexis de
Tocqueville, who spent six weeks in New York during his 1831 tour
of the United States with his friend Gustave de Beaumont. A German
duke who attended a party at the Chambers Street townhouse wrote in
his journal that the "very famous" Hosack was a "learned and pleasing man" whose recent marriage was rumored to have brought him an
annual income of $50,000 (more than a million dollars today).

Magdalena had also brought Hosack his first country house. In
1805, the Dutch-born Henry Coster had built a mansion on a piece of
land overlooking the East River at Kip's Bay, about three miles northeast of the Chambers Street townhouse. All around the Coster mansion, flower gardens and groves of trees ornamented the grounds so
beautifully that a later resident described the estate as a "remnant of
paradise." (Today this land is around Thirtieth Street.) Hosack began
spending summers on the Kip's Bay estate with Magdalena and the children, while also keeping up with his patients and attending meetings

in town. Although his idyllic new situation in no way quelled his longing to wrest control of Elgin from Columbia, he was finally surrounded by plants again and could resume some of his botanical work. The first summer after his marriage, he contributed twelve bunches of white muscadine grapes from the Kip's Bay estate to the annual banquet of the New-York Horticultural Society.

HOSACK MISSED WORKING among his Elgin collections, but eight months after his wedding to Magdalena, he brightened at the possibility of thousands of novel specimens flowing to New York—a development he owed in large part to his friend Clinton. The Erie Canal was finished. It had taken eight years, but by dint of political will, brilliant engineering, and hard labor, New York had carved out a watery path approximately three hundred fifty miles long. Goods, crops, settlers, and soldiers could now glide on mule-drawn barges between the Atlantic Ocean and the Western states.

By eight in the morning on November 4, 1825, all of New York seemed to be jammed into the park at the Battery, where the canal celebrations would begin with a boat procession from the East River to the edge of the ocean. It was a perfect autumn day, the air so still that the water reflected the lines and colors of the boats, as if they were floating on a lake rather than in a harbor bordering the choppy Atlantic. Bands played rousing tunes on the decks of vessels bearing the names of political heroes and historic events. Inevitably, the lead steamboat was called the *Washington*; she was followed by the *Fulton*, the *Chancellor Livingston*, and the *Constitution*, among others. But the sixteenth boat in the procession bore a different kind of name: *Linnaeus*—as clear a sign as any that Hosack's circle was excited about the completion of the canal. The opening of the Western route held the promise of new species of plants and animals, just as the journey of Lewis and Clark had done two decades earlier. The flotilla moved away from Manhattan toward the edge of the Atlantic, where Clinton—governor once again—poured water from Lake Erie into the ocean and spoke a few words to mark the occasion.

An estimated one hundred thousand spectators lined the city streets that day—an impressive showing, given that the most recent United States census, taken five years earlier, had put New York's population at around 124,000, about twice that of Philadelphia. Just before eleven o'clock, four trumpeters on horseback set off from the Battery, with the parade's Grand Marshal and his four aides riding behind them, serenaded by a band. Then came the main body of the parade. The New-York Horticultural Society held the signal honor of walking at the head of more than six thousand members of the city's professions and occupations. It was a vivid testament to Hosack's influence as well as to New Yorkers' newly keen interest in their flora. Each member of the horticultural society wore a sprig of flowers pinned to his left breast. Some of them had woven garlands around their hats. The men marched up Greenwich Street carrying a banner of blue silk that symbolized the sky; an embroidered sun spread its rays over a lush landscape filled with trees, cascading waters, and ripening fruit.

The men walked for hours, leading the procession on a circuit around a city that had in recent years spread so far north as to swallow the village of Greenwich. That night, more than two thousand candles and lamps lit up the façade of City Hall, and pillars of fireworks rocketed skyward from its rooftop. The most magical moment arrived when some of the rockets rained down "brilliant sparks in the form of a willow decorated with stars," while others "resembled the poplar, each being accompanied with showers of gold and silver rain." A forest of light was growing atop City Hall. Three miles north, the night obscured Hosack's ruined garden.

⸻

DEWITT CLINTON'S CANAL TRIUMPH signaled that the post-Revolutionary generation—Hosack's generation—was confidently taking the reins from the nation's elder statesmen. The deaths of Jefferson and Adams on July 4, 1826, the fiftieth anniversary of the Declaration of Independence, sharpened the impression of a solemn passage. Hosack, who had so admired Jefferson's dedication to botany, sent condolences to his family on behalf of the New-York Historical Society.

It may well have been this sense of old figures fading and new possibilities emerging that spurred Hosack to visit Washington that same year. Magdalena accompanied him on a trip combining visits to friends with scientific business. A letter Hosack wrote to John Francis from the road suggests his intention was to sound out the possibility of congressional support for Elgin. In Washington, Hosack met with some of the men involved in the Columbian Institute for the Promotion of the Arts and Sciences, a society organized in 1816 in large part by Dr. Edward Cutbush, a former student of Benjamin Rush with whom Hosack had overlapped in his Philadelphia days. Another moving spirit in the Columbian Institute was Thomas Law, Hosack's old acquaintance from the *Mohawk* voyage. In 1820, with the blessing of Congress and President Monroe, the institute had secured five acres at the boggy bottom of Capitol Hill for the creation of a botanical garden, although little progress had been made on the collections by the time Hosack visited in 1826. This benevolence may have encouraged Hosack to dream of securing federal funds to regain and restore Elgin. His Washington friends, however, had no interest in jeopardizing their own source of support, and they were quick to disabuse Hosack of the idea that Congress might take responsibility for New York projects.

Before leaving town, Hosack and Magdalena paid a visit to President and Mrs. John Quincy Adams and also went to the Capitol to hear Senator Martin Van Buren of New York give a speech. Then they went to see Mount Vernon and pay their respects to George Washington's memory. On their way back to New York, Hosack and Magdalena stopped to see friends in Philadelphia. Here Hosack paused in his social rounds long enough to send an update to John Francis in New York: "In Philadelphia we have seen everybody and everything." They visited the Rush family, toured the ingenious waterworks on the Schuylkill River, and went out to The Woodlands—which was no longer in the glorious state of William Hamilton's days but still lovely.

All around Hosack on this trip was the bittersweet sense of long lives well lived, in the service of natural history, the arts, and the nation itself. Not only the towering Washington and Jefferson, but also men such as Rush, McMahon, and the Bartrams. His old friend Peale was still around and still painting the occasional portrait, but he was now in his mideighties and no longer running his museum, having placed it in the hands of his sons Rubens, Rembrandt, and Titian. As usual, the museum was not doing well financially, and Peale was spending a good deal of time in New York, making and selling false teeth on the strength of long personal success with his own set. A widower once again, he was also trying to court Mary Stansbury, a single woman in her fifties who taught at the New-York Deaf and Dumb Asylum (an institution that Hosack's nephew John Eddy had helped found). Having no wife, Peale wrote in his diary, made him feel "like a fish out of water."

In May 1826, while on one of his New York visits, Peale was knocked to the ground by a horsecart near City Hall. Hosack happened to be riding by and helped the old man into his carriage. Peale insisted to Hosack that he was absolutely fine, but his trips to New York were beginning to take their toll, as did Mary Stansbury's eventual rejection. In February 1827, a grueling stagecoach journey back to Philadelphia left Peale bedridden with exhaustion and chest pains. Looking back at his life, he mused to Rubens, "Happiness is certinly a common Plant but the cultivation of it requires no little skill." It was during this illness that Peale suddenly became so frightened for his life that he hammered on the bedroom floor with his cane to rouse his children. They summoned a doctor, who bled him. He lingered a few weeks, then died on February 22.

Peale didn't live long enough to learn of the bequest of James Smithson, a wealthy British naturalist who died little more than two years later, in 1829. In a sign of admiration for the American democratic experiment, Smithson had left instructions for the money to be used "to found at Washington, under the name of the Smithsonian Institution, an establishment for the increase and diffusion of knowledge among men." President Andrew Jackson sent Benjamin Rush's son Richard to London to sail back with Smithson's bequest, and in the autumn of 1838 Rush arrived in New York Harbor on a ship that carried more than a hundred thousand gold sovereigns. By the middle of the nineteenth century, the Smithsonian Institution would fulfill Peale's eighteenth-century dream of "a great National Museum" for the United States.

In the three decades following Peale's death, his children struggled valiantly to maintain his museum, but many Americans seemed more interested in the freakish entertainments hawked by showmen masquerading as curators and professors—the kind of men who attached a human skull to the skeleton of a large fish and then sold the public tickets to see a mermaid. Around 1850, two of these men, P. T. Barnum and his partner Moses Kimball, acquired the Peale collections—in part to display pieces from it, but mostly to kill the competition to their new Philadelphia museum of wonders.

A FEW MONTHS after Peale's death, Hosack wrote to DeWitt Clinton saying he had heard a rumor of *Clinton's* death. Hosack knew the governor was alive, but he also knew he was in terrible health. The once-handsome Great Apollo, the man whose Erie Canal triumph had unleashed speculation that he would be the next president of the United States, had become, according to a cruel description left by John Francis, "a mass of obesity, unwieldy, and . . . loaded with adipose deposits." Hosack fretted about Clinton during the fall and winter of 1827, sending him medical advice that included exercise and a high-protein diet but also the retrograde measure of bloodletting to ease excess pressure in his blood vessels. Early in February 1828, Hosack warned him in person of the danger of his situation, but Clinton assured him he was "not afraid to die." His words reminded Hosack of George Washington's reported equanimity in the face of death in 1799. This turned out to be the last time Hosack ever spoke with his oldest, dearest friend. A few days later, Clinton died at his desk, in the presence of two of his sons.

Hosack immediately began writing to Clinton's family, friends, and associates, compiling memories and documents for a biography. At the memorial service for Clinton, Hosack read from these biographical materials at such length that many people in the audience gave up and filed out before he was even half finished, according to one of his medical students. When Hosack published his biography of Clinton the following year, it ran to more than five hundred pages. He sent copies to everyone he could think of, including James Madison, who in his thank-you note paid a lovely compliment to Hosack himself. "Permit me Sir, on this occasion to express the particular esteem, I have long been led to entertain for the Endowments, intellectual & ornamental, which distinguish your character; and to join in the public tribute which the use you have made of them, must always command."

After Clinton died, Hosack lost yet another old comrade. This time, the news came from across the Atlantic. On March 17, 1828, Sir James Edward Smith succumbed to a seizure at the age of sixty-eight. For thirty-five years, Hosack's friendship with the king of British botany had been a source of inspiration and pride. It had been Smith who had masterminded the greatest honor of Hosack's life—election to the Royal Society. Bolstered by Smith's moral support and his steady example, Hosack had never ceased fighting for the cause of botany in the United States.

In fact, Hosack had recently launched into a new botanical campaign with all his customary vigor, after receiving an intriguing letter from Secretary of the Treasury Richard Rush. Rush, as one of Benjamin Rush's sons, knew Hosack well, but on this occasion he was writing on behalf of President John Quincy Adams. "The President has it at heart to cause to be introduced into our country all such trees and plants, (not heretofore known among us,) from other countries, as may give promise, under proper cultivation, of flourishing and of becoming useful in any part of the U. States." President Adams was seeking advice from American men of science about which plants his diplomats around the world should be sending home. Hosack was so excited about this sign of presidential interest in botany that at the next meeting of the New-York Horticultural Society, he created a committee to draft a report for the president. When it was ready, Hosack sent it to Rush, who promised to show it to President Adams.

Two decades earlier, in 1806, Hosack had tried to persuade President Jefferson to create a national network of botanical gardens so that American naturalists could conduct research on foreign and native plant species in different climates, with the results shared systematically across the country. He had failed on that occasion—and then he had been forced to watch helplessly as Elgin fell apart, as well. Now

Hosack wrote Secretary Rush pointedly that the United States had "no public botanic gardens of any consequence, nor experimental gardens, which are deemed of so much importance, that every part of Europe, even places of minor notice, can display one." Hosack and his committee were recommending the creation of a federally funded network of at least three gardens: one in New York City, with its excellent access to national and international shipping routes and its northern climate; one in the South (Florida, Georgia, or South Carolina), where warmer-weather plants could be tested; and one in the nation's capital, where, as Hosack knew, a botanical garden was already taking shape on the Mall. But even twenty years after he first floated the idea to Jefferson, Hosack was still ahead of his time. Not until 1887 would Congress enact legislation creating a national network of plant research facilities—today's agricultural experiment stations.

In proposing his idea to Secretary Rush, Hosack couldn't resist pointing out that Elgin could have been "the pride and ornament of our state and country." Still, the new signs of interest in botany and horticulture coming from Washington seem to have increased his appetite for tussling with Columbia over the Elgin land. In April 1828, Hosack told the New-York Horticultural Society that he had reopened negotiations with the college. He allowed himself to feel optimistic about the outcome.

With all these encouraging developments, that summer should have been a relaxed and happy one for Hosack, when he could retreat from teaching to putter among his grapevines and flowers at the Kip's Bay estate and work on his plans for Elgin. Instead, it was a season of miseries. In July, his nephew Caspar Wistar Eddy died of an inflammation of the brain at the age of thirty-eight. Caspar's place of death was officially recorded as Bloomingdale. Perhaps when he had fallen ill of a brain malady he had been placed in the Bloomingdale insane asylum, with some of the familiar old Elgin trees ornamenting the grounds around him. Or maybe he had simply been staying with friends in the pretty countryside of northern Manhattan, where he had as a boy roamed through the meadows with his cousin John and collected rabbit's-foot clover to carry back to their uncle for his garden. Caspar had been a constant fixture in Hosack's life for more than twenty years. His death severed one of the few remaining links to the happy early days at Elgin.

One week after Caspar's death, Hosack's oldest son, twenty-five-year-old James Edward Smith Hosack, died in London, where he had gone to study medicine in his father's old haunts. James was buried in the graveyard of St. Martin-in-the-Fields, the cause of death unrecorded. Hosack knew the church from his own days in London, having attended the funeral of the famous anatomist John Hunter there more than thirty years earlier. Ten days after James's death—but before Hosack had received this devastating news—he was forced to report to the horticultural society that his negotiations with Columbia to lease the land back so he could rebuild the garden had failed. The trustees had bigger plans for the property: leasing it to builders of apartments. Hosack gave up on Elgin forever.

For three decades he had tended his garden—first in his imagination, then in the earth, and finally in his aching heart. "I fear New York is destined to be what Napoleon called England, a nation of shopkeepers," he wrote to a friend. "As I once said the Tontine Coffee House is her university and the Insurance companies her colleges." Thanks to a chance encounter the following spring, Hosack found a poignant way to mark the loss of the garden. On a Saturday afternoon in late May 1829, he ran into a friend named Thomas Cole in the street. Cole, a young painter, was running errands around town as he prepared to set sail on his first artist's trip to Europe, where he planned to tour the beautiful old cities and landscapes with his sketchbook in hand. It was a dream itinerary, but for the past year Cole had been wrestling with depression and anxiety over his faltering fortunes. By the time Hosack encountered him in the street, he was panicking.

One year earlier, at an exhibition held in May 1828, New Yorkers had been dazzled by two biblical paintings by Cole. They were large works, each more than three feet by four feet. One was entitled *The Garden of Eden*. Cole had poured his love of nature into every inch of its canvas, imagining Eden as a lush and colorful valley where the tiny figures of Adam and Eve were almost lost to the viewer's eye in a glorious profusion of plants. Here species from disparate climates flourished in a single garden—hollyhocks, strawberries, irises, tulips, roses, aloe, poplars, palms, pines, and more. The closest thing to this botanical fantasy that New York had ever seen before was the Elgin Botanic Garden. A reviewer raved that *The Garden of Eden* was "a picture on which the eye could banquet for hours."

Cole's other biblical painting was jarringly different. On the right half of the canvas, a swath of the Garden of Eden glowed in the sunlight, while on the left half, Adam and Eve traversed a frightening terrain of jagged rocks and twisted tree trunks. The title of this painting was *The Expulsion from the Garden of Eden*. Cole was justifiably proud of both works, and over the next months, he received a great deal of attention for them. Nonetheless, a year after the exhibition had opened, they still hadn't sold. Cole even went so far as to try to organize a raffle in New York. He grew resentful about the paintings, fretting that he had wasted his time on them while blaming their failure to sell on the "apathy" of "this wholly commercial city." Cole desperately needed the proceeds from their sale in order to travel to Europe, where he knew he could improve his technique and fire his artistic imagination. It was the best way to draw more American buyers for future works, but he couldn't afford the trip without selling his current works. He felt trapped.

One year after Cole had first exhibited the two paintings, he put them on display in New York once again—this time at a bookstore on Broadway. As an ad ran in the newspaper touting these "fine productions of American talent," he started planning his European trip in earnest. In early May 1829, he dashed off to see Niagara Falls for the first time, writing to his patron Robert Gilmor in Baltimore, "I wish to take a 'last lingering look' at our wild scenery. I shall endeavor to impress its features so strongly on my mind that in the midst of the fine scenery of other countries their grand and beautiful peculiarities shall not be erased." He found the falls grander by far than he had anticipated. When he returned to New York City, he learned that a local banker had agreed to pay $400 for *The Garden of Eden*. Cole still didn't have enough money for his voyage, however, and he was scheduled to sail for Europe on June 1, now just days away.

Then, on the afternoon of May 30, Cole ran into Hosack in the street. Hosack made Cole an offer then and there for *The Expulsion from the Garden of Eden.* There was no more suitable owner in the United States than the man now permanently exiled from his American Eden. But if this painting captured Hosack's stormy past, a second painting he bought from Cole that day heralded a more serene future. This was a smaller work, *The Subsiding of the Waters of the Deluge,* which depicted the first moments of ethereal calm after the Great Flood. Cole had painted a single white dove flying toward Noah's Ark across waters that shimmered like mother-of-pearl in the sun. Hosack swore Cole to secrecy about the amount of his offer for the two paintings; the latter complained the next day to Gilmor that it was far lower than he had wanted. Still, the sale enabled Cole to leave for Europe, where he saw the ancient ruins that would inspire *The Course of Empire,* the series of five monumental paintings he created after his return to New York in 1832.

Hosack, meanwhile, collected *The Expulsion from the Garden of Eden* and hung it in his townhouse on Chambers Street. Around this time, he went to survey the remains of his own lost garden. A specimen from a persimmon tree (*Diospyros virginiana*), held today by the herbarium of the New York Botanical Garden, is labeled in his hand: *"Elgin Garden 1829."*

BY NOW, THOUGH, Hosack had an exciting new distraction. In the autumn of 1828 he and Magdalena had purchased Samuel Bard's Hyde Park estate from Bard's son William, a lawyer in New York. For Hosack, the Bard mansion was filled with happy memories of a mentor who had cherished the time he spent there, surrounded by his family. All around the mansion lay hundreds of acres of breathtaking Hudson River scenery. Thanks to Samuel Bard's boyhood studies with Jane Colden in upstate New York, he had loved botany and horticulture his whole life and had made many improvements to the Hyde Park farms and gardens after inheriting them from his father, Dr. John Bard. Now Hosack, drawing on his two decades of experience at Elgin, began to form his own new vision for the estate. All through the winter of 1828 to 1829, he daydreamed down in New York City about what kinds of gardens he would lay out, which fences he would remove, and which hayfields he would mow to feed the livestock he planned to import from Britain. In March 1829, he told William Bard that he was longing for the frozen Hudson to thaw so he could take a steamboat up to Hyde Park and get to work.

By summertime, Hosack, Magdalena, and some of their younger children had settled into an old farmhouse on the property. Magdalena fell in love with the estate, too—a good thing, because it was largely her fortune that Hosack now began spending on building materials, plants, labor, and animals. John Pintard sniped in a letter to his daughter that Hosack had become so obsessed with the estate that he was no longer donating to charitable causes. This was a bit unfair, given all the time and money Hosack had poured into Elgin and the many other institutions he had been involved with over the previous thirty years. But it was true that people were taking note of Hosack's new spending habits. Jacob Harvey, who had recently married Hosack's daughter Mary, wrote to his father in Ireland, "My father-in-law has laid out a great deal of money at Hyde Park which will probably never be got again, but he enjoys the change very much."

Thanks to Magdalena, Hosack could now dream on as grand a scale as almost any man in New York City who suddenly had a country estate for his plaything. He summoned one of New York's foremost architects, Martin Thompson, to Hyde Park. Thompson designed two new wings for the Bard mansion. They contained a billiard room, a music room, a library, and a gallery for Hosack's large collection of European and American paintings—among which were two fourteen-foot panoramas of Niagara Falls by John Trumbull and a portrait of George Washington by Gilbert Stuart. A Philadelphia physician who visited Hosack's Hyde Park gallery concluded that "as a connoisseur of fine art, none of his contemporaries excelled him."

The façade of the mansion now stretched for almost one hundred forty feet, much of it two stories high. Hosack had a coal furnace installed in the basement to help keep the family and guests cozy in the sprawling house on cool nights. Piazzas graced both the front and the back of the mansion, and the view from the back piazza had one guest after another swooning. A friend of Hosack named James Thacher was awed by what he saw when he stood there: "The noble Hudson, which is nearly a mile in width, speckled at all times with the white spreading canvas, or the more formidable Fulton steamers. A richer prospect is not to be found, a more varied and fascinating view of picturesque scenery is scarcely to be imagined." The British writer Harriet Martineau, who visited in the fall of 1834, loved the way the land behind the house made its way down toward the Hudson—"not square and formal, but undulating, sloping, and sweeping, between the ridge and the river." Although Martineau was in her early thirties, this grassy hillside tempted her to "run up and down the slopes, and play hide-and-seek in the hollows."

She found these enticing effects repeated throughout the estate. They were the signature technique of André Parmentier, the gifted young horticulturist Hosack had hired to redesign the estate grounds. Parmentier was a nurseryman and fellow member of the New-York Horticultural Society who had grown up in Belgium in a family of celebrated horticulturists. In 1824 he had emigrated to Brooklyn, where he opened a nursery and began writing about and designing American gardens. He brought with him a philosophy of landscape design that was honed in reaction to the stiffness of French formal gardens. He urged wealthy New Yorkers to embrace the British predilection for soft lines and ever-changing vistas. He thought straight lines "ruinous" to

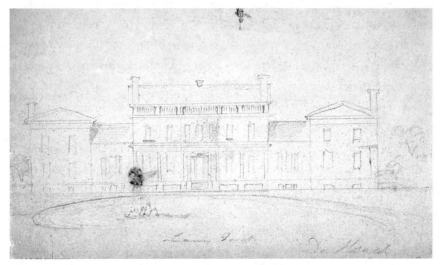

Hosack's mansion at Hyde Park

a lovely view and argued "how ridiculous it was, except in the public gardens of the city, to apply the rules of architecture" to a landscape.

Parmentier's ideas were new to most New Yorkers, and indeed to most Americans. Just before he began his work for Hosack at Hyde Park, Parmentier laid out his revolutionary philosophy of what he called the "modern style" of landscape design. Through the painterly use of plants and trees to give color and structure to a landscape, a skillful designer "presents to you a constant change of scene," thereby engaging the eye and the imagination. At the edges of an estate, the trees should have "thin and light foliage," while those grouped closer to the mansion should be deep green. The viewer's gaze is thus subtly directed toward the central subject of the designer's "landscape-picture"—the mansion. Hosack gave Parmentier all seven hundred fifty acres of Hyde Park to use as his canvas. The results were so spectacular that the landscape architect Andrew Jackson Downing pronounced the estate "one of the very finest in America." The writer Frances Trollope raved after a visit, "It is hardly possible to imagine anything more beautiful than this place."

Hosack and Magdalena felt the same way. They began spending most of each year on the estate, moving back to their New York townhouse only for the winter season. At Hyde Park, their neighbors were friends from the city, including the Roosevelts, the Livingstons, the Van Nesses, and the Pendletons. Members of these families often came over to socialize with the Hosacks, sometimes squiring guests onward for visits to their own estates. Hosack wrote to his friends in New York and elsewhere inviting them to visit him at Hyde Park and urging them to invite *their* friends.

Thanks to the lavish Chambers Street parties and the obvious joy that Hosack took in throwing them, he had earned an international reputation as a gracious host. Now he presided over an endless house party composed of an ever-changing array of guests both foreign and domestic. In September 1830 Philip Hone, a former mayor of New York who was Magdalena's cousin, was "walking and riding over Dr. Hosack's splendid grounds" during a stretch of lovely early autumn weather. He noted in his diary the arrival of Joel Poinsett, a diplomat who had recently served as President Adams's minister

View from across the Hudson to Hosack's Hyde Park estate,
sketched by his guest Thomas Kelah Wharton

to Mexico; Poinsett had brought the governor of Mexico with him to
meet Hosack.* And in 1832, Washington Irving warmly accepted an
invitation to Hyde Park, "of which I have heard the most delightful
accounts as well as the hospitality of the owner." When guests arrived
by steamboat and disembarked at a wharf at the western edge of the
estate, they often found Hosack himself waiting for them in his carriage. Parmentier believed in impressing visitors with the vastness of
a property before they caught their first glimpse of the residence, so
guests approached Hosack's house via a winding carriage drive that
Hosack's friend Thacher found "truly enchanting." The white mansion rising among the trees at the western edge of the lawn looked to
Thacher like a "palace."

Hosack finally had a grand country seat of his own, and he happily played the laird of the castle. He showed guests his art collection,
among them some vases that he told Harriet Martineau had belonged
to Louis XVI before being smuggled out of France during the Revolution. He also loved sharing books from his exceptional library. The
library, more than eight hundred square feet, was ornamented with

---

* Poinsett was also an amateur botanist, and a striking flower he found growing in Mexico
still bears his name: *Poinsettia*.

matching mantelpieces of black-veined marble perched above grates
that funneled in heat from the basement coal furnace. Five windows
illuminated the room during daylight hours. On the mahogany reading table in the middle of the room, candlesticks stood ready for the
evenings. The table also held bronze inkstands and a convex reading
lens on its own mahogany stand. One guest noted that the "carpet,
rugs, sofa, chairs &c. are in accordance with the sumptuous style of
the rest of the house." The room was so gorgeous, in fact, he thought
it would be hard to focus on reading there. Still, he couldn't help gravi-
tating toward the mahogany bookcases lining the walls of the library.
They were filled with thousands of beautifully bound volumes, many
inscribed to Hosack from their authors, including some from Sir
Joseph Banks. In one bookcase, Hosack kept extra copies of the biography he had written of Clinton, together with the many other works
he had authored.

Hosack was very proud of his library. He told more than one visitor that there were between four and five thousand volumes and that
he had spent at least $20,000 assembling them over the decades. It
was indeed a spectacular collection; some contemporaries thought it
was probably the finest in the United States, especially in the fields
of botany and medicine. When Andrew Jackson Downing wrote his
friend John Torrey in July 1834 requesting a letter of introduction to
Hosack, he explained that he was "anxious to get a moment's peep at
a book or two." Yet for all Hosack's pride in his art and his books, it
was the landscape of Hyde Park that seemed to delight him most. He
walked with his guests down to the Hudson and around a promontory
he called Cape Horn, and then he took them to rest in a pretty pavilion near the water. It was in this pavilion that Hosack sat with Harriet
Martineau one day and reminisced about his old friend DeWitt Clinton. At another point during her visit, Hosack told her about the summer day thirty years earlier when he had rowed to Weehawken with
his friend Hamilton. Hosack had also recently served on a committee
to erect an elegant sculpture of Hamilton in the rotunda of the new
Merchants' Exchange building in lower Manhattan.

Hosack banned the firing of guns anywhere near the house and gardens at Hyde Park so that the air would always be filled with birdsong.

The whole estate was, as one visitor wrote after a summertime ramble, nothing short of "a terrestrial paradise." The perceptive Martineau understood what this landscape meant to a man as passionate about nature as Hosack: "I felt that the possession of such a place ought to make a man devout, if any of the gifts of Providence can do so. To hold in one's hand that which melts all strangers' hearts is to be a steward in a very serious sense of the term."

Hosack was now in his midsixties. He had spent forty years driven by an intense feeling of stewardship—of people, nature, New York, and the nation. When he had fallen in love with medicine as a teenager in the 1780s and pushed himself so intensely in his studies, it was to become a capable steward of his fellow citizens' health. When he had created the Elgin Botanic Garden and then labored lovingly over it year after year, it was because he saw himself as a steward of a parcel of land whose cultivation would save lives and bring new foods to farmers and city dwellers alike. When he had devoted his time and expertise to Columbia College, the New-York Hospital, the Lying-In Hospital, the quarantine hospital at Bellevue, the College of Physicians and Surgeons, the almshouse, the New-York Historical Society, the Academy of Fine Arts, and the New-York Horticultural Society—in all these ways and places, he was trying to steward his young nation and especially his city toward a future securely anchored in the kind of charitable and cultural institutions that would keep its residents healthy, thriving, and as well educated as Europeans. It was an exhausting, vital struggle to create and sustain these institutions.

At Hyde Park, though, Hosack was free to make his own world. He faced no legislative committees, no sneering stockjobbers, no stingy trustees. He liked to stroll down the gravel path that led south from the mansion between beautifully planted borders. Right next to the path, on each side, Parmentier had placed small herbaceous plants. Then he had angled the profile of the border upward with low shrubbery planted just outside the smaller plants, and finally he had culminated the whole design in tall shrubbery and small trees. When Hosack reached the end of this path, he entered a garden. He was surrounded on all sides by flowers, flowering shrubs, and flowering trees. At the heart of this garden stood a tall greenhouse flanked by two smaller hothouses.

Hosack was creating his own new Elgin. The complete façade of his conservatory was about one hundred ten feet long, and its design was strikingly similar to the one he had built on the Middle Road twenty-five years earlier. Hosack took his friend Thacher on a tour of the conservatory in 1830, only a year and a half after purchasing the estate. He had already managed to fill it with plants from around the world—magnolias, bird-of-paradise flowers, acacias, an eight-foot-tall India rubber tree, and pineapples that Hosack served his guests for dessert. Hosack soon added a fringe tree, Mexican tiger flowers, lemon trees heavy with fruit, and "a thousand other beauties." Harriet Martineau thought the conservatory was "remarkable for America," while when Downing later visited he found "a handsome and well filled range of hothouses" and got to sample a tropical fruit called cattley guava. Hosack's new Elgin would be incomplete without medicinal species, of course, and there were many of these among the plants he was collecting in his conservatory. He was also working on plans for dedicated medicinal beds outdoors, according to his head gardener, Edward Sayers, who later recalled that when Hosack tended to the medical complaints of local villagers, he often advised them to "apply simples and herb tea, such as wormwood, horehound, &c."* Sometimes the villagers also picked his exotic specimens as if they were wildflowers. When Hosack saw his plants in their gardens, he asked for them back, and they were returned politely.

At Elgin, Hosack had had fewer than twenty acres to farm. Now he had five hundred. He loved driving guests to see his agricultural operations, which lay on the opposite side of the Albany Post Road from the mansion and the conservatory. Anyone who had ever visited Elgin and now toured Hosack's Hyde Park farm could see that he had outdone himself. By the autumn of 1830, he already had hundreds of acres of grasses and grains under cultivation. He had also ordered his gardeners to plant a kitchen garden filled with every kind of vegetable that would grow locally. A visitor who took a summer-

---

* In 2015, the Nobel Prize in Medicine was awarded in part to the Chinese researcher Tu Youyou for her discovery that an active compound in sweet wormwood (*Artemisia annua*) was highly potent against the malaria parasite.

time tour of Hosack's farm saw carts piled high with ripe watermelons, while still another enjoyed the fresh citron melons served for dessert at the mansion. Hosack was so proud of his kitchen garden that he sometimes took specimens to the city and presented them to the horticultural society.

One secret to his excellent produce was in the adjacent barnyard, where his men had installed a manure pit forty feet in diameter, with drains that captured the liquid runoff and channeled it into the kitchen garden. Here, among stables, barns, pens, and sheds, was the beating heart of Hosack's farm. Hogs lived in pens separated by breed (with a "cooking apparatus" located nearby), while Hosack's six hundred or so sheep—Merino and three other kinds—had a yard of their own between the barn and the kitchen garden. Hosack showed Harriet Martineau his flock of poultry—"a congregation of fowls [that] exceeded in number and bustle any that I had ever seen"—and his herd of cows, whose milk went to a dairy located in the cellar of the old farmhouse. When Thacher toured this dairy, he was struck by two novel sights: windows covered with wires that kept out the flies, and a dog running on a flat wheel attached to the butter churn, evidently trained to anticipate a reward of butter.

Hosack seemed captivated by every aspect of farming and country living. One of his guests "found the Doctor sagacious about long horns and short legs in a degree which impressed me with a due consciousness of my ignorance." Hosack made sure that the stream running through the farm was dammed in several places and that the resulting ponds were stocked with pickerel and trout. Hosack had a cider house ready that could hold one hundred barrels, and Jacob Harvey noted that "my father-in-law is very anxious to excel in fruits at Hyde Park." Hosack also began keeping bees. According to Thacher, Mitchill had given Hosack a colony of stingless bees from Mexico, and Hosack had let them loose in his greenhouse so they wouldn't die of cold. He wanted to raise native bees, as well, so Thacher—an expert beekeeper—designed Hosack a thirty-foot apiary containing forty hives, each of which was to be fitted with glass-fronted drawers that could be pulled out from behind to remove the honey. Several years after Thacher's visit, however, Hosack confessed to him that "my bees have not succeeded," because neither "my farmer nor his wife appear to understand the subject—they require another lesson from you."

Over and over in his life, Hosack had approached all the failures and setbacks as so many opportunities for self-improvement and renewed optimism. Now his enormous wealth and his complete control over the Hyde Park estate had removed all constraints. Thacher was amazed at how completely Hosack threw himself into his new projects each day, rising early and leaving the house for whatever spot on the estate his attention was most urgently needed.

Hosack told another of his guests, the artist Thomas Kelah Wharton, that he was writing a book about his improvements at Hyde Park, and he asked whether he could include Wharton's beautiful sketches of the estate as illustrations. Wharton was delighted at the proposal and insisted he didn't want any payment, but while he was out of the room fetching his latest picture, Hosack wrote out a large check from his Bank of New York account and pressed it on him when he returned. Hosack had developed a keen interest in Wharton's artistic future, and one evening they sat talking for a long time on the piazza as the Hudson faded into the gathering dusk. All around them lay the evidence of Hosack's undimmed interest in the natural world. Forty years earlier, in the eighteenth century, with the help of Curtis, Smith, and Banks, he had begun his journey with a magnifying glass in one hand and a walking stick in the other, straining to discern the plant structures that would teach him how to organize his collections and his mind. He had never stopped trying to catalogue the natural world in this way. But now, as he sat on his Hyde Park piazza with the fireflies sparkling against the dark trees below, his views of nature had been enriched by the romantic nineteenth-century visions of artists like Cole, Durand, and Parmentier, who labored to reveal nature's sublime, soul-stirring beauty.

In 1804, when Hosack had sent a sample of the young John Eddy's herbarium work to the botanist Martin Vahl in Denmark, he had also confided that "in a few years I hope to be enabled to withdraw from the *labour* of the Profession and to fix my residence in the country in the neighborhood of this city." In 1834, Hosack finally retired from his New York medical practice and began living on his estate all year.

It was around this time that John Francis began work on an article about Hosack for an encyclopedia of notable Americans. When he had completed a draft, he sent it to Hosack at Hyde Park for his corrections. Upon Hosack's retirement, some of his friends had tried to persuade him to run for public office. Hosack had declined with an eloquent reflection on the great passions of his life, which Francis quoted:

> If a party could be formed favorable to the interests of *education*, of *agriculture*, and the *commercial* character of our state; to the development of its natural resources and promotive of internal improvements; to such a party I could not hesitate to avow my allegiance, and to devote the best exertions of which I am capable to advance the interests of my native state and country: but under the existing dissentions, I must decline all connexion with our political institutions, and devote myself to the cultivation of the vine and the fig-tree, as more conducive to my own happiness and that of my family.

The "vine and the fig-tree" was a familiar Old Testament phrase, one that George Washington had famously used in a letter he wrote to Lafayette in 1784 describing how happy he was to return to Mount Vernon after the Revolutionary War. Hosack undoubtedly knew Washington's letter; it had been published in a number of local papers not long before Hosack wrote his own statement.

Washington had had his Mount Vernon, Jefferson his Monticello, Hamilton his Grange. Hosack now had his Hyde Park.

Chapter 17

"LIKE A ROMANCE"

IN THE AUTUMN OF 1835, HOSACK PRESENTED EACH OF HIS CHILDREN with a ring containing a lock of his hair, a common memento of a deceased loved one. Hosack was certain he would die soon.

It was around this time that Hosack took his oldest son, Alexander, aside for a private conversation. Alexander was now thirty years old and practicing medicine in New York. Hosack told Alexander that he expected to suffer a stroke before long and that he was trying to learn to write with his left hand, so he would still be able to communicate with his family if his right side became paralyzed. As Alexander watched, his father tried writing something and showed him the results. Hosack was as fascinated as ever by the afflictions of the human body, but the conversation made Alexander uncomfortable and he changed the subject. Hosack was as healthy as Alexander had ever seen him.

Not long after this talk, Magdalena's youngest daughter, who was seventeen, married a scion of the wealthy Schermerhorn family. The wedding was held on December 9 at the Hosacks' townhouse on Chambers Street, with an Episcopal minister from the Hyde Park village church performing the service. At the wedding supper afterward, one end of the table was laden with "superb" fruits and vegetables Hosack had grown in the Hyde Park greenhouse, as Magdalena's cousin Philip Hone reported in his diary. One week later, on December 16, Hosack and Magdalena hosted a second party, this time so that Magdalena's daughter could officially receive her family and friends as Mrs. Peter Augustus Schermerhorn. The thermometer had been below zero for days, and Hone and the other guests braved frigid winds as they made their way to the Hosacks' townhouse for the celebration.

Later that same night, as Hone sat at home in his library writing, the city's bells begin to peal an alarm—fire. He rushed outside and joined a stream of people surging toward the southernmost tip of the island, where he could see flames exploding into the sky "like flashes of lightning." Hosack also raced to the scene, prepared to offer medical assistance. The blaze had broken out at a warehouse near the docks, and now the whipping wind was spreading it rapidly to neighboring structures. The city's volunteer firemen were powerless to stop it, because the water had frozen in the wells and pipes. As the flames ripped through the buildings, people began dragging what goods they could out of the warehouses and shops, but many of these piles caught fire, too. Even the East River was aflame with burning turpentine slicks.

In New Jersey and Connecticut, people saw the red sky and knew that disaster had befallen New York. Documents from the businesses of lower Manhattan blew out over Long Island. A resident of Flatbush, more than five miles from Manhattan, later reported finding an insurance claim in his garden. By the end of the night, more than six hundred buildings had been destroyed, and virtually all of Manhattan's financial and commercial district was in ruins. The statue of Alexander Hamilton that Hosack and others had commissioned for the rotunda of the Merchants' Exchange now lay shattered under a pile of marble, just a few blocks south of Hamilton's tomb in the Trinity Church graveyard.

As dawn broke over the city on December 17, soot-covered firemen, nearly asleep on their feet, staggered home. Some of them were wrapped against the cold in fine imported blankets once destined for tranquil nights in wealthy homes. One company of firemen had chanced upon a cache of artificial flowers, and they trudged through the charred ruins wearing bright blooms in their caps.

Hone went home and struggled to record in his journal what he had seen that night. "I am fatigued in body, disturbed in mind, and my fancy filled with images of horror which my pen is inadequate to describe." It was, Hone thought, "the most awful calamity which has ever visited these United States." Hosack now faced a calamity of his own. Since retiring from his medical practice, he had supported his large family partly through investments of Magdalena's money. Hosack now owned stock in at least nine insurance companies, which were swamped with claims in the days after the fire. "The fire . . . has singed almost everybody," Washington Irving wrote his brother. Irving himself saw $3,000 of his own investments evaporate because of the fire, but Hosack's losses were on another scale entirely. The newspapers reported that of his $200,000 in insurance stock (more than $5 million today), he had lost all but $20,000. Even his daughter's dowry, not yet transferred to her new husband, was slashed away.

Less than forty-eight hours after the fire, on the morning of Friday, December 18, Hosack arose and went down to his study to sort through his business affairs. As he sat there working, the air around him was thick and smoky, and he felt a little short of breath. He ate his breakfast and ordered his wagon be brought around so he could take care of an errand in town. Putting on his overcoat, he started for the front door but abruptly sat down on a sofa. He assured his family that he would be fine in a moment. Then he fell over.

Out in the city, Philip Hone was just leaving for a walk with Alexander Hosack. The two men had made plans to survey the destruction together. Before they got far, they encountered a mutual friend who told them Hosack had just collapsed in a fit of apoplexy—a stroke. The men rushed to Chambers Street, where they found Hosack lying on a sofa in the same room where they had all gathered to celebrate the newlyweds two days earlier. He was paralyzed. He seemed unable to see or hear, and he could no longer speak.

The day after Hosack's stroke, while he lay semiconscious at home, many of his friends and fellow citizens gathered just across Chambers Street at City Hall to organize the rebuilding of the destroyed district. It was exactly the kind of civic project Hosack would have loved to lead, had he been well. Instead, John Francis and some of his other former medical students hovered over his makeshift sickbed as bulletins about his dangerous situation were rushed to newspaper offices from New Hampshire to South Carolina. After a few more days, Hosack rallied. On December 22, his condition visibly improved over the course of the day, and he tried to speak once or twice.

He died at eleven that night. The news ran the next day in the New York papers and began making its way to other cities. "A great man in the profession has fallen," announced the *New York Daily Advertiser*. "It is impossible that the death of such a man could occur in this community without producing a sensation, even at a moment when the mind of every citizen is absorbed with the awful dispensation with which we have been so recently visited," the *Evening Star* wrote. "There have been few men so extensively known in our own country and by the professional world abroad, as Dr. Hosack."

On Christmas Eve, local nurseryman Michael Floy Jr. noted in his diary that his beautiful scarlet camellia was finally in bloom. This was Floy's *Camellia hosackia*, named a decade earlier for the man who had done so much for the city's nurserymen through the New-York Horticultural Society. Hosack's funeral was held the next day at Grace Church. "Christmas Day, but not by any means a merry Christmas," Philip Hone wrote in his diary. Hosack's sprawling circle of family and friends jammed the church for the service. One of his former students broke down in sobs as he tried to read the eulogy. Among the pallbearers who bore the coffin out of the church were Edward Livingston, the painter John Trumbull, and Morgan Lewis—the former governor of New York who had championed the Elgin Botanic Garden three decades earlier. John Francis noted afterward that Hosack's "remains were followed to the grave by the eminent of every profession and by the humble in life whom his art had relieved." Hosack was buried in the New York Marble Cemetery on Second Street.[*]

---

[*] Hosack's remains were later transplanted, along with those of many other people, to Trinity Church's uptown cemetery, located by the Hudson River at 153rd Street. According to the headstones in the Hosack family plot, Hosack is buried there with his second wife, Mary, and their son David Hosack Jr. Nearby are their three daughters—Mary, Eliza, and Emily—and another of their six sons, Nathaniel Pendleton Hosack. In 1831, Nathaniel, named for Hamilton's second in the duel, had married a granddaughter of Angelica Schuyler Church, Hamilton's beloved sister-in-law.

The American writer Freeman Hunt reflected on what it meant to lose Hosack. "As a physician and man of science, his name was universally honoured as the first; as a citizen, his many virtues and excellences of character have made a deep impression upon the hearts of thousands." Hosack's death, Hunt wrote, had "left a blank in the scientific and social world." And of Hosack's work at Hyde Park, Hunt noted that his "extensive and practical knowledge" of horticulture, "connected with wealth and a refined taste, has rendered his garden second to none in the union." Second to none—it was a lovely piece of praise that could have doubled as a gentle obituary for Elgin. Six months after Hosack died, Magdalena and the children put the Hyde Park estate up for sale. John Jacob Astor soon bought it for his daughter.* Hosack's greenhouse and hothouse plants were put on sale at a shop in New York City.

Days after Hosack's death, the painter Thomas Cole published an essay celebrating the beauty of the Hudson River Valley—and sounding the alarm for future generations about the destruction of American landscapes. "We are still in Eden," Cole wrote. "The wall that shuts us out of the garden is our own ignorance and folly." In 1825, the inventor John Stevens—one of Hosack's patients—had built and tested the first American steam locomotive at Hoboken; when Hosack died a decade later, the rural world of old New York was starting to fall away before the oncoming steam train. Tracks were laid along the river's edge through Hosack's former estate, and his friend Washington Irving, who had done his own part to romanticize these landscapes, moved out of his bedroom overlooking the Hudson to a dark little room on the other side of his villa. "If the Garden of Eden were now on earth, they would not hesitate to run a railroad through it."

---

* Hosack's former mansion burned down in the 1840s, and a new one was built. In the 1890s, Frederick W. Vanderbilt, a grandson of Cornelius Vanderbilt, bought the estate and hired the firm of McKim, Mead and White to design a larger mansion. Today the Vanderbilt Mansion National Historic Site is run by the National Park Service, and although the Vanderbilts made alterations to the grounds, the sweeping views and the curved drives of Hosack's era are intact. Some of the surviving trees are believed to have been planted by André Parmentier.

———— ∞∞∞ ————

AND WHAT OF ELGIN?

Hosack was still alive when, in 1834, Columbia's latest tenant farmer on the property, William Shaw, sold many of the remaining trees, shrubs, and plants at a seed shop on Liberty Street. Meanwhile, at the Bloomingdale asylum in northern Manhattan, patients strolled among trees and shrubs that Columbia had stripped from the garden site twenty years earlier, and they also enjoyed a conservatory on the grounds that contained some of Hosack's exotic specimens. By the 1840s, the city was laying out cross streets through the old Elgin land from Fifth Avenue to Sixth Avenue. Forty-Seventh Street now bordered it on the south, Fifty-First Street on the north. Forty-Eighth and Forty-Ninth ran through Hosack's former fields of grain and past the pond where he had raised aquatic plants. Fiftieth Street was laid down across the top of Hosack's hill, running right past the spot where his conservatory had stood. Despite these changes, the area remained bucolic into the second half of the nineteenth century. In the 1850s, when Catholic officials were deciding on a site for the new St. Patrick's Cathedral, they chose Fifth Avenue between Fiftieth and Fifty-First Streets in part because it fronted the peaceful old fields of Elgin.

Columbia College finally abandoned the packed southern tip of Manhattan. In 1857, the college moved into a building one block to the southeast of its Elgin property, at Forty-Ninth Street and Madison Avenue. The same year that Columbia moved uptown, it demolished the century-old College Hall at Park Place, where Hosack, Clinton, and Hamilton had all studied and Hosack had taught for years. As for the Elgin land, after decades of paying taxes and assessments on it, the college trustees finally concluded that, with the city spreading up the island, they could now turn a profit from it. They sold off sixteen lots in 1857 and began slicing up the rest for lease to developers. New generations of immigrant laborers built townhouses and apartment buildings where their forebears, working for Hosack, had once rolled out garden paths and planted flowerbeds. By 1870, all of the Elgin land, now divided into more than two hundred separate lots, was blanketed with buildings. As the eastern edge of Central Park became the most desirable frontage in Manhattan, one Vanderbilt after another built American palaces along Fifth Avenue to the north of Elgin, and other wealthy New Yorkers joined them in the area. In the 1880s, Andrew Carnegie bought a townhouse on the northern side of Hosack's old land and gave it to his bride as a wedding present. Columbia, meanwhile, used its profits from the Elgin leases to help turn itself into a world-class university, creating new graduate faculties (including a law school) and taking over the independent College of Physicians and Surgeons—which Hosack had helped found—to form the medical school for which Columbia is still known today. As it expanded, Columbia outgrew its Forty-Ninth Street campus, and in 1892, the trustees bought the insane-asylum property on the beautiful northern heights of Manhattan. There, overlooking the Hudson at 116th Street, the eminent architectural firm McKim, Mead and White designed a large new campus for Columbia. In 1893, Columbia authorities consulted the landscape architect Frederick Law Olmsted (as well as his son Frederick Law Olmsted Jr.) about their plans for the grounds. Olmsted pointed out that some of the trees now thriving on the Mall in Central Park had been brought from the countryside north of Manhattan. He recommended they transplant the mature trees on the property—some of which had come from Elgin—to spots where they would ornament the new buildings. The focal point of the campus was a domed library whose front steps were flanked by two English yews that had been among the trees and shrubs moved from Elgin to the asylum grounds in the 1820s. The yews, which Olmsted reportedly pronounced the "oldest and finest" in North America, survived until 1914—one century after the state had turned Elgin over to Columbia.

It was not long after this that Hosack's old Elgin land caught the attention of the Metropolitan Opera. The Met's nineteenth-century opera house at the corner of Broadway and Thirty-Ninth Street was feeling cramped and antiquated to both its performers and its stockholders. Among the opera's stockholders, John D. Rockefeller Jr. took a particular interest in the matter of a new opera house. He had grown up on Fifty-Fourth Street, three blocks directly north of the old Elgin

Hosack's old Elgin yews flanking the steps to
Columbia's new library in 1897

site, and he was still living there in the 1920s with his own family. The neighborhood was now punctuated by speakeasies and a rattling elevated train that ran along Sixth Avenue. Rockefeller began dreaming of creating a monumental complex of cultural and commercial buildings, with a new opera house at its heart. In the fall of 1928, as Rockefeller's representatives were negotiating with Columbia for the Elgin property, President Nicholas Murray Butler sent Rockefeller a cordial note saying that he wanted to tell Rockefeller in person about the history of the land. "That history reads like a romance."

Columbia agreed to lease the Rockefellers eleven acres for more than $3 million a year. The United States hurtled into the Great Depression nine months after the deal was signed. Plans for a new opera house were abandoned, but Rockefeller decided to go ahead with a revised version of his center, which would eventually include a different kind of theater: Radio City Music Hall. In 1931, construction officially began on the largest private building project the country had ever seen. Men desperate for jobs swarmed the site. They razed tenement buildings and carted away the earth in which Hosack had planted his seeds. They hauled in blocks of Indiana limestone and dangled at lunchtime from beams in the sky. By the end of the decade, they had built fourteen buildings, all centered on a sunken skating rink. The tallest one, 30 Rockefeller Plaza, shot seventy stories into the air.

On November 1, 1939, in the lobby of the new building at the corner of Sixth Avenue and Forty-Eighth Street, Rockefeller pulled on a pair of workman's gloves and drove in the ceremonial last rivet of Rockefeller Center while three hundred people watched. Via a radio broadcast, the sharp rhythm of the pneumatic hammer reverberated into homes from New York to California. Rockefeller's son Nelson next gave a brief speech praising his father's perseverance in the face of economic catastrophe, and then he introduced President Butler of Columbia to talk about the land. "Probably no other piece of land on Manhattan Island, and few other pieces of land anywhere in this Western World, have had so few owners over a period of three hundred years," Butler told the audience, noting that it was thanks in part to a visionary American doctor that they were all gathered here.

The former Elgin land, looking from Fifth Avenue toward 30 Rockefeller Plaza

Radio City Music Hall was built over the footprint of Hosack's conservatory.

In 1985, the Rockefeller Group finally purchased the land on which their buildings had stood since the 1930s. Columbia received $400 million in the deal. Fifteen years later, the Rockefellers and their partners sold Rockefeller Center to a consortium of developers for nearly $2 billion. On one of the low walls lining the Channel Gardens, facing toward the old Middle Road, hangs a plaque that is easily overlooked in the crush of tourists from around the world.

> *In memory of David Hosack*
> *1769–1835*
> *Botanist, physician, man of science and citizen of the world*
> *On this site he developed the famous Elgin Botanic Garden.*

EPILOGUE

AARON BURR DIED ABOUT NINE MONTHS AFTER HOSACK, IN the wake of a stroke that had left his legs paralyzed. With his father gone, it fell to Alexander Hosack to care for Burr during his last illness. A friend of Alexander's later noted that he had once asked Burr whether he regretted having shot Hamilton. Burr reportedly had told Alexander, "No, sir; I could not regret it. Twice he crossed my path. He brought it on himself."

Sometime in 1836, the year Burr died, Alexander Hosack went to Europe. During a tour of the South of France, he stopped in the city of Montpellier. At five in the morning the day after he arrived, Alexander walked from his hotel to Montpellier's botanical garden—the oldest in France. He went to a house on the grounds and stated his business to a servant, who conducted him to a room where an aged man was bent over a microscope, examining a flower. Hearing them enter, the man turned around and stared at Alexander for a surprised moment. Then he exclaimed, "I know you, sir; you are the son of Dr. Hosack." It was Alire Raffeneau Delile, Hosack's former student at Elgin and Columbia. He embraced Alexander in tears.

After returning to France in 1807, Delile had spent twelve years in Paris, where he had collaborated with François André Michaux, Aimé Bonpland, and Pierre-Joseph Redouté on a series of beautiful botanical volumes. In 1819, he had been appointed a professor of botany at the Montpellier medical school as well as director of the botanical garden—posts he still held when Alexander Hosack arrived in 1836. Delile now took Alexander to a chest in the corner of the room and pulled out stacks of handwritten pages. He had saved the notes he had taken as a young man in Hosack's lectures in New York, along with the letters he had received from Hosack since then.

Scholars have sometimes wondered why Delile was an anomaly among French botanists in that he insisted on using Linnaeus's sexual system of plant classification rather than the prevailing natural system developed by Jussieu. It was likely thanks to Hosack's influence that Delile continued to use the Linnaean system, and he was also paying tribute to Hosack in another way. When Delile lectured to his medical students, he told them that it was critically important that aspiring doctors learn about plants. Three decades after he had studied in New York, Delile was still telling his students that a doctor "must know his food from his poison." It had been one of Hosack's favorite maxims. He had learned it from William Curtis forty years earlier.

If Hosack's name was well known in France at the time of his death—he even garnered a mention from Tocqueville—in Britain he was more famous still. In the last two decades of his life, his constant exchange of plants and ideas with British scientists had inspired one tribute after another. The greatest came from David Douglas, the Scottish botanist who had met Hosack in 1823. The following year Douglas had sailed once again from Britain to North America, this time landing on the Pacific coast at the mouth of the Columbia River. There he discovered a majestic tree species he thought was "one of the most striking and truly graceful objects of nature," a conifer known today as the Douglas fir (*Pseudotsuga menziesii*). He also discovered a new genus of wildflower with multicolored blooms that he decided to name in honor of his favorite American: *Hosackia*. It was the greatest possible gesture of respect from one botanist to another, and Douglas thought it was an entirely fitting tribute for David Hosack, to whom "the scientific men of North America owe the same gratitude as those of England did to Sir Joseph Banks."

Hosack deserved all the praise and recognition. From founding

Elgin, to teaching young botanists, to presiding over the New-York Horticultural Society, to turning his Hyde Park estate into the most inspiring new designed landscape on the continent—Hosack had done more than any man of his generation to foster in his fellow Americans a fascination with plants. He had lost his battle to save Elgin, but his student Amos Eaton later observed that it was Hosack's work there that ignited "the first spark of zeal for Botany." As Hosack and his students explained the importance and pleasures of botany in newspapers and in speeches during the first two decades of the nineteenth century, Americans who had previously paid little attention to the scientific study of plants began taking notice. At first, the botany guides available to amateur American botanists were by British naturalists such as Curtis and Smith, but by the 1820s Hosack's former students and associates from Elgin had joined him in lecturing and publishing on botany, taking their knowledge to audiences in villages and towns around the United States.

Thanks to Hosack, Eaton, Torrey, and others in the Elgin circle, Americans caught a fever for botany like the one Curtis had helped launch in England fifty years earlier with his *Flora Londinensis*. Hosack himself had taught only young men, but in the last decade of his life, he called for both boys and girls to be educated in botany, and his students followed suit. Budding botanists began traversing the countryside all over the United States with their new manuals in hand. They attended public lectures and arranged for botany to be taught to their children in the schools. It was not long before the botany craze helped fuel another that continues to this day—small-scale home gardening. Hosack's head gardeners at Elgin and Hyde Park both published home gardening guides, and in 1846, Andrew Jackson Downing, who had been so inspired by Hosack's work with André Parmentier on the Hyde Park estate, founded a wildly successful magazine called *The Horticulturist*. In the decades following the Civil War, new companies such as Burpee and R. H. Shumway began including color lithographs in their seed catalogues. For the first time, Americans could sit indoors in the dead of winter and hold the colors of summer between their hands, thrilling to the promise hidden in the homeliest little seed— just as they do today.

This national passion has never waned. In American cities today, in fact, interest in gardens seems stronger than ever. Farmers' markets and the farm-to-table movement are connecting city dwellers with rural farmers, helping the latter survive in an age of factory farming. People are turning abandoned city lots into urban farms to help eradicate "food deserts"—neighborhoods where the best food on offer sits in plastic packaging. Schoolchildren are learning in their classrooms and at botanical gardens how to grow their own food. Hotels and restaurants have begun keeping bees and growing herbs, vegetables, fruits, and flowers on their rooftops. Seed exchanges are helping Americans work together to preserve seeds from nonhybridized species, commonly known as heirlooms, so these species won't disappear from the world completely—as hundreds or possibly thousands already have.

Hosack's name has largely been forgotten, but his influence lingers in these efforts to keep plants at the forefront of city life and to cultivate and treasure every single species on earth. His legacy can also be found in some of the most beautiful public spaces in the United States today. This path of influence runs from Hosack's work at Hyde Park with Parmentier to Downing, who in turn launched the landscape architects Calvert Vaux and Frederick Law Olmsted on their brilliant American careers. In 1858, Vaux and Olmsted submitted an early version of what became the winning entry to the competition for Central Park, and Olmsted in particular would go on to design many other stunning landscapes that still shape the lives of American cities. The conversations about nature that Parmentier had with Hosack as they walked through the grounds of Hyde Park still echo, ever so faintly, through these landscapes.

Hosack had an even greater impact on American science. The year after his death, the New-York Horticultural Society elected John Torrey, the last of his prize students, to its presidency. Torrey was also pursuing another cause long dear to Hosack's heart. With a young botanist named Asa Gray, he was compiling a *Flora of North America*, which they published in multiple volumes from 1838 to 1843. It was by far the most ambitious work on American plants ever to appear, and it established Torrey's and Gray's international reputations. Gray was hired by Harvard, and over the next decades he assembled an herbarium that remains one of the most important in the world today. He also helped create Boston's famous Arnold Arboretum. Torrey further honored Hosack's legacy by advancing botany in New York. Like Hosack before him, he attracted a circle of young men devoted to the study of plants. They got in the habit of holding regular meetings in Torrey's office at Columbia, calling themselves the Torrey Botanical Club. After Torrey died in 1873, the group continued to meet. They welcomed women members and started a botanical journal that remains in circulation today. Although they had no botanical garden, they tried to make do with dried specimens from Torrey's herbarium. Then, in the 1880s, two members of the Torrey Botanical Club, Nathaniel Lord Britton and Elizabeth Knight Britton, went to England and toured the Royal Botanic Gardens at Kew. Like Hosack a century before them, they saw beauty and science mingled in an English garden and thought of their beloved New York. It was Elizabeth Britton who suggested to her husband that the city should have a botanical garden as glorious as Kew.

While Hosack had had to forge ahead at the dawn of the nineteenth century on his own, the Brittons lived in another age entirely. They shared New York with a group of stupendously wealthy Americans who were as interested in founding cultural and scientific institutions as Hosack and his friends had been—people such as the Rockefellers, Morgans, Carnegies, and Vanderbilts. As Nathaniel Lord Britton took the lead in the campaign for a new botanical garden, he made sure to harness this wealth and civic concern. Carnegie, Morgan, Rockefeller, and Cornelius Vanderbilt II each contributed $25,000, and they lobbied their friends and associates to contribute, too. Vanderbilt agreed to serve as the garden's first president, Carnegie as its first vice president, and Morgan as its first treasurer.

The New York Botanical Garden was built on two hundred fifty acres of Bronx parkland graced by a virgin hemlock forest and a river. Britton chose Calvert Vaux to lay out winding carriage drives through formal gardens to a palatial conservatory sheltering an entire acre of exotic plants under glass. A huge building was constructed to hold exhibits about medicinal, commercial, and agricultural plants and to allow botanists to work in state-of-the art laboratories. Garden officials began collecting thousands of botanical and horticultural volumes. Among them were more than two hundred from Hosack's personal library, including those he had brought back from London on the *Mohawk* a century earlier. Britton persuaded Columbia to donate its herbarium, which contained Torrey's collections as well as specimens gathered for Elgin by Hosack, Delile, and John and Caspar Eddy.*

Today, this herbarium contains more than seven million specimens. The New York Botanical Garden recently celebrated its one hundred twenty-fifth anniversary. Its botanical and horticultural achievements place it in the company of only one or two other gardens in the world. In 1897, when the garden was under construction, a local newspaper argued that Hosack deserved to be immortalized with a statue there. Nothing came of the idea, but in a way it doesn't matter. Hosack's spirit is visible in every blossom and under every microscope. The love of nature he brought home from Britain and nurtured among generations of his countrymen helped give rise to this national treasure. Today there are more than four hundred botanical gardens and arboreta across the United States.

In the late nineteenth century, the search for new medicines migrated from botanical gardens to universities and pharmaceutical laboratories. But this search has never ceased to depend on botanists who roam the planet collecting and studying plants in the wild and learning about the uses of medicinal plants from local experts. In the face of an alarming rise in antibiotic resistance, some pioneering medical researchers are now going into the field themselves in a race to identify new plant-based drugs. As they travel back and forth between the laboratory and the field, they are reconnecting the scientific domains between which Hosack once moved so easily.

We like our heroes to stand alone, so we can easily discern and celebrate their achievements. But Hosack's greatest legacy is perhaps the one that is the hardest to see. He showed his fellow citizens how to build institutions. Over and over, in the face of criticism and mis-

---

* The Linnaean specimens that James Edward Smith gave to Hosack in 1794 are now lost, perhaps destroyed in a fire at the Lyceum of Natural History in 1866. There is a small but tantalizing possibility that they are still out there somewhere, hiding among other uncatalogued specimens in some American museum or botanical garden.

fortune, he rallied people around him to create the charitable, medical, and cultural institutions that make cities worth inhabiting and that educate a nation for generations to come. Philanthropic work is hard and complex. The daily lives of civic organizations—full of meetings, bylaws, elections, and the like—strike many people as dull and unheroic. Because this work and its results are collective, we can't easily single out one hero to celebrate. Yet they take just as much patience, ingenuity, and money as any discovery or invention. Perhaps today more than ever, Hosack's quieter sort of heroism deserves emulation. He dreamed from boyhood about what his generation could do to improve the lives of others. Acting on those generous dreams for half a century, he helped build a new nation.

ACKNOWLEDGMENTS

This book is my effort to bring David Hosack into living relief. For many years now, I have been traveling in his footsteps, seeking to recapture the vanished sights and sounds of the places he frequented—such as the neighborhood in lower Manhattan where he lived just blocks from Hamilton, Jefferson, and Washington, and the cow pastures he coaxed into bloom where Rockefeller Center now rises. There are even a few places remaining today that he would recognize, among them Hamilton's country house in northern Manhattan and the unspoiled view of the Hudson River from the high bluff at Hyde Park where Hosack's own mansion once stood. On the northern coast of Scotland, near the town of Elgin, I found the little castle where Hosack stayed with friends for two weeks in the spring of 1793, and I saw why he fell so in love with the wild landscape there that he would later name his American garden for Elgin.

Above all, though, I found Hosack in the archives; his papers are scattered across more than thirty collections in the United States and Europe. As I worked my way through thousands of pages of letters, plant lists, medical and botanical books and essays, weather diaries, travel diaries, newspapers, and more, my constant companion was Christine Chapman Robbins's 1964 biography, *David Hosack: Citizen of New York*. I want to express my admiration for her rigorous scholarship, which provided me with a sort of treasure map as I explored the archives further. I also want to thank Peter Mickulas for his writing about Hosack in his wonderful book about the New York Botanical Garden, *Britton's Botanical Empire*. It was when I stumbled across Mickulas's description of the Elgin Botanic Garden that my fascination with Hosack began.

Many dozens of curators, archivists, and librarians helped me as I tracked down Hosack's papers and those of his correspondents and contemporaries, and I'm indebted to each of them. My special thanks to Susan Fraser, Vice President and Director of the LuEsther T. Mertz Library at the New York Botanical Garden; Vanessa Sellers, Humanities Research Coordinator at Mertz; Stephen Sinon, Head of Archives; and the entire Mertz staff, all of whom aided me on so many occasions. I gratefully acknowledge the support of the Andrew W. Mellon Foundation and the Humanities Institute at Mertz for supporting my research with a Mellon Visiting Scholar Fellowship in the summer of 2016. My deep gratitude also to Lisa O'Sullivan, Vice President and Director of the Library of the New York Academy of Medicine, and her staff, especially the inimitable Arlene Shaner; to Tammy Kiter and her colleagues at the Patricia D. Klingenstein Library of the New-York Historical Society, where I always felt Hosack's presence vividly (and not only because I got to visit the regal bust of him in the lobby); to Elaine Charwat and Lynda Brooks at the Linnean Society of London; to Keith Moore at the Royal Society of London; to Hector Rivera at the Manhattan Borough President's Office; to Don Pfister, Curator of the Farlow Library and Herbarium of Cryptogamic Botany and Asa Gray Professor of Systematic Botany at Harvard University; to Lisa DeCesare and Michaela Schmull of the Harvard University Herbaria; and to Marika Hedin, Director of the Linnaeus Museum in Uppsala, Sweden, and Jesper Kårehed, Scientific Curator of the Linnaean Gardens, Uppsala University. Naomi Kroll Hassebroek, Senior Conservator with the National Park Service, gave a wonderful tour of the Grange and helped me find beautiful photographs of the house; thanks also to Minerva Anderson for those. Archie Drummond gave me a memorable tour of Brodie Castle in Scotland, and on a thrilling day at the BNY Mellon archives in lower Manhattan, Christine McKay showed me Hosack's subscription certificates to the secret Hamilton family fund.

I'm also grateful to the many archivists and librarians who helped me at the following institutions: the American Philosophical Society Library; the Botanisk Centralbibliotek, Copenhagen Botanical Garden; the Bristol (England) Record Office; at Columbia University, the Augustus C. Long Health Sciences Library, Avery Architectural & Fine Arts Library, and the Rare Book & Manuscript Library; the Historical Society of Pennsylvania; the Municipal Archives of the City of New York; the Mystic Seaport Museum; the New York Society Library; the Library of Congress; the New York State Library; Princeton University's Firestone Library; the Université de Montpellier; and at Yale University, the Manuscripts & Archives division at Sterling Memorial Library and the Beinecke Rare Book & Manuscript Library.

I began this research as a professor at the University of Michigan in Ann Arbor, where I taught for thirteen years on philanthropy and the natural environment. For their enthusiasm, hospitality, and/ or readings of the book proposal while I was in Ann Arbor, I would like to thank: Doug Anderson, Sue Ashford, Wayne Baker, Dave Barger, Jerry Davis, Jane Dutton, Melissa Eljamal, Betsy Ellsworth, Adam Grant, Bob Grese, Rebecca Henn (to whom I owe the original idea of studying botanical gardens), Andy Hoffman, Bruce Judge, Carole Kirby, Greta Krippner, Peggy McCracken, Mark Mizruchi, Jason Owen-Smith, Scotti Parrish, Mary Price, Jenni Steiger, Katie Van Dusen, Ari Weinzweig, and Denise Yekulis. I still draw daily on what I learned from Rick Price about patience, mentorship, and how to run a meeting. Nathan Harris brightened many wintry Michigan days. To Andrew Port and Sylvia Taschka—and Rebekka and Hannah—thank you so many times over. When I first began studying botanical gardens a decade ago, Richard Piacentini, Executive Director of the spectacular Phipps Conservatory and Botanical Gardens in Pittsburgh, made me feel so welcome and generously allowed me to interview his staff and study the garden's institutional records, as did Gregory Long, CEO and The William C. Steere Sr. President of The New York Botanical Garden. Directors at more than a dozen other gardens across the country were similarly welcoming. My thanks to all, especially Susan Lacerte, Executive Director of the Queens Botanical Garden, and Scot Medbury, President of the Brooklyn Botanic Garden. My undergraduate students in three classes embraced the study of botanical gardens with energy and dedication (Tigerlilies, Snapdragons, and Botanicals!). Patrick Cullina shared his extraordinary botanical and horticultural expertise with us. Kate Brierty, Mitch Crispell, Brian Pogrund, Caroline Rooney, Rebecca Sunde, and Lea Wender joined me on joyful adventures in Hosack's New York. These travels, along with much of my archival research, were made possible by very generous funding from the Organizational Studies Program, the Barger Leadership Institute, and the Erb Institute for Global Sustainable Enterprise.

Several years ago, I moved back to New York to join the Department of Urban Policy and Planning at Hunter College. I would like to acknowledge President Jennifer Raab for welcoming me so warmly to Hunter, Professor Joseph Viteritti for his constant moral support, and all my colleagues for their kindness to me and their exemplary devotion to our students. Miriam Galindez and Marisol Otero-Morales helped me settle into my new academic home. I wrote most of this book during a dreamy fellowship year at the Cullman Center for Scholars and Writers of the New York Public Library in 2015–2016. Director Jean Strouse presided over a group of Fellows from whom I learned daily about writing, politics, and history. Vanessa Schwartz prodded me to celebrate Hosack's larger accomplishments; Paul Yoon kindly paused in his own work to help me unlock a closed door in my writing; and Nick Wilding deciphered a Latin document for me (and also shared his wicked sense of humor all year). I am especially grateful to my fellow Fellows Larry Rohter and Edward Ball for many conversations about the writing of biographies, a topic Jean Strouse also kindly discussed at great length with me. Lauren Goldenberg, Paul Delavardac, and Julia Pagnamenta were cheerful guides throughout the year and afterward; Lauren also read a chunk of the manuscript. Upstairs at the Manuscripts, Archives and Rare Books Division, I received aid from Thomas Lannon, John Cordovez, Tal Nadal, Katie O'Connell, and Jessica Pigza, among others. Mark Boonschoft read my entire manuscript and gave me detailed comments. Sam Swope, Melanie Locay, and Elizabeth Denlinger showed the kind of interest in the project that keeps a writer writing, as did Kristin McDonough, Director of the Science, Industry and Business Library. Jason Baumann took time out of his day to help me with a logistical challenge, and I received additional help from staff members in the Prints and Photographs Division and the Map Division. I would also like to acknowledge the many other people, including those working in maintenance and security, who make the New York Public Library such a welcoming place. I feel certain Hosack would have understood exactly how much labor and largesse go into this institution every day.

As I pieced together Hosack's life, many writers, historians, and botanists were extremely generous with their expertise and ideas, especially Ron Chernow, Heather Ewing, Eric Foner, Eric Hilt, Beth Hyde, Maya Jasanoff, Mark Laird, John Merriman, Kaitlin Mondello, Amy Meyers, David Nasaw, Dr. Michael Nevins, Therese O'Malley, and Penelope Rowlands. (Any errors are not theirs, but mine, of course.) I am grateful to Barbara Thiers, the Patricia K. Holgrem Director of the William and Lynda Steere Herbarium and Vice President for Science Administration at the New York Botanical Garden, for convening a meeting about Hosack's specimen lists with a fantastic team of NYBG experts, including Daniel Atha, Director of Conservation Outreach; Brian Boom, Vice President for Conservation Strategy; Todd Forrest, Arthur Ross Vice President for Horticulture and Living Collections; Robbin Moran, Nathaniel Lord Britton Curator of Botany; and Robert Naczi, Arthur J. Cronquist Curator of North American Botany. I am grateful to them for their time and expertise—especially to Daniel Atha, for his excitement about this book and his continued work with me on Hosack's plant lists. Joel T. Fry, Curator of Bartram's Garden in Philadelphia, shared his truly encyclopedic knowledge of plants and botanists of the early Republic and took me on a fascinating tour of the Bartram family's beautiful house and garden. Lisa Vargues, Rob Naczi, and Neil Snow helped me track down the current name of one of Hosack's herbarium specimens, and I'm also grateful to Nancy Slowick for her course on native flora at the NYBG. Julie Sakellariadis attended a talk I gave at the NYBG and kindly put me in touch with botanists at Harvard. At lightning speed, David Hosack Barnes located a painting that had been passed down in the Hosack family, and Alice Lloyd generously granted me permission to reproduce it in these pages.

David Larabell whipped an early draft of the proposal into shape, and George Gibson gave me both precious time and the benefit of his editorial insights on the whole manuscript. I am so grateful to you both. My thanks also to Heather Brooke and Katherine Ibbett for hosting me in London, to Neil Brenner for hosting me in Boston (and for all those years of encouragement), and to Christopher Merriman for critical help with the plant catalogues. Thank you to Dr. Laura Corio for the chance to try smelling salts for myself. Guenther Roth taught me to look for Columbia's history when I first arrived on campus as a graduate student. David Stark, a mentor and friend for nearly two decades, read drafts of chapters as fast as I wrote them, and his early enthusiasm kept me going. Woody Powell joined me in studying Hosack for an article, which led to fascinating conversations and walks through lower Manhattan, Central Park, and Rockefeller Center. Mauro Guillén, who loves New York City and its history, has also been a steadfast source of encouragement.

I am deeply indebted to the following people who took time out of their busy lives to read and comment on the entire draft: Dr. John Hornby, Dr. Edward Huey, Byron Knief, Christine Laporte, Dr. Patrick Shin, Roy Tsao, and Tracey Van Dusen. Thomas Woltz read a draft on trains and in airplanes as he crisscrossed the globe; his comments and our conversations have taught me much about landscapes, gardens, and the art of friendship. Andrea Wulf began sharing her scholarly contacts and her vast knowledge of early Republic science and politics within minutes of our first meeting (in a New Jersey sleet storm in January 2015)—and has been doing so with enormous generosity ever since. Markley Boyer allowed me to see the vanished Mannahatta and also solved many a challenge involving spreadsheets, websites, and digital images. He also read a draft of the manuscript; his ideas and enthusiasm have shaped this book in so many ways. Vanessa Sellers has a genius not only for the history of landscapes and gardens, but also for recognizing which scholars should meet one another. Dan Kevles shared my fascination with medical botany treatises and created camaraderie in the archives; I have learned so very much about writing and about the history of science from our conversations. I'm grateful to Simon Lipskar for connecting me with my fantastic literary agent, Geri Thoma of Writers House. My editor at Liveright, Katie Adams, is endowed with some kind of X-ray vision into the bones of a book.

It has been a joy to work with her—and with the rest of the Liveright team, especially Steve Attardo, Gina Iaquinta, and Peter Miller. I'm deeply grateful to the meticulous Kathleen Brandes for her hard work as copy editor on the manuscript.

My oldest friends have been incredibly patient and supportive while I wrote . . . and wrote and wrote. My love and gratitude especially to Nitsan Chorev, Joyce Robbins and Alan Glickman, Clare Johnson and Dino Valaoritis, Justin Bischof, Laura Wolfson, and the Merriman family: John, Laura, Christopher, and the much-missed Carol. My family is full of history-lovers, among them my uncles, and I am grateful for their interest and inspiration. My cousin Drew Stephens, a GIS oceans expert, helped me pinpoint Hosack's 1794 location on the Atlantic. Thank you to Steve Honigberg and Georgi Kostov for making me your little sister, and to my five "neffuses" for being the delights of my life. My sister Elizabeth Kostova read each chapter almost as I soon as I had written it and gave me just the right combination of criticism and encouragement. My sister Jessica Honigberg trained her artist's eye and her playful wit on the book proposal, the manuscript, and my obsession with Hosack. My sister Betsy Sheldon is my sunshine.

The wellsprings of this book are my parents, David and Eleanor. They are passionate gardeners, city planners, citizens, philanthropists, and lovers of literature and history. They have shown nothing but excitement about my research from start to finish—even through hundreds of hours of phone calls about what I found in the archives *this* week. They also read each chapter in turn and cheered me on through the writing of the next one. Thank you for making me weed the garden and look at old maps and old buildings all those years. This book is dedicated to you, Mom and Pop, with profound love and profound admiration.

It is also dedicated to a dear friend who has been on this whole adventure with me. If you were to cross an artist, a computer scientist, a careful historian, and a loving daughter, and then added the patience and dignity of a saint, you might—if the stars were aligned—get Rebecca Sunde. Without her, no *American Eden*.

Thank you, my beloved R.

NOTES

The following abbreviations are used in these notes. See Sources and Bibliography for more complete information.

ABBREVIATIONS: PEOPLE

AB	Aaron Burr
AEH	Alexander Eddy Hosack
AH	Alexander Hamilton
ARD	Alire Raffeneau Delile
BR	Benjamin Rush
CWB	Catharine Wistar Bache
CWE	Caspar Wistar Eddy
CWP	Charles Willson Peale
DC	DeWitt Clinton
DH	David Hosack
FAM	François André Michaux
JE	John Eddy
JES	James Edward Smith
JWF	John Wakefield Francis
MH	Mary [Eddy] Hosack
SLM	Samuel Latham Mitchill
TB	Theodosia Burr (Aaron Burr's daughter; married name, Theodosia Burr Alston)
TJ	Thomas Jefferson
WC	William Curtis
WH	William Hosack

ABBREVIATIONS: INSTITUTIONS

APS	American Philosophical Society
CPS	College of Physicians and Surgeons, Columbia University
CU	Columbia University
HSP	Historical Society of Pennsylvania
LC	Library of Congress
LS	Linnean Society of London
NYAM	New York Academy of Medicine
NYBG	New York Botanical Garden
N-YHS	New-York Historical Society
NYPL	New York Public Library
RS	Royal Society of London

ABBREVIATIONS: ARCHIVES AND SOURCES

AB Memoirs	*Memoirs of Aaron Burr: With Miscellaneous Selections from His Correspondence*. Edited by Matthew Livingston Davis.
AB Journal Davis	*The Private Journal of Aaron Burr, During His Residence of Four Years in Europe; with Selections from His Correspondence*. Edited by Matthew Livingston Davis.
AB Journal Bixby	*The Private Journal of Aaron Burr, Reprinted in full from the original manuscript in the library of Mr. William K. Bixby of St. Louis, MO*. Edited by William H. Samson.
AEH 1861	Hosack, Alexander Eddy. "David Hosack." In *Lives of Eminent American Physicians and Surgeons of the Nineteenth Century*, 1861.
APS-CWB	Catharine Wistar Bache Papers, American Philosophical Society
AvH *Briefe*	Humboldt, Alexander von. *Briefe aus Amerika, 1799–1804*
AvH VS	Humboldt, Alexander von. *Alexander von Humboldt und die Vereinigten Staaten von Amerika: Briefwechsel*
Bot. Mag.	Curtis, William. *Botanical Magazine*
CPPJJ	*The Correspondence and Public Papers of John Jay*

CU-AB	Papers of Aaron Burr, Columbia University
CU-CC TM	Trustees' Minutes, Columbia College
CU-CPS TM	Trustees' Minutes, College of Physicians and Surgeons, Columbia University
CU-DC	DeWitt Clinton Papers, Columbia University
CU-JJ	Papers of John Jay, Columbia University
DH-LB	David Hosack Letter Book, New York Academy of Medicine
DH-MB	David Hosack Memorandum Book, New York Academy of Medicine
Douglas Journal	*Journal Kept by David Douglas during his Travels in North America, 1823–1827*
FCAB	Fuller Collection of Aaron Burr, Princeton University
HSP-BR	Rush Family Papers, Series I: Benjamin Rush Papers, Historical Society of Pennsylvania
JE Diary	John Hartshorne Eddy Diary, 1810, New York Public Library
Flora Lond.	Curtis, William. *Flora Londinensis*
FO-TJ	Thomas Jefferson Papers, Founders Online
Hone Diary	*Diary of Philip Hone*
INPS	Stokes, I. N. Phelps. *The Iconography of Manhattan Island, 1498–1909*
LC-AHP	Alexander Hamilton Papers, Library of Congress
LC-TJ	Papers of Thomas Jefferson, Series 1, General Correspondence, Library of Congress
LPAH	*The Law Practice of Alexander Hamilton: Documents and Commentary*
LS-JES	Correspondence of James Edward Smith, Linnean Society
LS Minutes	Minutes of the Linnean Society
N-Y Hort. Soc. Min.	Minutes of the New-York Horticultural Society, New York Botanical Garden
PAH	*The Papers of Alexander Hamilton*
RS-TJ	*Papers of Thomas Jefferson, Retirement Series*
SPCWP	*The Selected Papers of Charles Willson Peale and His Family*
Wharton Diary	Thomas Kelah Wharton Diary and Sketchbook, 1830–1834

PROLOGUE

3 island Eden: See Prest 1981 on the influence of the Garden of Eden story and the European discovery of America on the rise of botanical gardens.

3 build the civic institutions: Appleby 2000; Chaplin 2003.

5 "perhaps no one person": [no author, 1904] "Dr. David Hosack and His Botanical Garden," 517. On Hosack's reputation and contributions, see also Harnagel 1959; Robbins 1960; Robbins 1964; Jeffe 2004; and Hoge 2007.

5 more than a million: This and all subsequent financial comparisons are based on purchasing power in 2016 dollars (sourced at measuringworth .com).

9 "land of botanists": Milbert 1968 [1828], 25.

9 Emerson's epiphany: see esp. Brown 1997 and Walls 2003.

9 "natural alphabet": Emerson quoted by Brown 1997, 116.

10 now lies dormant: Sanderson 2009 reconstructs in detail the ecosystems and landscapes on which New York was built and has begun to inspire new urban-design projects that are more sensitive to the city's natural environment. Rogers 2016 offers a field guide to the intersections of the built and natural environments of New York. On the environmental history of New York State, see esp. Stradling 2010.

CHAPTER ONE: *"Tear in Pieces the Doctors"*

15 DH birthplace and father: Robbins 1964, 5–8.

15 volunteer fireman: Common Council Minutes, vol. 1, 203.

15 piles of volumes: circulation records for Alexander Hosack, 1789–1805, New York Society Library.

16 Washington Irving's birth: Jones 2008, 2.

16 "naturally very dull": AEH 1861, 300.

16 "gloomy reflections": AEH 1861, 300.

16 three awards: AEH 1861, 291.

17 "Is my son Dewitt Dead": quoted by Cornog 1998, 17.

17 "study—study—study": DC to David Hosack Jr., 15 May 1824, CU-DC.

17 later become famous: Hosack 1829, 41.

17 needed more doctors: Starr (2008 [1982], 40) has estimated that there were about two hundred physicians with medical degrees in the American colonies on the eve of the Revolution.

18 causes of death in late eighteenth-century New York: see, e.g., Klein and Reader 1991; *New-York Gazette & Weekly Mercury*, 17 January 1780 and 17 July 1780; *Independent Gazetteer*, 7 June 1787; *New-York Journal*, 11 May 1786.

18 "How just is yr observation": Sarah Livingston Jay to Susannah French Livingston, 17 April 1788, CU-JJ.

18 relationships among these humors: On humoral theory in the early Republic, see, e.g., Rosenberg 1979, Gronim 2006, and Wood 2014.

19 the main precursors: Sonnedecker, ed., 1986, 181ff.

19 plant trade routes in eighteenth century: see esp. Schiebinger 2004.

19 Beekman's shop: *New York Daily Advertiser*, 25 May 1789.

19 Peruvian bark: see esp. Achan et al. 2011; Maehle 1999, chapter 4. On the dosing of Peruvian bark, e.g., Andrew Duncan Jr. 1805, 189–94.

20 cold cream: *New York Daily Advertiser*, 5 January 1790.

20 *sal ammoniac*: Lewis 1791, 271.

20 calomel's side effects: Rothstein 1985, 50–51.

21 yellow fever: Blake 1968, 674–75.

21 smallpox epidemic: Wood 2014, 43.

21 smallpox symptoms, epidemics, and inoculation: Gronim 2006; Rothstein 1985, 29–32.

21 smallpox inoculation in Europe: Wootton 2006, 154–58.

21 "good harvests and good times": *New-York Packet*, 27 April 1786.

22 Bayley's career: Anderson 2004, 28.

22 "ardently attached": AEH 1861, 292.

22 barely tolerated: Anderson 2004, 99.

23 pillaging of a fresh grave at Trinity: *New York Daily Advertiser*, 26 February 1788.

23 Poughkeepsie in June 1788: Burrows and Wallace 1998, 291.

23 drafting of the *Federalist Papers*: Chernow 2004, 246–67.

24 "different parts of bodies": "Extract of a letter from New York dated April 18, 1788," *Charleston City Gazette*, 13 May 1788.

24 beat on the doors: *Litchfield Monitor*, 12 May 1788.

24 "tear in pieces the Doctors": Sarah Livingston Jay to Susannah French Livingston, 17 April 1788, CU-JJ.

24 "two large holes in his forehead": Sarah Livingston Jay to Susannah French Livingston, 17 April 1788, CU-JJ.

24 attributed to his exposure: Sarah Livingston Jay to Susannah French Livingston, 17 April 1788, CU-JJ.

24 one estimate: *Charleston City Gazette*, 13 May 1788.

24 at least three people were killed: Wilf 1989, 513.

25 remained calm: Chernow 2010, 586–87.

25 Dutch façade: Brissot de Warville 1792, 160; Webster 1886 [1786], iv.

25 Harlem name: Burrows and Wallace 1998, 69–70.

25 one Frenchman who visited New York: Brissot de Warville 1792, 156.

25 gilt-edged mirrors: *New York Argus, or Greenleaf's New Daily Advertiser*, 18 January 1796.

26 English tailor: *Diary or Loudon's Register*, 25 May 1792.

26 harpsichords imported from London: *New-York Packet*, 26 June 1787.

26 "bosoms very naked": Brissot de Warville 1792, 169, quoted by Burrows and Wallace 1998, 301.

26 poured into the streets: Burrows and Wallace 1998, 293; Chernow 2004, 268–69.

26 owed their well-appointed townhouses: Burrows and Wallace 1998, 290.

26 only a powerful central government: Chernow 2004, 243.

26 AH and Eliza Hamilton: Brissot de Warville 1792, 165–66.

27 "good nature": quoted by Chernow 2004, 130.

27 Hamiltoniana: Chernow 2004, 269; Burrows and Wallace 1998, 293.

27 it was feared Congress would pack up: Burrows and Wallace 1998, 292.

27 "safe in this place": quoted by Burrows and Wallace 1998, 292; see also Chernow 2004, 268.

27 "eternal buzz": quoted by Meacham 2012, 257.

28 would merge in 1791: see editor's note in Elihu Hubbard Smith 1973, 7.

28 "looks of tenderness": quoted by Brodsky 2004, 88.

29 "perpetual stream of eloquence": Hosack 1814, 52.

29 "garden of Eden": quoted by Gaudio 2003, 57.

29 "mere spectator": quoted by Hosack 1814, 50.

29 "all philosophical Experiments": Van Doren 1943, 280.

30 "daily pitted": quoted by Meacham 2012, 262.

30 botanical tour: see Wulf 2011, 89–99.

30 "What a field": TJ to Joseph Willard, 24 March 1789, LC-TJ.

30 "Toppan Sea": TJ, "Journal of the Tour," entry for 21 May 1791, FO-TJ.

30 "vast abundance": TJ to Thomas Mann Randolph, 5 June 1791, LC-TJ.

30 "innocent maple sugar": Rush 1793, 76.

31 coffee sweetened: https://www.monticello.org/site/house-and-gardens/sugar-maple.

31 "This organ": Hosack 1791, 9.

31 Snow and Koch: Porter 1997, 413, 437.

32 "painful": DH in AEH 1861, 294.

32 Robert Burns: DH to John Bostock, 27 August 1815, DH-LB, 123r.

32 twelve hours a day: AEH 1861, 294.

33 "Old Spasm": Porter 1997, 260.

33 novel structure: Wilson 2010, 442–44.

33 "while Astronomy claims": quoted by Brodsky 2004, 49.

34 DH wrote Rush: DH to BR, 12 December 1792, HSP-BR, vol. 27, 55.

34 jutting chin and squashed nose: Szatkowski 2007, 118–19.

34 "very much mortified": AEH 1861, 297.

35 were still struggling: Wilson 2010, 441. On the glacial pace of medical progress by the late eighteenth century, see esp. Wootton 2006.

35 four compounds: Quave 2016, 4.

36 botanical gardens: A growing scholarly literature explores the critical role of botany and botanical gardens in producing and maintaining the political, cultural, and scientific domination of European imperial powers over colonial lands and peoples. On Britain, see esp. Drayton 2000; on France, Mukerji 1997 and 2005. On botanical gardens in North America prior to the Revolution, see esp. O'Malley 1998 and Kevles 2011. On the role of Linnaean binomial nomenclature in the creation and maintenance of these networks of power, see Müller-Wille 2005.

37 professors and students: Fletcher and Brown 1970, v. On the intellectual life of Edinburgh at the time of Hosack's studies, see esp. Buchan 2003 and Uglow 2002.

CHAPTER TWO: *"An Endless Source of Innocent Delight"*

38 Wolf of Badenoch: Byatt 2005, 16.

39 "really unhappy": quoted by Fisher 1986: http://www.historyofpar liamentonline.org/volume/1790-1820/member/brodie-james-1744 -1824.

40 breakneck pace, shy: W. Hugh Curtis 1941, 100–101.

40 "innocent delight": Curtis 1783, 11; WC studies: Thornton 1805, 3.

40 "putrid air": Curtis 1803, 2.

41 His job: Field 1820, 72–76. On the history of the Chelsea Physic Garden, which can still be visited today, see Minter 2013.

41 garden in Brompton: Curtis first founded a garden in the London neighborhood of Lambeth, south of the Thames, but he soon realized that the wealthy patrons whose support he needed lived too far away to take much notice of him. He had also placed his garden right in the path of London's deadly coal smoke, so in 1789 he moved his plants across the river to the fresher air of the Brompton neighborhood.

41 "Botanic Garden, open to subscribers": quoted by W. Hugh Curtis 1941, 86. The layout of the Brompton Botanic Garden is given in W. Hugh Curtis 1941 and Curtis 1805.

42 "bouquets of the flowers": Curtis 1792, 11.

42 three and a half acres: W. Hugh Curtis 1941, 86; there were ten more acres for crops.

42 "frightened": Linnaeus 1775 [1750], iv, v.

42 "pompous expressions": Linnaeus 1775 [1750], 271.

43 illustrated guide: Curtis 1777.

43 James Sowerby: Laird (2015, 338) notes that William Kilburn did 32 plates for the first three fascicles.

43 "attract the notice": *Flora Lond.*, vol. 2, plate 63.

43 "poor dear *Fl. Londinensis*": Samuel Goodenough to WC, 8 January 1791, quoted by W. Hugh Curtis 1941, 90; two thousand subscribers: W. Hugh Curtis 1941, 74.

44 DH at Brompton: The following account of the Brompton Botanic Garden is based on the catalogues Curtis published (Curtis 1790b, 1792), as well as on the garden plan given in Thornton 1805 and reproduced in W. Hugh Curtis 1941, 87. For Curtis's descriptions of particular species, I have drawn on his *Flora Londinensis* (1777–1798), *Botanical Magazine* (1787–1800), and *Assistant Plates to the Materia Medica* (1786). For the medicinal properties of specific plants, I have drawn on contemporary editions of the *Edinburgh New Dispensatory* and on William Woodville's *Medical Botany* (1790–1793), both of which DH consulted regularly.

44 purgative effects: Linnaeus 1775, 408. Morning glories are now classified under the genus *Ipomoea*.

45 only in a botanical garden: Curtis 1783, 12; Curtis 1786, iv.

45 help the students commit to memory: Curtis 1778, 15.

45 reduce inflammation: Lewis 1786, 129. This plant is now classified in the genus *Polygonatum*.

45 *Cichorium intybus*: Lewis 1786, 125.

45 "uterine obstructions": Lewis 1786, 110.

46 "lax state of the solids": *Flora Lond.*, vol. 5, plate 32; Lewis 1786, 71.

46 *Lobelia siphilitica*: Lewis 1786, 173.

46 "powers of the imagination": *Flora Lond.*, vol. 1, plate 41.

46 sound farming practices: Curtis 1783, 13.

46 "his vegetable foes": Curtis 1792, 1.

46 "a much neglected tribe": Curtis 1783, 67.

46 "preferred the *Poa procumbens*": Curtis 1790a, 62.

47 *Aethusa cynapium*: *Flora Lond.*, vol. 1, plate 18.

47 *Pulmonaria maritima*: *Flora Lond.*, vol. 6, plate 18.

47 "skin of a cat": *Bot. Mag.*, vol. 2 (1788), 45.

47 room to room: W. Hugh Curtis 1941, 88.

47 hours in the little library: AEH 1861, 298.

47 "a thousand miles from London": Thornton quoted by W. Hugh Curtis 1941, 89.

48 books in library: listed in Curtis 1790b, various pages.

48 "E. Libris David Hosack": inscription in Hosack's copy at NYBG of Tomus II, Part 1 of Linnaeus's *Systema Naturae*, edited by Johan Friedrich Gmelin and published in 1791.

48 "Do not however": Rousseau 1787, Letter I, 21.

48 *Antirrhinum triste*: *Bot. Mag.*, vol. 3 (1788), 74; *Cineraria lanata*: *Bot. Mag.*, vol. 2 (1788), 53.

49 "small roll": *Bot. Mag.*, vol. 3 (1790), 97.

49 "sunk in spirits": Thornton 1805, 29.

49 "pursue the foxtail-grass": Curtis 1792, 1.

49 inns and taverns where the class met: Curtis 1792, 2.

49 heavy woolen vest: Thornton 1794, 22.

50 *New Illustration of the Sexual System*: Thornton 1807.

50 "read in that book": Curtis 1783, 11; face lit up: Thornton 1805, 29.

51 "rage for building": Curtis 1777–1798, vol. 1, n.p.

51 *Anagallis arvensis*: Curtis 1777–1798, vol. 1, plate 12.

51 "glowed with youthful fire": Thornton 1805, 29.

CHAPTER THREE: *"Ripping Open My Belly"*

53 John soon surpassed: My account of Hunter's work is based on Moore 2005.

53 Edward Jenner: Porter 1997, 276.

53 "every peculiarity": Sawrey 1815, xxx.

54 "boy's belly" and other quotations: Marshal 1815, 43, 46, 180–83.

55 "curved trocar": Savigny 1800, 23.

55 "exquisitely painful" and "horror": Earle 1791, 39, 41.

55 "than his head": Earle 1796, 27.

56 "clear straw coloured fluid": Earle 1796, 48.

56 WC nominated DH: On 15 October 1793, WC signed a letter nominating DH as Foreign Member; nomination put forward same day (LS Minutes 1788–1802, 135). Hosack was elected a Foreign Member on 17 December 1793 (LS Minutes 1788–1802, 139).

56 DH elected Foreign Member: 17 December 1793 (LS Minutes 1788–1802, 139).

56 erstwhile foes: DH to BR, 8 September 1794, HSP-BR, vol. 27, 56.

56 "Voice from the grave": quoted by Gage and Stearn 1988, 178.

57 "King of Botany": Samuel Goodenough to JES, 31 March 1794, LS-JES.

57 Banks at the Chelsea Physic Garden: Wulf 2008, 174. On the voyage and the launching of Banks's scientific career in Britain, see esp. O'Brian 1987 and Holmes 2008.

57 "totally unworthy": DH to Banks, 27 August 1815, DH-LB, 123v.

57 "imperfect dried specimen": Hosack 1824b, 23.

58 Tuesdays, Thursdays, and Saturdays: AEH 1861, 298.

58 taller: according to [no author] "The Linnean Herbarium," *Hooker's Journal of Botany and Kew Garden Miscellany* (vol. 4 [1852], 219), each cabinet was seven feet six inches tall.

58 shoved loose: Wulf 2008, 57.

58 Linnaeus's own hand: Benjamin Daydon Jackson 1912, 7.

58 thousands of plants: For the countries and regions from which the specimens came, see Benjamin Daydon Jackson 1912.

59 Kalm's travels and Swedish Academy: Koerner 1999, 117.

59 letter "K": On Kalm's North American specimens, see Juel and Harshberger 1929 and Jarvis 2007.

59 Catesby in North America: Meyers and Pritchard, eds., 1998; Parrish 2006.

59 "different distances": Hosack 1794, 3.

60 Banks suggested: DH in AEH 1861, 299.

60 listened to Hosack's paper: proceedings recorded in entry of 1 May 1794, Journal Book of RS, vol. 35 (1792–1795), 291.

60 "widely known": Gross 1893, 87.

60 promotion to Fellow: 20 May 1794, LS Minutes 1788–1802, 151.

61 fashioned of pine and iron: *Lloyd's Register* (1794), n.p.

61 Copies of Jay's initial dispatches: *Baltimore Daily Intelligencer*, 1 September 1794.

61 "Kiss our little ones for me": John Jay to Sarah Livingston Jay, 12 May 1794, CPPJJ, vol. 4; cannonfire and cheers: *American Minerva*, 12 May 1794. On Jay's mission, see Stahr 2005, 313ff.

61 "all Vessels of War belonging to Foreign Nations": *American Minerva*, 13 June 1794.

61 when word arrived: John Jay to Sarah Livingston Jay, 21 June 1794, CPPJJ, vol. 4.

62 letter to President Washington: Jay to GW, 23 June 1794, in CPPJJ, vol. 4; letter to Secretary of State Edmund Randolph: Jay to Randolph, 23 June 1794, CPPJJ, vol. 4.

62 Law traveling with Duncanson: Clark 1911.

62 families making for Newhaven: *Oracle and Public Advertiser*, 4 July 1794.

63 "oxygene": DH to BR, 3 September 1795, HSP-BR, vol. 37, part 3.

63 news of a British attack: *London Morning Chronicle*, 6 August 1794.

64 "embellishing every story": DH to BR, 8 September 1794, HSP-BR, vol. 27, 56.

64 not until after: see, e.g., *New-York Daily Gazette*, 19 August 1794.

64 steer clear of politics: Hosack 1826, 17.

64 typhus symptoms: Raoult, Woodward, and Dumler 2004.

64 "frothy and offensive discharges": Hosack 1815, 8. Hosack distinguished in this treatise between typhus in particular and "the typhoid state of fever" (a contemporary term for the advanced stage of a number of diseases, including scarlatina). Hosack argued that both typhus and "the typhoid state of fever" result in symptoms such as diarrhea and delirium, and should be treated by doctors with the same medicines. It would be 1849 before a British doctor, William Jenner, provided evidence of a distinction between typhus and typhoid fever, not to be confused with the earlier classification of the "typhoid state of fever" (Porter 1997, 349).

65 *typhos*: Porter 1997, 26.

65 "dangerous" and "indiscriminate" treatments: Hosack 1815, 7–10.

66 "western disturbances": *Hartford Gazette*, 1 September 1794.

66 special militia: Chernow 2004, 471.

66 TJ on right to assemble: Meacham 2012, 289.

66 "I could wish the experiments were repeated": DH to BR, 3 September 1795, HSP-BR, vol. 37, part 3.

66 old Continental Army fortifications: Yocum 2005, 20.

66 banded together in work teams: see, e.g., *New York Daily Advertiser*, 2 May 1794; *New York Diary or Loudon's Register*, 6 May 1794; *New York Columbian Gazetteer*, 8 May 1794.

68 "Typhus fevers": *Philadelphia General Advertiser*, 8 September 1794.

68 Captain Allen besieged: Allen's letter to *New York Daily Advertiser*, 29 August 1794.

68 men and horses: INPS, vol. 1, 401–2.

68 post office at 30 Wall Street: William Duncan 1794, 12.

68 "My d[ea]r Mr. Jay": Sarah Livingston Jay to Jay, 27 August 1794, CU-JJ.

68   "Having observed": William Hunter et al. to Allen, letter of 27 August 1794, reprinted in *New-York Daily Gazette*, 2 September 1794.

68   overjoyed to be home: DH to BR, 8 September 1794, HSP-BR, vol. 27, 56.

69   "blueish colour": Hosack 1798b, 507–8.

69   "their natural expression": Hosack 1798b, 509.

CHAPTER FOUR: *"He Is As Good As the Theatre"*

70   sixty thousand: The 1790 census counted about 33,000 people in New York City (Sanderson and Brown 2007, 547), and the 1800 census about 60,000 (Pomerantz 1938, 200).

70   goats and cows: e.g., Common Council Minutes, 8 September 1795, 176; *New York Daily Advertiser*, 2 February 1795.

70   it suited him: DH to BR, 8 September 1794, HSP-BR, vol. 27, 56.

70   LS election: Hosack elected 15 July 1794, LS Minutes 1788–1802, 157.

71   Not even Rush: On the meaning of membership in the RS to Franklin's generation of colonials, see Parrish 2006, 125–27.

71   "minions of despotism": *New York Daily Advertiser*, 5 July 1793.

71   not allowed to vote: on voting rights in New York, see Brooke 2013 and Harris 2003, 58.

71   Clinton's allies: Burrows and Wallace 1998, 317; Isenberg 2007, 106.

72   "from despotism": quoted by Meacham 2012, 224.

72   by mid-1793: on the shaping of American political cleavages by events in Revolutionary France, see Cleves 2009. On New Yorkers' responses to the Revolution, see Rapport 2017, chapter 12.

72   publicly insulted: *New York Diary or Loudon's Register*, 14 June 1793.

72   Tontine Coffee House: Burrows and Wallace 1998, 311. The name *Tontine* was for Lorenzo Tonti, a seventeenth-century Italian banker (Kamensky 2008, 90–91).

72   "scorn and hatred": *New York Diary or Loudon's Register*, 14 June 1793.

72   challenged the captain of *L'Embuscade*: Burrows and Wallace 1998, 318.

73   French anthems: Burrows and Wallace 1998, 321.

73   "so full of French": quoted by Wallace and Burrows 1998, 313. On French émigrés in the early American Republic, see Furstenberg 2014.

73   French consul on William Street: William Duncan 1794, 6.

73   "immediate evils" and following quotations: DH to BR, 8 September 1794, HSP-BR, vol. 27, 56.

74   Athens of America: *Boston Independent Ledger and the American Advertiser*, 5 December 1785.

74   "a NATIONAL CONCERN": quoted by Andrew J. Lewis 2011, 18.

75 "I neglect": quoted by Sellers 1980, 152.

75 TJ and AH at Peale's museum: Sellers 1980, 58, 61.

75 scientific knowledge: On Philadelphia's prominence in North American natural history, see Meyers, ed., 2011, esp. chapters by O'Malley and Fry.

75 CWP's portraits of GW: see Sellers 1951.

75 Peale daughters: Ward 2004, 142.

75 "The first Linnaeus": quoted by Sellers 1980, 74.

76 DH at 60 Maiden Lane: William Duncan 1795, 105.

76 DH's purchases at Posts' shop: entry for July 1800, DH-MB, n.p. (medical section).

76 dosing and effects of myrrh and camphor: Andrew Duncan Jr. 1805, 246, 264.

76 DH's purchases at Philips & Clark: 13 April 1801, DH-MB, 11 (medical section).

77 $1,500: AEH 1861, 301.

78 "suffocating anguish": SLM quoted by Hosack 1811b, 31.

78 Bard and Washington: Chernow 2010, 586–87, 624–26.

78 "all the Presidents": SLM 1826, 6–7.

78 outside of Syracuse: SLM opinion noted in anonymous review of "An Address Delivered Before the New-York Historical Society at its Fortieth Anniversary," *Knickerbocker* 25, no. 3 (1845), 254. On the uses and meanings of the Eden story in the early American Republic, see esp. Merchant 2004.

78 "zeal, Industry and Talents": SLM to Reverend Richard Provoost, 15 February 1795; quoted by Robbins 1964, 54. On 28 April 1795, SLM recommended to the Columbia trustees that DH be appointed Professor of Botany in his place, and the trustees moved to do so (CU-CC TM, vol. 2, part 1, 218).

79 DH waited: appointment announced in *American Minerva*, 27 May 1795.

79 founded in 1791 by SLM and others: Burrows and Wallace 1998, 376.

79 "performances intended" and manure: *Transactions of the Society for the Promotion of Agriculture, Arts, and Manufactures* 1792, vii.

79 cast iron: *Greenleaf's New-York Journal*, 2 February 1793.

79 brewers: *Albany Register*, 7 January 1793.

79 members: see *Transactions of the Society for the Promotion of Useful Arts* 1807; Robbins 1964, 49, 53; and Pomerantz 1938, 411.

79 DH proposal: DH to Robert R. Livingston, reprinted in *American Minerva*, 10 April 1795.

80 DH procedure: see Hosack 1798a; Thacher 1828, vol. 1, 59.

81 June 15: DH-MB, 2 (medical section).

81 DH portly and animated: Blatchford 1861, 20–22.

81 heavily underlined: e.g., Hosack 1804.

81 "read his lectures": Mott 1850, 8.

81 "black eyebrows" and "thunder-cloud frown": Blatchford 1861, 20, 22.

82 "vegetables": Hosack 1804, 1. This manuscript fleshes out DH's syllabus outline of 1795 (Hosack 1795).

82 "Dr. Smith": Hosack 1804, 20.

82 "in the vegetable economy": Hosack 1795, 12.

82 "want of food": Hosack 1804, 84.

82 "sexual machine": Hosack 1804, 78–79.

82 New Yorker observed: *Medley, or, Newbedford (MA) Marine Journal*, 14 August 1795.

82 Hosack checked a thermometer: DH to BR, 3 September 1795, HSP-BR, vol. 37, part 3.

83 Bayley on Broadway: William Duncan 1795, 10.

83 rind of mold: Bayley 1796, 52.

83 overcast day: Webster, ed., 1796, 28.

83 Hamilton assaulted: Wood 2009, 198; Chernow 2004, 490.

83 Jacobins: *American Minerva*, 11 July 1795.

83 "humiliating": *Greenleaf's New York Journal*, 1 July 1795.

83 burning a copy of treaty: Burrows and Wallace 1998, 322; Freeman 2001, xiii–xiv; Chernow 2004, 490–92.

83 two duels: Chernow 2004, 490–92.

83 dumped coffee: Bayley 1796, 14.

83 "Yellow Fever prevails": *Medley, or, Newbedford Marine Journal*, 14 August 1795.

83 "clot[t]ed": *Medley, or, Newbedford Marine Journal*, 14 August 1795.

84 weeded out thousands of people: Chernow 2004, 448.

84 victims often lay with their knees drawn up: Alexander Hosack Jr. 1797, 14.

84 coffee grounds: Alexander Hosack Jr. 1797, 15.

84 damaged shipment of coffee: Brodsky 2004, 326.

84 "I found Mr. Cochran": Rush 1796, vol. 4, 6.

84 loss of blood: Rush 1796, 188–92.

84 In a report he published: Rush 1796.

84 bled Sally Eyre nine times and John Madge twelve times: Rush 1796, 85.

85 "scarlet coloured sediment": Rush 1796, 84.

85 mean-spirited letter: TJ to Madison, 8 September 1793, FO-TJ.

85 AH, Eliza Hamilton, and yellow fever: Chernow 2004, 449–51.

85 Hamilton published open letter: AH to the College of Physicians, 11 September 1793, Alexander Hamilton Papers, Founders Online. Stevens's approach is detailed in Currie 1794, 54–57.

85 heard people praising: DH to BR, 8 September 1794, HSP-BR, vol. 27, 56.

85 held out hope: Dr. William Pitt Smith to Dr. Samuel Duffield, 1 September 1795, reprinted in *American Minerva*, 8 September 1795.

85 converged on City Hall: *Philadelphia Gazette and Universal Daily Advertiser*, 20 August 1795.

85 research on nitrous oxide: Mitchill 1795; Bergman 1985.

86 "no case of yellow fever": *Philadelphia Gazette and Universal Daily Advertiser*, 20 August 1795.

86 decamped: Mitchill signed the preface to his work on nitrous oxide on 20 August 1795 at Plandome (Long Island), the site of his country house (Mitchill 1795, 4).

86 "on the subject of fever": DH to BR, 3 September 1795, HSP-BR, vol. 37, part 3.

86 "found in Water Street": DH to BR, 3 September 1795, HSP-BR, vol. 37, part 3.

86 yellow from the contagion: Bayley 1796, 24.

86 worst in people: Bayley 1796, 10.

86 come down with a cold: *Philadelphia Gazette and Universal Daily Advertiser*, 11 September 1795.

86 "better to fear too far": *Philadelphia Gazette and Universal Daily Advertiser*, 28 August 1795.

87 "Raw head and bloody bones": *American Minerva*, 5 September 1795.

87 "they had trifling colds": *Albany Gazette*, 11 September 1795.

87 August 25 alone: Dr. William Pitt Smith to Dr. Samuel Duffield, 1 September 1795, published in *American Minerva*, 8 September 1795. Number of deaths by 8 September: John Broome to Governor John Jay, 8 September 1795, published in *New York Herald*, 30 September 1795.

87 "hearse monsters": Sarah Anderson to Alexander Anderson, 10 October 1795, Alexander Anderson Papers, MssCol 98, NYPL.

87 $300 fine: *Albany Gazette*, 11 September 1795.

87 "general Health prevails": Common Council Minutes, 8 September 1795, vol. 2, 177.

87 "Two hundred carcases": *Albany Register*, 11 September 1795.

88 "This destructive Terror": entry for 25 September 1795, Elihu Hubbard Smith 1973, 62.

88 washing limbs and torsos: DH to BR, 3 September 1795, HSP-BR, vol. 37, part 3.

88 tamarind: Alexander Hosack Jr. 1797, 31–32.

88 "When I find my patient sweating": Hosack 1797, 455.

89 "their Maker": Sarah Anderson to Alexander Anderson, 19 October 1795, Alexander Anderson Papers, MssCol 98, NYPL.

89 "no time for much reflexion": Hosack [n.d.], "Notes on Midwifery," 17.

89 "the soft parts": Hosack [n.d.], "Notes on Midwifery," 7.

89 "vital functions": Hosack [n.d.], "Notes on Midwifery," n.p., note inserted between 7 and 8.

89 "her friends": Hosack [n.d.], "Notes on Midwifery," 7.

90 two weeks after Kitty's death: 24 February 1796, CU-CC TM, vol. 2, part 1, 230.

90 a perfect combination: Hosack 1801, 41.

90 "dig their graves": Hosack 1830, 4.

90 "art of war": Hosack 1826, 12.

90 "after a hearty dinner": quoted by Pratt 1956, 28.

91 "ever anxious": quoted by AEH 1861, 305.

91 "Dr. H. is of the opinion": Bass 1817, n.p.

91 "most eloquent and impressive teacher": Francis 1858, 84.

91 "fine, manly" and "No reader": Henry W. Ducachet, Preface to Hosack 1838, vol. 1, xiii.

91 "good as the theatre": Moses Champion to Reuben Champion, 20 December 1818, CU-CPS, Misc. MSS, Box 1, Folder 32.

91 "Make frequent visits": Hosack 1801, 31.

91 "altogether incompatible": Hosack 1826, 17; on DH and alcohol, see also Wood 2009, 339.

91 humiliated any student: Mott 1850, 8.

91 kill a patient: Hosack 1826, 31.

91 "punishment was terrible": Blatchford 1861, 22.

92 "habits of inattention": Hosack 1801, 6.

92 "*Poppies* (Puppies?)": quoted by Pratt 1956, 28.

92 "charms and incantations": Hosack 1801, 25–26.

92 "fashion in medicine": Hosack 1801, 28.

92 prevailing humoral approach: e.g., Hosack 1801, 34.

92 found in the plant world: e.g., Hosack 1795, Hosack 1804.

92 "the beautiful science of botany": quoted by AEH 1861, 304.

92 agaric: Hosack 1804, 61.

92   "thrown aside": Hosack 1801, 40.

93   "immense treasures": Hosack 1801, 42.

93   "*Foxglove* for those of *Mullein*": Hosack 1801, 41; original in Curtis 1792, 1.

CHAPTER FIVE: *"The Grass Is Three Feet High in the Streets"*

94   shy man: Samuel Powell to George Washington, 11 June 1790, quoted by Peck 2010, xv.

94   cypress tree: Ewan and Ewan 2007, 111; Wulf 2011, 71.

94   vomited twice: Ewan and Ewan 2007, 197.

94   Collinson and Bartram partnership: Wulf 2008.

95   "his Idol Flowers": Garden to Cadwallader Colden, 4 November 1754, quoted by Fry 2004, 164; see also Wulf 2008, 130.

95   humble or plain: Wulf 2008, 113–14.

95   announcement: Fry 2012, 4. The announcement was published in the *Pennsylvania Gazette*, 20–27 July 1738. On John Bartram, see also Fry 2004, 2011, and 2014.

95   "Mountains and Swamps": "A Copy of the Subscription Paper, for the Encouragement of Mr. John Bartram," *Pennsylvania Gazette*, 17 March 1742, quoted by Van Doren 1943, 278.

95   British guide: Short 1746.

95   Bartram wrote little editorials: John Bartram 1751.

95   "Womens After-pains" and pleurisy root: John Bartram 1751, 2, quoted by Fry 2012.

96   Isaac and Moses as apothecaries: Baird 2003.

96   botanizing trips: Hallock and Hoffman, eds., 2010, 3–4.

96   "roaring terribly": William Bartram 1958 [1791], 76. For a chronology of William Bartram's travels, see Fry 2010.

96   "expected every moment": William Bartram 1958 [1791], 76.

96   "dangerous": William Bartram to Thomas, 15 July 1786, reprinted in Hallock and Hoffman, eds., 2010, 133.

97   "green meadows": quoted by Harper 1958, liii.

97   "wonder of Creation": Barton to William Bartram, 30 December 1792, quoted by Ewan and Ewan 2007, 277.

97   "vegetable Diuretick": William Bartram to Barton, 29 December 1792, reprinted in Hallock and Hoffman, eds., 2010, 167.

97   "Alexander, Caesar": quoted by Ewan and Ewan 2007, 137. On Barton, see also Andrew J. Lewis 2005.

97   Bartram's *pharmacopoeia*: Wilson 2010. There is some question as to the

authorship of this pharmacopoeia, but Wilson makes a persuasive case that it is by William Bartram.

99 opium test and "never buoy up": quoted by [no author, 1968] "Samuel Bard, Colonial Physician," 586.

99 "*Summus Perfectus*": quoted by Chaplin 2006, 96.

99 father's estate: on Cadwallader Colden, see Dixon 2016; on Jane Colden, see Gronim 2007, Paula Ivaska Robbins 2009, and Parrish 2006, 196–200.

99 "first lady": quoted by Beatrice Scheer Smith 1988, 1091.

100 "My dear Hosack": quoted by McVickar 1822, 207.

100 "an internal monitor" and following quotations: Hosack 1826, 9–10.

100 When Bard returned: AEH 1861, 301.

100 AB wrote from Philadelphia: AB to TB, 28 February 1797, quoted by Stone 1865, vol. 2, 456.

100 Peale's painting: Kelsay 1984, 577. AB's Senate term ended in March 1797 (Isenberg 2007, 155).

101 "court-like manners": Rev. Dr. Samuel Miller, quoted by Stone 1865, vol. 2, 457.

101 renting Richmond Hill: INPS, vol. 1, 359, 416–17; the city directory of 1792 indicates that Burr also had an office at 4 Broadway (William Duncan 1792, 22).

101 "red waistcoat": quoted by INPS, vol. 5, 1254.

101 "delicious": quoted by INPS, vol. 5, 1274; see also McCullough 2008, 412.

101 "golden harvest": Abigail Adams to Elizabeth Shaw, quoted by INPS, vol. 5, 1254.

101 Burr's lease: INPS, vol. 1, 417, and vol. 5, 1340; Isenberg 2007, 158.

101 Burr bought flowers: Plantsman's ledger [author unknown], 1793–1796, NYBG.

101 icehouse: INPS, vol. 5, 1352.

101 "Ma begs you": TB quoted in AB to TB, 14 January 1794, AB Memoirs, vol. 1, 374. Matthew Livingston Davis heavily edited Burr's letters (published in two volumes as *Memoirs of Aaron Burr* in 1836–37) as well as Burr's 1808–12 journal (published in two volumes as *The Private Journal of Aaron Burr* in 1838). I therefore rely wherever possible on the original manuscripts of the letters and on the Bixby/Samson edition of the original journal (published in two volumes as *The Private Journal of Aaron Burr* in 1903). My sources for Burr's original letters are the 27 microfilm reels of the Papers of Aaron Burr held by Columbia University; the Fuller Collection of Aaron Burr at Prince-

ton University; and *The Political Correspondence and Public Papers of Aaron Burr*, Kline, ed., 1983.

102 "*infusion* and *decoction*": AB to TB, 23 January 1794, AB Memoirs, vol. 1, 375.

102 "beautiful assortment": AB to TB, 31 March 1794, AB Memoirs, vol. 1, 378.

102 "How much of your taste": TB to AB, 10 December 1803, AB Memoirs, vol. 2, 252–53, quoted by INPS, vol. 5, 1415.

102 "narcotic powers of Opium": AB to wife Theodosia Burr, 24 December 1793, CU-AB, Series I, Reel 3.

102 "wonderful cures": AB to wife Theodosia Burr, 24 December 1793, CU-AB, Series I, Reel 3.

102 "Doctor Rush thinks": AB to TB, 23 January 1794, AB Memoirs, vol. 1, 375.

102 "my dear little girl": AB to TB, 16 January 1794, AB Memoirs, vol. 1, 374.

102 Senate chamber: see, e.g., AB to TB, 31 December 1793, AB Memoirs, vol. 1, 368.

102 week-old letter: AB to Pierpont Edwards, 24 May 1794, CU-AB, Series I, Reel 3.

102 "more pain": quoted by Isenberg 2007, 127.

103 "Dream on": AB to TB, 5 January 1795, AB Memoirs, vol. 1, 386.

103 By the spring of 1796: In an 1826 letter to a friend in Philadelphia, DH recalled having been on a house call to the Hamiltons when AH received from President Washington an outline "on several sheets of foolscap paper" of what would become known as the Farewell Address. This would thus have been sometime in or soon after mid-May 1796 (Chernow 2004, 505). DH wrote of this episode, "I shall never forget the gratification displayed by the General upon receiving this high compliment from his great chief" (DH to Thomas James, 9 July 1826, DH Collection, N-YHS; I am grateful to the New-York Historical Society for permission to quote from its collections here and elsewhere in this book). In 1833, DH recalled "having been the physician of [AH's] family from the year 1795" (DH to John Church Hamilton, 1 January 1833, LC-AHP, Reel 30), a claim borne out by AH's record of a payment on 1 February 1797 to "Doctors Bard & Hosack in full for accounts 95 & 96" (*LPAH*, vol. 5, 474).

103 AH age: It is not known precisely when AH was born. Chernow puts his birthdate at 11 January 1755 (Chernow 2004, 17); Eliza was born on 9 August 1757 (Chernow 2004, 130).

103 "leaving my dear family": AH to Eliza Hamilton, 12 September 1797, *PAH*, vol. 21, 294; quoted by Chernow 2004, 544.

104 "love of liberty": *Claypoole's American Daily Advertiser*, 19 September 1796.

104 AH and election of Adams: Chernow 2010, 510–11, 514; McCullough 2008, 463.

104 "superabundance of secretions": JA to BR, 11 November 1806, John Adams Papers, Founders Online; quoted in part by Chernow 2004, 522.

104 "singularly Critical": AH to Oliver Wolcott Jr., 5 April 1797, *PAH*, vol. 21, 22.

104 fell ill with a fever: It is not clear precisely what illness Philip had contracted. In an 1833 letter to AH's son John Church Hamilton recalling the events of this night, DH described Philip's illness as a "severe, bilious fever, which soon assumed a typhus character" (DH to John Church Hamilton, 1 January 1833, LC-AHP, Reel 30; quoted by Chernow 2004, 544). Given Hosack's usage of the terms *typhus* and *typhus character* in his medical writings, he could have meant by *typhus character* either typhus specifically or an advanced stage of one of several other illnesses; Hosack's son Alexander later recalled hearing that Philip had contracted scarlet fever (AEH 1861, 307).

104 "very anxious": AH to Eliza Hamilton, 12 September 1797, *PAH*, vol. 21, 294; quoted by Chernow 2004, 544.

105 Charlton sent for DH: DH to John Church Hamilton, 1 January 1833, LC-AHP, Reel 30. See also *PAH*, vol. 21, 294–95.

105 "overwhelmed with distress": DH to John Church Hamilton, 1 January 1833, LC-AHP, Reel 30.

105 letter to James Monroe: AH to Monroe, *PAH*, vol. 21, 200.

105 resurfaced in the press: Chernow 2004, 529–32.

106 AH statement: AH 1800 [1797]. On content and AH motivations, see Chernow 2004, 532–42.

106 "great distress": DH to JCH, 1 January 1833, LC-AHP, Reel 30.

106 reworked his lectures: Minturn Post, quoted by AEH 1861, 308. DH states in his 1833 letter to John Church Hamilton that he was still "personally unknown" to AH when AH came to his room to thank him for saving Philip, but this claim conflicts with DH's 1826 recollection to Thomas James (letter of 9 July 1826, DH Collection, N-YHS) that he was in AH's presence when the latter received Washington's draft of the Farewell Address, which would have been in the spring of 1796.

Also conflicting with this claim is AH's payment of 1 February 1797 to "Doctors Bard & Hosack in full for accounts 95 & 96" (*LPAH*, vol. 5, 474), as noted above.

106 *Aedes aegypti*: Porter 1997, 300.

107 "it is impossible for Hosack": entry for 13 October 1797, in Elihu Hubbard Smith 1973, 376.

107 "Parke & I are bleeders": entry for 13 October 1797, in Elihu Hubbard Smith 1973, 376.

107 medical board of New York recommended: Langstaff 1940, 194.

107 "mentioned with horror in some companies": quoted by Brodsky 2004, 334.

107 Columbia trustee: Langstaff 1940, 194; CU-CC TM, vol. 2, part 1, 250.

107 "the opposition to my appointment": BR to John R. R. Rodgers, 6 November 1797, quoted by Brodsky 2004, 335.

107 "brother in the republic of medicine": BR to DH, 7 June 1809, Robbins Mss. B.H.78, APS.

107 AH with his family: DH to John Church Hamilton, 1 January 1833, LC-AHP, Reel 30.

107 "highly endowed and cultivated mind": DH to John Church Hamilton, 1 January 1833, LC-AHP, Reel 30.

108 "great regret": reprinted in Hosack 1811a, 7; see also minutes for 3 November 1797, CU-CC TM, vol. 2, part 1, 250.

108 "effectual remedy": reprinted in DH 1811d, 8.

108 "Hosack has been some time at Phila.": Elihu Hubbard Smith to Levi Wheaton, 20 December 1797, Elihu Hubbard Smith 1973, 406.

108 Mary Eddy, the Wistars, and Franklin: Robbins 1964, 151.

108 *Wisteria*: Nuttall 1818; Ewan and Ewan 2007, 540.

109 "only lady here": John Torrey quoted by Rodgers 1942, 23.

109 neighbors of DH: Longworth 1797, various pages.

109 DH's cow: *New York Daily Advertiser*, 19 July 1803.

109 DH father and slaves: "United States Census, 1790," 77; slave population in NY: Harris (2003, 56) notes: "By 1790, the free black population in New York City had grown to an unprecedented 1,036 out of a total black population of 3,092." On slaveholding in eighteenth-century New York, see Lepore 2005.

109 antislavery advocates: DH was friendly with members of the New York Society for Promoting the Manumission of Slaves, founded in 1785 by Jay, Hamilton, and others (Harris 2003, 56). Some of the society's founding members owned slaves themselves but hoped to achieve grad-

ual emancipation in New York. Other members were Quakers whose New York congregation had already banned slavery (Harris 2003, 56).

109 DH and slaveholding: see United States Census, 1800: United States Census, 1800. *FamilySearch* (https://familysearch.org/ark:/61903/3:1:33SQ-GRZN-FF3?cc=1804228&wc=3V1X-9TB%3A1585148702%2C1585148902%2C1585149561: 10 June 2015), New York > New York > New York Ward 1 > image 16 of 22; citing NARA microfilm publication M32 (Washington DC: National Archives and Records Administration, n.d.).

For DH's slaveholding as of 1810, see United States Census, 1810: United States Census, 1810. *FamilySearch* (https://familysearch.org/ark:/61903/3:1:33SQ-GYBQ-9MXV?cc=1803765&wc=QZZZ-M4T%3A1588180303%2C1588181204%2C1588181203: 1 December 2015), New York > New York > New York Ward 1 > image 12 of 27; citing NARA microfilm publication M252 (Washington DC: National Archives and Records Administration, n.d.).

On slaveholding among DH and his Columbia colleagues, see Foner 2016 and Wilder 2013. DH freed a twenty-eight-year-old slave named Nelly in 1805 (Yoshpe 1941, 84).

109 warm, clear evening: Laight 1795–1803, entry for 14 September 1798.

110 DH ill: entry for 14 September 1798, in Elihu Hubbard Smith 1973, 464.

110 Dozens had died: see, e.g., Laight 1795–1803, various entries for September 1798.

110 "perpetual tears": entry for 6 September 1798, in Elihu Hubbard Smith 1973, 463.

110 conspiratorial silence: *Companion, and Commercial Centinel*, 1 September 1798.

110 "*three* feet high": *New-York Gazette*, 24 August 1798.

110 "North and East River": *New-York Gazette*, 24 August 1798.

110 "If they do not": *Greenleaf's New York Journal*, 4 September 1798.

110 Greenleaf died: Laight 1795–1803, entry for 14 September 1798.

110 Titian Peale died: *SPCWP*, vol. 2, 226.

110 "this *cankerworm*": CWP to A. M. F. J. Palisot de Beauvois, 16 June 1799, *SPCWP*, vol. 2, 244.

110 true Peale fashion: *SPCWP*, vol. 2, 232–33.

110 Smith died: editor's note in Elihu Hubbard Smith 1973, 464.

110 deserted properties: Common Council Minutes, 24 September 1798, vol. 2, 469.

111 "Sampson of the *materia medica*": Hosack 1824a, vol. 3, 431.

111 good results: Hosack 1824a, vol. 3, 427.

111 "without whose aid": quoted by Langstaff 1940, 196.

112 "trodden under foot": DH to Eaton, 30 August 1810, quoted by Robbins 1964, 69.

112 "My little boy": MH to CWB, 19 November 1799, APS-CWB.

112 named him Titian: *SPCWP*, vol. 2, 273, note 1.

112 bled him half-dry: Chernow 2010, 807.

112 "painful duty": *Alexandria Times*, 16 December 1799.

112 "behold the HERO": *Weekly Museum*, 21 December 1799.

113 bells rang daily: by order of the Common Council "from 12 to 1 every day for five days in a row," *New York Daily Advertiser*, 24 December 1799.

113 "drooping wings": *Spectator*, 4 January 1800.

113 "my heart sad": AH to Charles C. Pinckney, 22 December 1799, quoted by Chernow 2004, 600–601.

113 Clinton had gone to work: on Clinton's political career, see Cornog 1998.

113 Clinton back at Columbia: Cornog 1998, 32.

114 "formidable and dangerous diseases": DH to DC, 18 February 1800, CU-DC.

114 "cheerfully lend": DH to DC, 18 February 1800, CU-DC.

114 three assemblymen: *Journal of the Assembly* (23rd Session, 1800), 115–16.

115 endorsed SLM: *New York Daily Advertiser*, 19 April 1800.

115 mobilize voters: Chernow 2004, 607–8.

115 "destroy the Constitution": *Spectator*, 23 April 1800.

115 bribery or fraud: *New York Daily Advertiser*, 19 April 1800.

115 "prostitute": *New York Daily Advertiser*, 25 April 1800.

115 end of the French Revolution: Burrows and Wallace 1998, 328.

116 Hamilton's land and architect: Chernow 2004, 641–42.

116 "wicked enough": quoted by Chernow 2004, 633.

CHAPTER SIX: *"Doctor, I Despair"*

117 scarlet fever: Robbins 1964, 163.

117 "animal of uncommon magnitude": quoted by Sellers 1980, 113.

117 Jefferson and moose: Dugatkin 2009.

118 "true political carriages": Brissot de Warville 1792, 108.

118 "boy" and "pison the minds": quoted by Sellers 1980, 125.

118 "sullen as any hog": quoted by Sellers 1980, 125.

118 "wild fire": *SPCWP*, vol. 2, 334.

118 "Every body seemed rejoiced": *SPCWP*, vol. 2, 334.

119 "bones togather": *SPCWP*, vol. 2, 334.

119 Peale recounted: CWP to James G. Graham, 30 June 1801, *SPCWP*, vol. 2, 339.

119 "contrary sentiment": *SPCWP*, vol. 2, 334.

119 did not mention Hosack: CWP to TJ, 29 June 1801, LC-TJ; *SPCWP*, vol. 2, 338.

119 "bloody Panther": Broadside reproduced in Semonin 2000, 328, and see chapter 13 for an analysis of this episode.

121 Sale of confiscated land: Burrows and Wallace 1998, 267. On the departure of Loyalists from the American colonies, see Jasanoff 2011.

121 Absentee landlords: On the rise of rental markets in late eighteenth-century Manhattan, see Blackmar 1989.

121 "put a stop to your improvements": quoted by Hartog 1983, 93.

121 "sufficient quantity of ground": quoted by Hartog 1983, 93–94.

121 Lispenard's Meadow: called by contemporaries variously Lispenard Meadows, Lispenard's Meadow, or Lispenard Meadow.

121 "two majestic rivers": Brissot de Warville 1792, 152.

121 deep forests: New York was not alone in this; see Cronon 1983 on the widespread colonial and early Republic deforestation in the Northeast, especially in New England.

122 blue skies: Laight 1795–1803, entries for September 1801.

122 "fields of Long-Island": Hosack 1810a, 1.

122 viburnum and violets on Hosack's property: see notations of collection by CWE in copy of Hosack 1806 held by Mertz Library, NYBG.

123 "exceedingly rough": Hosack 1811d, 16.

123 sunshine: Laight 1795–1803, 12 September 1801.

123 DH signed a bill of lading: Entry of Merchandise, 12 September 1801, David M. Rubenstein Rare Book & Manuscript Library, Duke University.

123 DH lent books to Bard: DH-MB, 12 (medical section).

123 chilly, overcast: Laight 1795–1803, weather entry for 23 November 1801, N-YHS.

123 reportedly fainted: Chernow 2004, 653. My account of this episode is based on Chernow 2004 and DH to John Church Hamilton, 1 January 1833, LC-AHP, Reel 30; the latter document is partially reproduced in *PAH*, vol. 25, 437.

123 "little impertinences": *New-York Evening Post*, 10 December 1802.

124 smoking: *Mercantile Advertiser*, 19 November 1801.

124 "Republican Young Men": *American Citizen*, 19 June 1801.

124 "man of the people": quoted in *American Citizen*, 24 July 1801.

124 "persecuted patriot": quoted in *American Citizen*, 24 July 1801.

124 less stringently prosecuted in New Jersey: Chernow 2004, 700.

125 "agony of grief": DH to John Church Hamilton, 1 January 1833, *PAH*, vol. 25, 437, and partially quoted by Chernow 2004, 653.

125 lay down: Chernow 2004, 654.

125 "grave of his hopes": *PAH*, vol. 25, 436, note 1.

125 "much agitated": Robert Troup to Rufus King, 5 December 1801, quoted in *PAH*, vol. 25, 437, note 1.

125 "many tears": BR to AH, 26 November 1801, *PAH*, vol. 25, 435.

126 "precious to me": AH to BR, 29 March 1802, *PAH*, vol. 25, 584.

126 "Hamilton is more composed": Robert Troup to Rufus King, 5 December 1801, *PAH*, vol. 25, 437, note 1.

CHAPTER SEVEN: *"There Are No Informed People Here"*

127 "Amer'can Farmer": *New York Commercial Advertiser*, 21 October 1805.

128 stark inequality: see esp. Wilentz 2004 [1984]. Burrows and Wallace (1998, 351) note that "by 1800 the richest 20 percent owned almost 80 percent of the city's wealth. The bottom half owned under 5 percent."

128 "other than medical subjects": DH-MB, 26 (medical section).

128 docked their pay: DH-MB, 24 (garden section).

129 porter's lodge: *Columbian*, 3 October 1810.

129 Curtis's epitaph: quoted by Thornton in Curtis 1805, vol. 3, 33.

129 Clinton and botany: Hosack 1829, 35.

130 FAM's injury: Savage and Savage 1986, 104.

130 France had been sapped: Savage and Savage 1986, 33. On the relation of Michaux's mission to the French political context, see Hyde 2012, 2013. On the Jardin du Roi before and during the Revolution, see Spary 2000.

130 "I imagine myself": quoted by Savage and Savage 1986, 41.

130 a dollar: Robbins and Howson 1958, 353.

130 "no informed people": AM to Louis Guillaume Lemonnier, 26 January 1786, quoted by Fry 2011, 81.

131 "majestic" and "magnificent": e.g., François André Michaux 1819, vol. 1, 41 and 152.

131 "Cramberrie": quoted by Robbins and Howson 1958, 354.

131 "romantic" and "awful": *New-Jersey Journal*, 27 June 1787, quoted by Robbins and Howson 1958, 357–58.

132 offers for its purchase: Robbins and Howson 1958, 362.

132 DH reputation: François André Michaux 1805, 126.

132 DH on medicinal species: Hosack 1801, 42–43.

132 Linnaeus and Jussieu: Hosack 1806, 5.

133 size of Brompton: about ten and a half acres, according to W. Hugh Curtis 1941, 86.

133 "naturalize as soon as possible": DH to TJ, 10 September 1806, FO-TJ.

133 native to Africa: on American efforts to naturalize cotton species, see Beckert 2015.

133 *Aconitum napellus* and *Nicandra physalodes*: Hosack 1806.

134 "evergreen foliage": François André Michaux 1819, vol. 1, 297. See also Fraser and Leone, eds., 2017.

134 plea for conservation: Savage and Savage 1986, 357. Thoreau and FAM: Thoreau 2004 [1854], 241, note 60; Walls (2017, 308) notes that Thoreau consulted works by André Michaux.

135 FAM's opinion of DH's land: François André Michaux 1804, 15, and 1805, 11.

135 prison: Hosack was an "inspector" and physician at the state prison from as early as 1799 (Longworth 1799, 95).

135 nails for Hosack's garden: DH-MB, 8 (garden section).

135 old assembly room: *Commercial Advertiser*, 29 March 1802.

135 Rembrandt had raked in $340: Rembrandt Peale to CWP, 7 April 1802, *SPCWP*, vol. 2, 425; Semonin 2000, 331.

135 *"exact counterpart"*: *American Citizen*, 20 May 1802.

135 he announced: *American Citizen*, 20 May 1802.

135 Banks: Sellers 1980, 156.

136 lost macaw: *Poulson's American Daily Advertiser*, 10 September 1802.

136 Randall's will: *PAH*, vol. 25, 389.

136 "merciless and unfeeling quacks": *New York Daily Advertiser*, 30 August 1790.

136 Bellevue: Oshinsky 2016, 14–37. See also INPS, vol. 1, 388.

137 Livingston and Monroe's mission to Paris: Wood 2009, 367–69.

137 "climate will not be": *Morning Chronicle*, 22 January 1803.

137 DH founding member: American Academy of Fine Arts 1817, 3.

137 *Laocoön*, the *Apollo Belvedere*, and the *Dying Gaul*: *Poulson's American Daily Advertiser*, 1 November 1802; *Spectator*, 3 July 1802.

137 City Hall: *Poulson's American Daily Advertiser*, 1 November 1802; Bender 1987, 62.

138 elected Napoleon: Academy of Fine Arts 1817, 33.

138 lavish volumes: Mitchill 1807, 158.

138 Livingston sent seeds: Hosack 1806, 7.

138 "Philips funeral": *PAH*, vol. 25, 437.

138 Philip born 2 June 1802: Chernow 2004, 654.

138 completed by the summer of 1802: Chernow 2004, 642.

138 goldenrod: CWE collected *Solidago canadensis* on Manhattan, noted in NYBG copy of Hosack 1806.

138 tomes on trees and soil: on AH at the Grange, see Chernow 2004, 641–44.

138 "chapel in this grove": AH quoted by John Church Hamilton, "Chapter 168," n.p., Box 20, John Church Hamilton Papers, CU.

138 "refuge in a Garden": AH to Richard Peters, 29 December 1802, *PAH*, vol. 26, 69.

138 "horticultural or rural Persuits": Richard Peters to AH, 8 January 1803, *PAH*, vol. 26, 74.

138 "helm of the UStates": quoted by Wulf 2011, 70n.

139 AH stops at Elgin: James A. Hamilton 1869, 3; Chernow 2004, 642–43.

139 sketch inspired by Hosack: Chernow 2004, 643; AH, "Plan for a Garden," *PAH*, vol. 26, 182.

139 "a few waggon loads": AH, "Plan for a Garden," *PAH*, vol. 26, 182.

139 *ha-ha*: see O'Malley 2010, 343.

139 "a few dogwood trees": AH, "Plan for a Garden," *PAH*, vol. 26, 182, quoted by Chernow 2004, 643.

139 "interfere with the hot bed": AH, "Plan for a Garden," *PAH*, vol. 26, 183.

139 AH support of Pinckney: Chernow 2004, 616–18.

139 watermelon and muskmelon: AH to Charles Cotesworth Pinckney, 29 December 1802, *PAH*, vol. 26, 71; Pinckney to AH, 6 March 1803, *PAH*, vol. 26, 92.

140 *Erythrina herbacea*: Pinckney called it "coral shrub"; see Pinckney to AH, 6 March 1803, *PAH*, vol. 26, 92.

140 planted thirteen sweetgums: Chernow 2004, 643; the last tree was cut down in the early twentieth century (American Scenic and Historic Preservation Society 1909, 81).

141 "Flora does not receive": DH to Vahl, 1 March 1804, University of Copenhagen, Botanical Garden Library.

141 plants collected and locations: see annotations in copy of Hosack 1806, NYBG.

142 "an artificial alphabet": DH to Vahl, 1 March 1804, University of Copenhagen, Botanical Garden Library.

143 *Anemone quinquefolia* and other plants: DH-MB, 1–2 (garden section).

143 "very astonishing progress" and "delight in Botanic pursuits": DH to Vahl, 1 March 1804.

143 when he needed help: Beginning around 1798, DH made notations in his copy of *Systema Naturae* (held by NYBG), keying them to plants he found in Curtis's *Botanical Magazine*.

144 "quarters of the globe": *New-York Evening Post*, 29 June 1803.

144 "first attempt of the kind": *New-York Evening Post*, 29 June 1803.

144 description of greenhouse: DH to Thomas Parke, 25 July 1803, Ch.C.11.76.77, Boston Public Library, Rare Books and Manuscripts.

145 William Hamilton: on The Woodlands, see esp. O'Malley 1992[b], 1998, 2010, and 2011, and Chesney 2014. TJ on The Woodlands: Wulf 2011, 43.

145 FAM shot: Savage and Savage 1986, 104.

145 DH greenhouse nearly identical to Woodlands: DH to Parke, 25 July 1803.

145 "miniature hill": quoted by O'Malley 2010, 359.

145 "catalogue": on Bartram's catalogues, see Fry 1996.

146 new City Hall: the cornerstone was laid on 26 May 1803 by Mayor Livingston (INPS, vol. 1, 397).

146 McComb and Mangin: *Morning Chronicle*, 6 October 1802.

146 Clinton returns: Burrows and Wallace 1998, 330; Cornog 1998, 49–50.

147 Michaux's aborted trip: Ambrose 2005, 37.

147 recent British expedition: Meacham 2012, 370; Ambrose 2005, 74.

148 "whose virtue and talents": Meriwether Lewis to Ferdinand Claiborne, 7 March 1801; quoted by Ambrose 2005, 60.

148 TJ as gardener: see esp. Wulf 2011 and Hatch 2012.

148 "Were I to venture to describe": Jefferson 1787, 59.

148 "particular plants": TJ to Meriwether Lewis, 20 June 1803, LC-TJ.

149 first botany textbook: Barton 1803. Barton was also bringing out a medical textbook in parts (Barton 1801).

149 Barton's health: Ambrose 2005, 91.

149 Wistar had sent Jefferson: Caspar Wistar to TJ, 1 January 1802; TJ to Caspar Wistar, 14 July 1802, LC-TJ.

150 Wistar alerted TJ: Caspar Wistar to TJ, 8 January 1802, LC-TJ.

150 traversing territory: Meacham 2012, 387; Ambrose 2005, 101.

150 Lewis met up with Clark: Ambrose 2005, 117.

150 May 22, 1804: Ambrose 2005, 123.

150 two volumes: Miller 1779, 1789.

CHAPTER EIGHT: *"H—k Is Enough, and Even That Unnecessary"*

151 ostrich landed: *New York Morning Chronicle*, 7 May 1804.

151 his first name has not survived: *SPCWP* gives Delacoste's first name as Isaac, but another source gives it as "J.-B." (possibly for Jean-Baptiste). Delacoste signed publicity documents for his exhibits simply as "Monsieur Delacoste."

151 ostrich feathers on ladies' hats: *New York Daily Advertiser*, 11 May 1804.

151 first scientific museum: an earlier museum of art, history, and natural history, founded in 1791 by John Pintard, had, after 1795, shed its scientific focus in favor of sensational (and sometimes fraudulent) items (Burrows and Wallace 1998, 316).

151 "hair-ball": Delacoste 1804, 78.

152 "first institution of *this nature*": *New York Daily Advertiser*, 25 June 1804.

152 Humboldt visits Wistar: Ewan and Ewan 2007, xxiv. On Humboldt's visit to the United States, see Walls 2009 and Wulf 2015.

152 "most extraordinary traveller" and subsequent quotations: CWP to John DePeyster, 27 June 1804, *SPCWP*, vol. 2, 725.

153 English, Spanish, and French: *SPCWP*, vol. 2, 683.

153 "The native Indians": *SPCWP*, vol. 2, 685.

153 cupola of the Capitol: *SPCWP*, vol. 2, 692.

153 "I only lament": TJ to CWP, 19 August 1804, LC-TJ.

154 "sumtious" dinner: *SPCWP*, vol. 2, 693.

154 "changed my Teeth": *SPCWP*, vol. 2, 694.

154 "his ardent wish": Humboldt quoted by CWP, *SPCWP*, vol. 2, 694.

154 as quickly as three days: *SPCWP*, vol. 2, 725.

154 Peale felt tempted: CWP to John DePeyster, 21 June 1804, in *SPCWP*, vol. 2, 718.

154 "little Syren": Peters to AH, 8 January 1803, *PAH*, vol. 26, 76.

154 "pay for the Rattle": Peters to AH, 8 January 1803, *PAH*, vol. 26, 74.

154 "3 barrels full of the *clay*": AH to Eliza Hamilton, 14 October 1803, *PAH*, vol. 26, 159.

155 "You see I do not forget": AH to Eliza Hamilton, 14 October 1803, *PAH*, vol. 26, 160.

155 "his house a most joyous one": James A. Hamilton 1869, 3.

155 "variegated landscape of hill & dale": John Church Hamilton, "Chapter 168," Box 20, John Church Hamilton Papers, CU, 3.

155 "the last sunny days": quoted by Chernow 2004, 693.

155 AH at dinner party: Chernow 2004, 680. On the duel, see also Sedg-

wick 2015, Fleming 1999, and Ellis 2000, chapter 1. On dueling in the early Republic, see Freeman 2007.

155 "more despicable": Charles D. Cooper to Schuyler, 23 April 1804, *PAH*, vol. 26, 246; quoted by Chernow 2004, 681.

155 "a prompt and unqualified acknowledgment or denial": AB to AH, 18 June 1804, *PAH*, vol. 26, 243.

155 joined Pendleton's law practice: Longworth 1803, 234.

156 moved to 65 Broadway: *New-York Evening Post*, 11 August 1803.

156 "You can't think": AB to TB, 24 June 1804, AB Memoirs, vol. 2, 290.

156 "mistress of natural philosophy": AB to TB, 24 June 1804, AB Memoirs, vol. 2, 290.

156 drafted words of challenge: AB had not yet issued the final formal challenge to AH but had written a letter to him on 22 June 1804 that included these words: "Thus, Sir, you have invited the course I am about to pursue, and now by your silence you impose it on me" (quoted by Chernow 2004, 688). According to *PAH*, vol. 26, 256, note 1, this letter was never shown to AH.

156 CWP began painting on June 24: *SPCWP*, vol. 2, 728.

156 "alegator" and "*Rapacious paws of the War Hawks*": *SPCWP*, vol. 2, 727.

157 "so interesting": Humboldt to DH (in French), 25 June 1804, David Hosack Correspondence, APS, microfilm 842; reprinted in AvH VS, 98–99. Humboldt also mentions in this letter having read and appreciated essays by SLM.

157 "as good a portrait": *SPCWP*, vol. 2, 728. See also Walls 2009, 106.

157 sixty thousand plant specimens: Wulf 2015, 111.

157 "I love seeing a Secretary of State": Humboldt to Madison (in French), 27 June 1804, AvH VS, 102.

157 "the simplicity of a philosopher": quoted by Wulf 2015, 101.

158 "without enjoying the pleasure": Humboldt to AH (in French), 27 June 1804, *PAH*, vol. 26, 274.

158 AB challenged AH: Van Ness to Pendleton, 27 June 1804, *PAH*, vol. 26, 272–73; Chernow 2004, 689.

158 "progressive rise": *PAH*, vol. 26, 289.

158 AH paid DH: *LPAH*, vol. 5, 621.

158 "advantages of multiplying writings": *SPCWP*, vol. 2, 733.

159 "emotions of gratification": *SPCWP*, vol. 3, 22.

159 "Green & thickest foliage": *SPCWP*, vol. 2, 730.

159 wait for a gap: *SPCWP*, vol. 2, 730.

159 "well arranged Museum": CWP to Rubens Peale and Sophonisba Peale Sellers, 6 July 1804, *SPCWP*, vol. 2, 737.

159 "linian System" and "Ribons": *SPCWP*, vol. 2, 731.

159 previous two months: Delacoste had launched his appeal in the *New-York Evening Post* by 2 May 1804.

159 subscribers at $2: Delacoste 1804, 81–84.

159 "society of Gentlemen": *SPCWP*, vol. 2, 731–32.

160 "last critical scene": AH to Pendleton, 4 July 1804, *PAH*, vol. 26, 294.

160 "my very dear Eliza": quoted by Chernow 2004, 709.

160 subscribers at $50: *Commercial Advertiser*, 7 July 1804.

160 both principals to bring physicians: see Pendleton, 4 July 1804, *PAH*, vol. 26, 295.

160 "*but* get on": AB to Van Ness, 9 July 1804, *PAH*, vol. 26, 300.

160 "H——k is enough": AB to Van Ness, 9 July 1804, *PAH*, vol. 26, 301; quoted by Chernow 2004, 696.

160 throwing away his shot: on AH's plan to do this, see Ellis 2000, 28–30; Chernow 2004, 694. On Pendleton's communication of this plan to DH, see DH to William Coleman, 17 August 1804, *PAH*, vol. 26, 345.

161 Van Ness carried an umbrella: Chernow 2004, 703.

161 chestnut, hickory, and oak: Slowick 2004, 24.

161 these and following species on and near the Hudson's New Jersey shore: see annotations in Hosack 1806 (copy held by NYBG) and Torrey 1818–1820. See also Torrey et al. 1819.

161 DH left behind: Chernow 2004, 701.

161 DH notices that several seconds elapsed: Ellis 2000, 29, quoting Van Ness, *PAH*, vol. 26, 335.

162 "countenance of death": DH to Coleman, 17 August 1804, *PAH*, vol. 26, 344, quoted by Chernow 2004, 704.

162 "mortal wound": AH quoted in DH to Coleman, 17 August 1804, *PAH*, vol. 26, 344.

162 "My vision is indistinct": AH quoted in letter of DH to Coleman, 17 August 1804, *PAH*, vol. 26, 345.

162 "asked me": DH to Coleman, 17 August 1804, *PAH*, vol. 26, 345.

162 "alone appeared tranquil": DH to Coleman, 17 August 1804, *PAH*, vol. 26, 346.

163 "almost intolerable": DH to Coleman, 17 August 1804, *PAH*, vol. 26, 346.

163 "beloved wife and children": DH to Coleman, 17 August 1804, *PAH*, vol. 26, 347.

163 "*Remember, my Eliza*": AH quoted in DH to Coleman, 17 August 1804, *PAH*, vol. 26, 347; quoted by Chernow 2004, 706.

163 "not be very pleasant": AB to Van Ness, 11 July 1804, *PAH*, vol. 26, 341, note 1.

163 gathering on street corners: INPS, vol. 5, 1425.

164 "opened his eyes": DH to Coleman, 17 August 1804, *PAH*, vol. 26, 347.

164 "Mr. Burr's respectful Compliments": AB to DH, 12 July 1804, *PAH*, vol. 26, 312.

164 "Mr. Burr begs": AB to DH, 12 July 1804, *PAH*, vol. 26, 312.

164 Church bells tolling: *New-York Gazette*, n.d., quoted by Coleman 1804, 59.

164 "O America!": *Frederick Town-Herald*, n.d., quoted by Coleman 1804, 97.

164 "What do you think": *Republican Watch-Tower*, 28 July 1804.

164 "the ball struck": DH to Coleman, 17 August 1804, *PAH*, vol. 26, 344.

165 "Our Troy": *Boston Repertory* [1804], quoted by Coleman 1804, 236.

165 AH's funeral: Chernow 2004, 710–13; see also Coleman 1804, 29–46.

165 "Citizens in general": Coleman 1804, 37.

165 live sheepshead fish: *SPCWP*, vol. 2, 734–35.

165 "Alexr. Hambleton will be buried": *SPCWP*, vol. 2, 735.

166 "well digested plan": *SPCWP*, vol. 2, 735.

166 forty-eight minutes: Coleman 1804, 46.

166 Morris decided: Chernow 2004, 712.

166 "It seemed as if God": Morris quoted by Coleman 1804, 42.

166 "*Would Hamilton have done this thing?*": Morris quoted by Coleman 1804, 45.

167 "unlawfully wilfully wickedly": "The People v. Aaron Burr," 14 August 1804, *PAH*, vol. 26, 342.

167 "I attended him": *PAH*, vol. 26, 347, note 4.

167 "*When will incorruptible Faith*": translation given in *PAH*, vol. 26, 347, note 3, which points out that DH got a word wrong in the quotation, writing *quidem* instead of *bonis*.

167 "The design is": Oliver Wolcott Jr. to James McHenry, 16 July 1804, quoted by Steiner 1907, 531.

167 purchasers: Mayer and East 1937. I am indebted to Christine McKay of BNY Mellon for providing copies of the certificates purchased by Hosack.

167 fund kept secret: Chernow 2004, 725.

167 "even Burr himself": quoted by Coleman 1804, 213.

168 "a MONSTER, and an ASSASSIN": Coleman 1804, 214.

168 murder charge: Chernow 2004, 717–18.

168 AB sells Richmond Hill: AB to Joseph Alston, 5 November 1804, reported he has sold the house and furniture; quoted in INPS, vol. 5, 1427.

168 "lancet might be useful": AB to BR, 3 August 1804, CU-AB, Series I, Reel 5; AB arrived in Philadelphia by 24 July 1804 (Chernow 2004, 717). See also Kline, ed., *Political Correspondence and Public Papers of Aaron Burr*, vol. 2, 891ff.

168 "Procure and read it": AB to TB, 15 September 1804, in AB Memoirs, vol. 2, 342.

168 "bear all the cost": CWP to DH, 29 June 1805, *SPCWP*, vol. 2, 858.

168 *"None of us think we will die"*: CWP to DH, 17 July 1805, *SPCWP*, vol. 2, 865.

CHAPTER NINE: *"This Delicious Banquet"*

169 return of yellow fever to New York: *Albany Register*, 10 July 1804.

169 "reduced to skin & Bone": MH to CWB, 8 October 1804, APS-CWB.

170 Clinton promised: Vail 1954, 30.

170 commissioned Trumbull: Common Council Minutes, vol. 3, 636.

170 Hosack's portrait of Hamilton: On 15 November 1806, John Trumbull noted in his account book that he received "[$]140 from Dr. Hosack for a portrait" (Series III, "Personal and Family Papers, Account Book, 1804–1806, Folder 92), John Trumbull Papers, Yale University. Today the National Gallery owns a Hamilton portrait painted by Trumbull ca. 1806 and passed down from the Hosack family.

170 "Being unsetled": AB to TB, 5 November 1804, CU-AB, Series I, Reel 5.

171 Burr on the dais: Chernow 2004, 718.

171 "a citadel of law": quoted by Chernow 2004, 719.

171 "burst into tears": quoted by Isenberg 2007, 279.

171 "a zeal and ability": quoted by Hosack 1811d, 11.

172 Hosack's purchases and his hiring of men in spring and summer 1805: DH-MB, 29ff. (garden section).

174 botanical terminology: Hosack 1811d, 12.

174 "attendance &c": *PAH*, vol. 26, 347, note 4.

175 Clark inscribed his name: 3 December 1805, Lewis and Clark Journals (Moulton, ed., 2003, 296). Hosack's peas "to be uncovered when the weather is fine": DH-MB, 37 (garden section).

175 running ad: *New-York Gazette*, 28 December 1805.

175 "individual fortune": quoted by Hosack 1811d, 12.

175 DH selling produce: DH-MB, 16, 26 (garden section).

176 first comprehensive garden manual: O'Malley (1992[a], 425–26) notes,

however, that portions of McMahon's book were direct appropriations from British gardening publications.

176   McMahon's wife and seed store: Hatch 1993, n.p.

176   "Ceres, and Flora, and Pomona": Anon., *Medical Repository* 4 (1807), 178.

176   Jefferson loved McMahon's book: Hatch 1993.

176   Hosack bought his own copy: Vail 1900, 24.

177   "lively, warm, steamy quality": McMahon 1806, 4.

177   cabbage and parsley: DH-MB, 37 (garden section).

177   March 21, 22: DH-MB, 37 (garden section).

177   April planting: DH-MB, 37 (garden section).

178   names of fruits: DH-MB, 2 (garden section).

178   Elgin fruit trees: Hosack 1806.

178   "exhausted by moss": McMahon 1806, 38–39.

178   fruit trees pruned: DH-MB, 37 (garden section).

178   looking forward: Hosack 1810a, 3.

178   Hosack's correspondents: Hosack 1806, 6–7.

178   long flues: DH to Thomas Parke, 25 July 1803.

179   dozens of species: all greenhouse species names in this section are drawn from Hosack 1806.

179   "splendid petals": Hosack 1810a, 3.

179   "cool and unsteady": DH-MB, 37 (garden section).

179   McMahon's advice: McMahon 1806, 80, 83.

179   almost entirely of glass: Hosack 1806, 4.

180   snowdrops, crocuses, etc.: Hosack 1810a, 4.

180   "evil of great magnitude": McMahon 1806, 92.

180   "raise a comforting steam": McMahon 1806, 97.

180   cinnamon, ginger, pineapple, etc.: all hothouse plants mentioned in this section are from Hosack 1806.

180   "zest to this delicious banquet": Hosack 1810a, 4.

182   hot poultice: Coxe 1810, 382.

182   antispasmodic: Coxe 1810, 304–5.

182   "immoderate discharges": Lewis 1796, 145.

182   writings of explorers and traders: Hosack's own writings rarely mention Native American medical and botanical knowledge. Joyce Chaplin (2003, 87) argues that many early Republic natural historians avoided calling attention to Native American knowledge, "lest this contradict claims of cultural superiority to and political authority over native peoples." On uses of medicinal plants by Native Americans, see, e.g.,

Huron H. Smith 1928; Speck 1937; Taylor 1940; Turner and Bell 1973; Hamel and Chiltoskey 1975.

183 "prince of good fellows": Gross 1893, vol. 2, 91.

183 "Bodily & Mental Inferiority": Original manuscript of 27 February 1808 in the John W. Francis Papers, NYPL. On Francis's views and a court case involving DH, SLM, and Alexander Whistelo, DH's coachman, see Wilder 2013, chapter 7.

183 "member of our family": MH to CWB, 23 January 1817, APS-CWB.

184 "wealth of John Jacob Astor": JWF 1858, 86.

184 his exotics: DH-MB, 39 (garden section).

184 pleasant, mild days: [no author], "Sketch of the Weather and Diseases in the Summer and Autumn of 1806," *Medical Repository*, Hexade 2, vol. 4 (1807): 214.

185 Napoleon reportedly preferred: Motte 1956, 11. On the scientific life of Napoleonic France, see Gillispie 1980.

185 camel and plague: Motte 1956, 12.

185 *Ulva fasciata*: Delile 1813.

186 French commissioner: Ewan and Ewan 2007, 560.

186 Josephine's instructions to ARD: Joly 1859, 7.

186 American bullfrog: Ewan and Ewan 2007, 562.

186 ARD studying with DH: DH-MB, n.p. (medical section).

187 mortality records: see, e.g., *New-York Medical and Philosophical Journal*, vol. 2 (1810), 134–38.

187 "child-bearing is a disease": Rush 1803, 27.

187 Jacob Schieffelin's shop: *New York Daily Advertiser*, 2 January 1804.

187 Hosack's medicinal species: All plant names in the following section are drawn from Hosack 1806. Unless otherwise noted, the medicinal uses are drawn from Andrew Duncan Jr. 1805, Coxe 1806, and Thacher 1810.

188 "urtication, or whipping": *Flora Lond.*, vol. 6, n.p.

188 *Aristolochia serpentaria* and anthrax: Hosack 1812a.

188 "admit a goose-quill": Hosack, "A Case of Trismus Cured by Laudanum," *Medical Repository* 6 (4): 386–88.

189 "very plentiful in Jersey": Bartram 1751, 4.

189 *Verbena urticifolia* uses: e.g., Huron H. Smith 1928, 251–52.

190 Hosack often cared for consumptives: Delile 1807, 40.

190 "nails . . . curve inwards": Delile 1807, 9.

190 "it manifestly did harm": DH quoted by Delile 1807, 41.

191 "lovely library": ARD to Joseph Philippe François Deleuze, 22 September 1806, quoted (in French) by Robbins 1964, 75.

CHAPTER TEN: *"I Long to See Captain Lewis"*

192 poplars along Broadway: INPS, vol. 1, 400.

192 "general panic": [Mitchill] 1806, 98–100.

193 "violent internal pains": *Commercial Advertiser*, 3 July 1806.

193 "to observe the worm": John P. Van Ness to TJ, 5 July 1806, Coolidge Collection of Thomas Jefferson Manuscripts, Massachusetts Historical Society.

193 the term *arboretum*: O'Malley (2010, 93) notes an appearance of the term by 1832.

193 "How differently nations act!": [no author] Review of *History of the American Oaks*, by André Michaux, *Medical Repository* 6 (1) (1803): 64–70, quotations on 65–66.

193 "our native citizens": [no author] Review of *History of the American Oaks*, by André Michaux, *Medical Repository* 6 (1) (1803): 64–70, quotation on 65.

194 "above all the trees": ARD to Joseph Deleuze, 22 September 1806, quoted (in French) by Robbins 1964, 74. My account of Hosack's tree species in this section comes from Hosack 1806.

194 "trouble you for a few Bermuda Cedars": DH to Augustus Harvey, 13 August 1809, DH-LB, 33r.

194 "the application of the Ice": CWP to Caspar Wistar, 13 November 1804, *SPCWP*, vol. 2, 780–81.

195 soon reporting to Wistar: CWP to Caspar Wistar, 23 June 1806, *SPCWP*, vol. 2, 972–73.

195 "I readily detached the artery": DH to ARD, 9 April 1808, DH-LB, 13r.

195 "I certainly look with very different eyes": DH to Amos Eaton, 30 August 1810.

195 "my young gentlemen": DH to BR, 5 December 1809, DH-LB, 51v.

196 native New Yorkers: Hosack 1806.

196 "cost me": Francis 1858, 31; quoted by Robbins 1964, 72.

196 down at the harbor: Entry of Merchandise, 8 October 1806, MS 71, NYAM.

196 NYBG herbarium: Thiers et al. 2016.

196 *Festuca fasciculata*: some botanists now classify this species as *Diplachne fusca* (L.) P. Beauv. subsp. *fascicularis* (Lam.); see Peterson et al. 2012.

197 gentlemen in Cambridge: *Boston Repertory*, 17 May 1805.

197 Rush was having as little luck: On 6 January 1806, he was informed by Dr. John Syng Dorsey, "a member of the committee appointed by the

legislature," that "as regards the Botanic Garden I have little hope—the present legislature are not exactly the giving kind" (quoted by Ewan and Ewan 2007, 553).

197 sent his new Elgin catalogue to President Jefferson: DH to TJ, 10 September 1806, LC-TJ.

198 "rich in vegetable productions": DH to TJ, 10 September 1806, LC-TJ.

198 Bartram and Barton declined: Flores, ed., 1984, 57, 59.

198 "a Cross and bad man": quoted by Flores, ed., 1984, 194.

199 "the gentlemen who are at present": DH to TJ, 10 September 1806, LC-TJ.

199 "compliments to Mr. Hosack": TJ to DH, 18 September 1806, LC-TJ.

199 "arrival of myself": quoted by Ambrose 2005, 406.

199 "penitrated the Continent": quoted by Ambrose 2005, 406.

199 "unspeakable joy": quoted by Ambrose 2005, 417.

200 "Robinson Crusoes": *New-York Herald*, 29 October 1806. On Lewis's reception in Washington, see esp. Ambrose 2005, 417–23.

200 "I long to see Captain Lewis": CWP to TJ, 10 February 1807, *SPCWP*, vol. 2, 1001.

200 "I have animals": quoted by Ambrose 2005, 433.

200 Lewis and Clark plant specimens: TJ to McMahon, 22 March 1807, quoted by Wulf 2011, 169.

201 "young man boarding": McMahon to Lewis, 5 April 1807, quoted by Cutright 1969, 359.

201 Pursh worked for Barton: Ewan and Ewan 2007, xxiv.

201 TJ and Barton: Ewan 1952, 606.

201 Lewis works with Pursh: Ambrose 2005, 433.

201 around one hundred fifty specimens: Pursh 1814, xi. For plants discovered by Lewis and Clark, see Cutright 1969, Appendix A.

202 Silliman on Pursh: quoted by Ewan and Ewan 2007, xiii.

202 "if he dont die drunk": quoted by Ewan and Ewan 2007, 46.

202 Lewis left Philadelphia in July 1807: Ambrose 2005, 441.

202 McMahon worried about Pursh: e.g., McMahon to TJ, 24 December 1809, FO-TJ.

CHAPTER ELEVEN: *"Strange Noises, Low Spirits"*

203 "everything that concerns New York": 8 April 1807, MH to CWB, APS-CWB.

203 Eliza Hamilton and orphanage: Longworth 1807; Chernow 2004, 728–29.

204 Mitchill's guidebook: Mitchill 1807.

204 descriptions of environs: Mitchill 1807, 180–87.

204 Gotham nickname: Burrows and Wallace 1998, 416.

205 "gods and goddesses": [no author] "The Stranger at Home; Or, a Tour in Broadway," *Salmagundi*, 24 June 1807, 268.

205 Irving and Knickerbocker: Irving 1809.

205 "Botanic Garden": Delile 1807, n.p.

205 "at Dr. Hosack's, N.Y.": ARD to TJ, 10 May 1807, LC-TJ.

205 "the objects which will employ him": TJ to ARD, 24 May 1807, LC-TJ.

205 By Hosack's estimate: Hosack 1811c, 6.

206 charter and garden: CU-CPS TM, 12 March 1807, 52.

206 DH dual positions: Robbins 1964, 77.

206 Elgin plants advertisement: *Commercial Advertiser*, 4 June 1807.

206 "grisly Bears": SLM to Catherine Mitchill, 9 November 1807, SLM Papers, Clements Library, University of Michigan; on the bears, see also Meacham 2012, 411; Sellers 1980, 207.

207 Clinton and deposition: *American Citizen*, 28 April 1806.

207 "electrical shock": quoted by Toll 2008, 279.

207 Fort Jay name: Yocum 2005, xix.

207 Hosack in the military: Jones 1805, 89.

208 effect on imports and exports: Burrows and Wallace 1998, 411.

208 "Not a box, bale, cask": quoted by Burrows and Wallace 1998, 411.

208 "Our embargo": quoted by Meacham 2012, 431; see also Beckert 2015, 158–59.

208 useful Elgin plants: species from Hosack 1806.

209 "I will not be neglected": Hosack to Romayne, 10 February 1808, Stookey 1967, 589–90.

209 DH resigned: Robbins 1964, 78–79.

209 Clinton visits to Elgin: e.g., DC Diary (July 1, 1802–November 6, 1810), N-YHS, entry for 24 August 1809.

209 Clinton's efforts and "mortifying": DC to John Stevens, 29 February 1808, CU-DC.

209 "highly honourable": quoted by Hosack 1811d, 16.

210 "great medical school": DH to John Warren, 22 June 1808, DH-LB, 16r.

210 "I wish every gentleman": DH to William Darlington, 27 June 1808, Jane Loring Gray Autograph Collection, Harvard University.

210 monument completed: *American Citizen*, 16 October 1806.

211 AB on ship: he left New York on 9 June 1808 and arrived in London in mid-July (Isenberg 2007, 371).

211 on AB's political activities and arrest: see esp. Chernow 2004, 720; Isenberg 2007, 307–16; Stewart 2011, 193–203; Lomask 1982.

211 "afflicting as unexpected": TB to Dr. William Eustis, 3 October 1808, FCAB.

212 "Amongst a collection": CWP to John Isaac Hawkins, 28 March 1807, *SPCWP*, vol. 2, 1010.

212 prolapsed uterus: TB to Eustis, 3 October 1808, FCAB. Unless otherwise noted, subsequent quotations on TB's health are from this memorandum.

213 DH advice to TB of May 1808: TB to Eustis, 3 October 1808, FCAB.

213 "general state of her health": DH to Alston, 12 June 1808, DH-LB, 12r.

213 "spring waters": AB to TB, undated, AB Journal Davis, vol. 1, 12.

213 AB on DH: In an undated note from AB to TB written shortly before he left New York for Europe, he told her to "order Dr. Hosack to be with you this evening" (AB Journal Davis, vol. 1, 13).

213 DH lent AB money: Lamb 1880, vol. 2, 540; Brown 1908, 4.

213 Bentham's hospitality: AB Journal Bixby, vol. 1, 3–4.

213 "dear little creature": Bentham quoted by AB in AB Journal Bixby, vol. 1, 5.

213 AB on WH: AB to "Mr. Gram," 17 September 1809, AB Journal Davis, vol. 1, 307.

214 TB to Ballston: DH to Alston, 12 June 1808, DH-LB, 12r.

214 "Our atmosphere": DH to William Currie, 13 July 1808, DH-LB, 16v.

214 "let me have": DH to TB, 20 August 1808, DH-LB, 28r.

214 TB turned elsewhere: TB writes from Ballston Spa to Eustis in Boston, 20 July 1808, FCAB. TB's appeal to Eustis thus predates DH's letter to TB of 20 August 1808. See also TB to AB, 1 February 1809, AB Journal Davis, vol. 1, 159: "You may recollect that I gave Hosack's preparation a trial. He is ignorant of my application to Eustis (so let him remain, I entreat), and advises me to continue taking his medicine."

214 AB and Eustis: Isenberg 2007, 162.

214 TB waiting: see TB to Eustis, 5 October 1808, FCAB.

214 "the cruelty": TB to AB, 31 May 1809, AB Journal Davis, vol. 1, 243.

214 "constant source of distress": AB to "Mrs. Prevost," 19 November 1808, AB Journal Davis, vol. 1, 97.

215 Lettsom: AB spelled it "Lettsome" in his letters and journal.

215 "Show this letter": AB to DH, 10 November 1808, AB Journal Davis, vol. 1, 81.

215 "Favour me also": AB to DH, 10 November 1808, AB Journal Davis, vol. 1, 81.

215 Tower of London: 27 November 1808, AB Journal Bixby, vol. 1, 20.

215 "What a feast": TB to AB, 31 May 1809, AB Journal Davis, vol. 1, 242.

215 "friend Bentham": AB to TB, 21 November 1808, AB Journal Davis, vol. 1, 100.

216 "American savages": AB to "Mr. Gram," 17 September 1809, AB Journal Davis, vol. 1, 307.

216 "Moheigungk": AB to Bentham, 23 January 1809, AB Journal Davis vol. 1, 167.

216 TB and DH hadn't received AB's letters: see TB to AB, 1 February 1809, AB Journal Davis, vol. 1, 158: "Your letters to the 24th November reached me two days ago. A voyage to join you at any season, and through any danger, would be a most delightful party of pleasure to me; but it is now impracticable."

216 DH visits TB daily: TB to AB, 3 January 1809, AB Journal Davis, vol. 1, 129.

216 "*I will not*": TB to AB, 1 February 1809, AB Journal Davis, vol. 1, 159.

216 AB on mercury: AB to TB, 15 February 1809, AB Journal Davis, vol. 1, 175.

216 late January: see TB to AB, 1 February 1809, AB Journal Davis, vol. 1, 158: "Your letters to the 24th November reached me two days ago."

216 "I presume": DH quoted by TB to AB, 31 May 1809, AB Journal Davis, vol. 1, 240. (AB appears not to have received this letter until much later; see Bentham to AB, 19 January 1811, AB Journal Davis, vol. 2, 436.)

216 "ignorant of my application": AB to TB, 1 February 1809, AB Journal Davis, vol. 1, 159.

216 okra and benne: TB to AB, 19 February [1809], AB Journal Davis, vol. 1, 163.

217 "Pray show": AB to TB, 5 May 1809, AB Journal Davis, vol. 1, 239.

217 "Many medicinal articles": Hosack 1811d, 18.

217 "with pain": Hosack 1811d, 16.

217 unfair to his family: Hosack 1811d, 25.

217 available organizational models: Johnson and Powell 2017. Chaplin (2003, 83) argues that the political structure of the new nation "guaranteed the slow development in the sciences" because of "the lack of wealthy and prestigious patrons for scientific enterprises, and the federal nature of American politics and society."

217 "first establishment of its kind": Hosack 1811d, 19.

218 "increase in value": Hosack 1811d, 19.

218 "erroneous impressions": Hosack 1811d, 22.

218 "public animosity": quoted by Robbins 1964, 81, note 11.

CHAPTER TWELVE: *"Such a Piece of Downright Imposture"*

219 "ornament of Manhattan": *Medical Repository,* April 1809, 400.

219 anxious letter: McMahon to TJ, 17 January 1809, LC-TJ.

219 Pursh's situation: McMahon to TJ, 17 January 1809 and 24 December 1809, LC-TJ; Pursh to Benjamin Smith Barton, 13 December 1808, Barton-Delafield Papers, APS. See also Cox 2004, 124.

220 "rash work": CWP to Rembrandt Peale, 17 November 1809, *SPCWP,* vol. 2, 1238.

220 "I never yet": McMahon to TJ, 24 December 1809, LC-TJ.

220 "bring us closer together": DH to BR, 3 November 1809, DH-LB, 51v; see also DH to Noah Webster, 29 August 1809, DH-LB, 36r.

221 "troublesome correspondent": DH to Thomas Dancer, 9 September 1809, DH-LB, 42r.

221 "purity of the atmosphere": DH to Thomas Schley, 19 July 1809, DH-LB, 31r.

221 wild pigeon, fish, and succotash: N-YHS Minutes, 4 September 1809, 24.

221 "our Dutch ancestors": DH quoted by Vail 1954, 366. On New Amsterdam, see Shorto 2004.

221 "an Indian cornfield": quoted by Vail 1954, 366.

221 "sensations of regret": Hosack 1811d, 22.

222 "botanic garden": quoted by Hosack 1811d, 23.

222 statement of support: Hosack 1811d, 25.

222 "If we suffer": Hosack 1811d, 30.

222 "pulse feeling": DH to [no first name] Inskeep, 30 November 1809, DH-LB, 54r.

222 AB's expulsion announced in South Carolina papers: e.g., *Carolina Gazette,* 7 July 1809.

222 "remained stupified": TB to AB, 1 August 1809, AB Journal Davis, vol. 1, 283.

223 early May: AB entry 6 May 1809, AB Journal Bixby, vol. 1, 105.

223 "perfectly wild and picturesque": AB entry 3 August 1809, AB Journal Davis, vol. 1, 254.

223 "New-York to Harlem": AB to TB, 13 October 1809, AB Journal Davis, vol. 1, 316.

223 Afzelius demonstrator at Botanic Garden: Skuncke 2014, 212.

223 Thunberg's travels: Skuncke 2014, 15. AB called him "Turnberg" in many journal entries.

223 "excited my curiosity": AB entry 12 August 1809, AB Journal Davis, vol. 1, 261.

223 "ten thousand things" and following quotations: AB entry 12 August 1809, AB Journal Davis, vol. 1, 260–61.

223 "grown from a nut": AB entry 22 August 1809, AB Journal Bixby, vol. 1, 206.

223 Linnaeus's house and garden: AB entry 16 August 1809, AB Journal Davis, vol. 1, 264.

224 "Where have I laid": AB entry 29 August 1809, AB Journal Bixby, vol. 1, 221.

224 *Views of Nature*: Wulf 2015, 132–33.

224 AB hoped to meet Humboldt: Isenberg 2007, 382.

224 "threw off my coat": AB entry 8 September 1809, AB Journal Bixby, vol. 1, 232.

224 AB to meet Humboldt: AB to "Mr. Hauterive," 10 September 1809, AB Journal Davis, vol. 1, 305.

224 "to see the people": AB to "Mr. Hauterive," 10 September 1809, AB Journal Davis, vol. 1, 305.

224 WH and Robinson: AB to "Mr. Gram," 12 September 1809, AB Journal Davis, vol. 1, 306.

224 Humboldt in Paris: Wulf 2015, 135–43. Humboldt had moved there in December 1807.

224 political reasons: Isenberg 2007, 381–82; Lomask 1982, 324–31.

224 Goethe and Wilhelm von Humboldt: AB entry 4 January 1810, AB Journal Bixby, vol. 1, 352.

225 "I suffer and freeze": AB entry 21 February 1810, AB Journal Bixby, vol. 1, 416.

225 ARD sees Humboldt and FAM: ARD to Wistar, 24 November 1808, Caspar Wistar Papers, APS.

225 AB sees Humboldt: AB entry 24 April 1810, AB Journal Davis, vol. 1, 447–48; slightly different language for same date in AB Journal Bixby, vol. 1, 438.

225 great lengths: e.g., AB entry 11 August 1810, AB Journal Davis, vol. 2, 24; AB entry 9 September 1810, AB Journal Davis, vol. 2, 38.

225 mild, wet days: Hosack 1824c, 334.

225 "a constant gale": DH weather observations for January 1810, in Hosack 1824c, 355.

225 Fulton's steamboat launched 1807: Stiles 2010, 42.

225 "perfect success": Hosack 1812b, 196.

226 "never ceasing jar": JE Diary, entry for 30 June 1810, 3.

226 "cannot by any means recommend it": JE Diary, entry for 30 June 1810, 4.

227 description of Capitol: Roseberry 1964, 14–19.

227 "Mrs. Hosack is recovering": JWF to DH, 7 February 1810, John W. Francis Papers, NYPL, Reel 2. Emily Hosack was born 31 January 1810.

227 botanical garden "inferior": 27 January 1810, quoted by Robbins 1964, 81.

228 description of interior: Wheeler 1993, 98.

228 "first establishment of this kind": Hosack 1811d, 32.

229 "relief of Doctor Hosack": *Columbian*, 6 October 1810.

229 "mere *science of hard* words": *Columbian*, 3 October 1810.

230 "*monstrum Americanum*": *Columbian*, 15 October 1810.

231 "nothing is more ridiculous": *Columbian*, 15 October 1810.

231 "vegetable riches": *Columbian*, 15 October 1810.

231 "valuable vegetables": *Columbian*, 15 October 1810.

232 Great Apollo: Clinton was known by the Latin version, *Magnus Apollo* (Cornog 1998, 6).

232 DC's interest in presidency: Cornog 1998, 88.

232 Clinton speech: see Hosack 1811d, 33.

232 DH on Clinton politics: DH to John Vaughan, 16 March 1811, DH-LB, 82r.

232 managers of lottery: New York State Comptroller Memorandum Book, NYPL, 23.

CHAPTER THIRTEEN: *"You Know, Better Than Any Man"*

233 clover seed: Hosack 1810b, 216.

233 oak and shrubbery: Hosack 1810a, 2.

233 appraisal and DH feeling: Hosack 1811d, 43.

234 Elgin lottery: Fox 1986.

234 hundreds more specimens: the following description of Hosack's expanded collections is drawn from Hosack 1811a.

236 Canada thistle: DH to SLM, 21 July 1810, reprinted in Hosack 1810b.

236 the Sharpleses' visit to Elgin and environs: Sharples 1810–1836 (entries for summer 1810).

237 lines quoted by DH: DH to Coleman, 17 August 1804, *PAH*, vol. 26, 347, note 3. Part of this monument is owned by the New-York Historical Society. A description of the monument published in the *New-York Gazette* on 2 December 1806 reads: "The Monument is in the form of

an obelisk on a pedestal 4 feet square, and nearly 3 feet above the base. The obelisk itself is composed of four pieces of white marble, 8 feet in length, and is surmounted by a flaming urn. . . ." The verse from Horace was inscribed on the back of the monument; on the front were recorded AH's name, the date of his death, and the notation that the monument was a gift from the St. Andrew's Society (*New-York Gazette*, 2 December 1806).

237 DH's membership in the St. Andrew's Society: Morrison Jr. 1906, 199.

237 "Tell Dr. Hosack": AB to Henry Gahn, 26 March 1810, AB Journal Davis, vol. 1, 442.

237 "style of magnificence": AB entry 9 August 1810, AB Journal Davis, vol. 2, 23.

238 "useful trees": AB entry 17 September 1810, AB Journal Davis, vol. 2, 43.

238 "a plant and a tree": AB entry 17 September 1810, AB Journal Davis, vol. 2, 43.

238 "You know": AB to DH, 20 December 1810, AB Journal Davis, vol. 2, 116; the request came to AB via his friend Madame Fenwick, a friend of Calmelet (AB entry 20 December 1810, AB Journal Davis, vol. 2, 107).

239 "estragore": This misspelling may have been introduced by Davis, editor of AB's journal, who occasionally made such mistakes (along with some intentional changes to the text meant to protect AB's reputation). AB to DH, 20 December 1810, AB Journal Davis, vol. 2, 116.

239 French merchant: *New York Diary or Loudon's Register*, 25 December 1793.

239 Hosack and Thouin shipment: Hosack 1811a, addendum, and Francis 1858, 30.

239 recipients of catalogue: DH-MB, n.p. (garden section).

240 first volume: DH to unknown recipient, 4 August 1810, DH-LB, 68v.

240 "much praise is due": DH quoted by McAllister 1941, 104.

240 "young gentleman of great genius": quoted by Robbins 1964, 66.

241 "American continent his GARDEN": No author, untitled review of Hosack 1811a and 1811d, *London Medical and Physical Journal*, vol. 26 (1811).

241 DH medicines for trip: noted in Clinton's 1810 diary for this trip, reprinted in Clinton 1849. My account of this trip is based on Clinton's canal journal and the diary of John Eddy (JE Diary).

241 Clinton recently removed: *Independent American*, 13 February 1810; Cornog 1998, 89.

244 "make DeWitt Clinton a member": DH to John Vaughan, 16 March 1811, DH-LB, 82r.

244 description of the weather: Hosack 1824c, 358.

244 no cases of fever: DH to Currie, 10 September 1810, DH-LB, 76v.

244 "frippery of a botanic garden!": *Columbian*, 11 June 1810.

244 "vile attempt": quoted by Robbins 1964, 83.

244 "blowing [on] coals": DH to Samuel Bard, 9 April 1811, DH-LB, 85v.

245 "very despicable opinion" and Mitchill's response: DH to unknown recipient, 4 August 1810, DH-LB, 68v.

245 "some precious things": DH to William Wilson, 7 October 1810, quoted by Robbins, 83.

245 estimates on garden: Hosack 1811d, 44–45.

245 "N. York cannot": DH to unknown recipient, 4 August 1810, DH-LB, 68v.

245 "immense importance": Hosack 1811d, 52.

245 Pursh leaves for West Indies: Pursh 1814, vol. 1, xiv.

246 DH accepts lower appraisal: Hosack 1811d, 46–48.

246 "too great an annual expenditure": quoted by Robbins 1964, 97.

246 DH move and motives: see Hosack 1811c.

247 "duties he owes to Columbia": Post et al. to CU trustees, 2 May 1811, Columbia College Records, Box 6.

247 "tigers concealed in their jungles": Hosack 1826, 4.

247 "continue to offend": quoted by Robbins 1964, 90.

247 DH resigns from Columbia: 6 June 1811, CU-CPS TM, vol. 1, 249–50.

247 "zeal for the propagation": 23 May 1811, CU-CPS TM, vol. 1, 309ff.

248 Pearl Street quarters and meeting at DH house: June–July CU-CPS TM, vol. 1, 244–53.

248 Tough and Dennison proposals: 3 June 1811, CU-CPS TM, vol. 1, 244ff.

248 DH on the drought: Hosack 1824c, 370.

248 "not to be sold": 1 July 1811, Inventory of Plants, Columbia College Records, Box 6.

248 needed blankets: Hosack 1824c, 370–71.

248 cracked panes: CU-CPS TM, vol. 1, 311.

249 "unfavorable to the publication": Pursh 1814, vol. 1, xiv–xvi; Ewan and Ewan 2007, xxv.

249 "least foundation": TJ to Bradbury, 21 March 1812, LC-TJ.

CHAPTER FOURTEEN: *"Instead of Creeping along the Earth"*

250 "I would gladly smoke": AB entry 2 April 1812, AB Journal Davis, vol. 2, 380.

250 true identity: AB entry 25 April, 1812, AB Journal Davis, vol. 2, 392.

250 "eat off the floor": AB entry 2 April 1812, AB Journal Davis, vol. 2, 380.

250 ink blotched: AB entry 4 April 1812, AB Journal Davis, vol. 2, 381.

250 "though half sick": AB entries 9–10 April 1812, AB Journal Davis, vol. 2, 384–85.

251 "J. Madison & Co.": AB entry 26 March 1812, AB Journal Davis, vol. 2, 368.

251 "their war-prattle": AB entry 26 March 1812, AB Journal Davis, vol. 2, 368.

251 Boston in early May: AB entry 4 May 1812, AB Journal Davis, vol. 2, 402.

251 AB on *Rose*: AB entry 30 May 1812, AB Journal Davis, vol. 2, 423.

251 "two vagabonds": AB entry 8 June 1812, AB Journal Davis, vol. 2, 434.

251 new forts: Yocum 2005, 47.

251 "a fierce little warrior": quoted by Yocum 2005, 49.

251 "battered down": *New-York Gazette*, 15 April 1812.

252 Proponents of war: Taylor 2011, 126–27.

252 "devastation and carnage": *New-York Spectator*, 20 June 1812.

252 "every once and a-while": *Public Advertiser*, 27 April 1812.

252 Clinton's political aspirations: Burrows and Wallace 1998, 424–25; Cornog 1998, 7, 96.

252 send his grandson: AB to TB, 9 May 1812, AB Journal Davis, vol. 2, 395–96.

252 "no more joy": TB to AB, 12 July 1812, AB Journal Davis, vol. 2, 439; quoted in part by Isenberg 2007, 387. Aaron Burr Alston was ten years old at the time of his death (Chernow 2004, 721).

253 AB after TB's death: Isenberg 2007, 387–88; Chernow 2004, 721.

253 "These animals": College of Physicians and Surgeons 1812, 18.

253 mild October days: On 15 October 1812, Randel recorded measurements for the new Fifth Avenue between Forty-Seventh and Forty-Eighth Streets, thus along the Middle Road on the east side of Elgin (John Randel Jr. Field Books, Pkge 61.2, Box 1, N-YHS). On Randel's mapping of Manhattan, see Holloway 2013. The weather in New York City on and around 15 October 1812 is recorded in DC's diary (November 6, 1810–November 8, 1815), N-YHS.

254 the Fifth Avenue: Klein states that "the year 1824 marked the official debut of the [Fifth] avenue with the acquisition of title to 13th Street" (Klein 1939, 288).

254 Rush's ministrations: Gordon-Reed and Onuf 2016, 318–19.

254 "Another of our friends of 76. is gone": TJ to Adams, 27 May 1813, LC-TJ.

254 "afflicting bereavement": DH to James Rush, 26 April 1813, DH-LB, 91v.

254 "whole odium": BR to DH, 20 June 1812, quoted by AEH 1861, 314–15.

254 "animates the face": DH to Sully, 16 April 1812, DH-LB, 90r.

254 "your city Hospital or University": DH to Sully, 16 April 1812, DH-LB, 90r.

255 Hosack confessed: DH to Julia Rush, 10 May 1813, DH-LB, 92v.

255 Rush's dying words: quoted by Hosack 1814, 53.

255 Irving's request for DH's eulogy: DH to Richard Rush, 21 December 1813, DH-LB, 104v.

255 bust of BR: Robbins 1964, 205.

255 CWE surgeon in 97th Regiment: Eddy 1930, 1188.

256 "execution in a nautical way": quoted by Robbins 1964, 93.

256 DH in debt: DH to Archibald McIntyre, 18 April 1816, DH Collection, N-YHS.

256 dissertation on boneset: Anderson 1813.

256 discovery was quickly picked up: see, e.g., Thacher 1813, xxii.

256 DH replaced by SLM: June 1812, CU-CPS TM, vol. 1, 353.

256 SLM and DH: Blatchford 1861, 30.

257 "exhausting my youth" and "a man drowning": 14 July 1813, CU-CPS TM, vol. 1, 395–96.

257 DH Elgin visit: 1 September 1813, CU-CPS TM, vol. 1, 400–405.

258 "mask of friendship": 8 September 1813, CU-CPS TM, vol. 1, 409.

258 "authorized Dr. Hosack": 8 September 1813, CU-CPS TM, vol. 1, 406–13.

258 *Camellia japonica*: [undated, but after August 1813] Dennison to JWF, John W. Francis Papers, NYPL, Box 3.

259 "You will find": quoted by Ewan and Ewan 2007, 547.

259 Pursh's new work: Pursh 1814.

259 scooping Barton: Ewan and Ewan 2007, 545; see also Cutright 2001, 20.

259 Henry Muhlenberg: Muhlenberg 1813; Eustis and Andrews 2015, 159–60.

260 "my remuneration": DH to JES, 28 August 1815, DH-LB, 125r.

260 "best work": DH to JES, 14 June 1817, LS-JES.

260 preserving Lewis and Clark's botanical discoveries: I am indebted to Joel T. Fry of Bartram's Garden for this observation.

260 officers of the new society: DH to James Kent, 8 February 1814, DH-LB, 105r. Clinton reappointed mayor: *Columbian*, 26 February 1814.

261 "where are the philosophers?": quoted by Cornog 1998, 68.

261 Morris and other members: *Transactions of the Literary and Philosophical Society*, vol. 1 (1815), v–viii, xiii–xiv.

261 "our next Governor": DH to Barton, 18 March 1815, DH-LB, 114v.

261 "master spirits" and following quotes from speech: Clinton 1815.

262 "He was indefatigable": AEH 1861, 311–12.

262 "the more a man": [Francis] 1835, n.p.

262 confined her to bed: MH to CWB, 15 May 1812, Bache Family Papers, Series 1, APS.

263 "dear little babe" and following quotations: DH to Bard, 4 June 1814, DH-LB, 107v.

264 war's effect on city: Burrows and Wallace 1998, 426.

264 almshouse figures for April: *Commercial Advertiser*, 7 May 1814.

264 new almshouse: Burrows and Wallace 1998, 502.

264 "*TO ARMS!!*": *Columbian*, 27 August 1814; quoted in part by Burrows and Wallace 1998, 428.

265 Plattsburgh: Taylor 2011, 402; impact of the defeat of Napoleon on British forces in North America: Taylor 2011, 413.

265 gained little: Wilentz 2005, 166–67.

265 news of Treaty of Ghent reaches New York: *Mercantile Advertiser*, 8 February 1815; Wilentz 2004 [1984], 23.

265 "Let me intreat": quoted by Vail 1954, 44.

265 Common Council: Common Council Minutes, 13 June 1815, vol. 8, 233.

266 SLM bragged: CWP diary entry 4 June 1817, *SPCWP*, vol. 3, 513.

266 "Citizens of New York": Common Council Minutes, 13 June 1815, vol. 8, 233.

266 "splendid zenith": Common Council Minutes, 13 June 1815, vol. 8, 235.

266 "City Hall": Pintard to Fitch Hall, 7 December 1812, John Pintard Papers, N-YHS.

266 Linnaean specimens: N-YHS Minutes, 8 July 1817, vol. 1, 100.

266 "one peppercorn": Robbins 1960, 300.

266 Peale's son Linnaeus: Sellers 1980, 222.

266 Inderwick's ship lost: *Analectic Magazine*, vol. 6 (1815), 23.

267 "Mr. Washington Irving": DH to JES, 27 May 1815, LS-JES.

267 "great men": DH to JES, 28 August 1815, LS-JES.

267 JES knighted: Kennett 2016, 280.

267 "I really rejoice": DH to JES, 28 August 1815, LS-JES.

268 "most ardent": JWF to DH, 15 November 1815, JWF Papers, NYPL.

268 "I can assure you": Townsend to JWF, 14 March 1816, JWF Papers, NYPL.

268 Banks on Lewis and Clark: JWF to DH, 26 May 1816, LC-TJ.

268 "own worst enemy": JWF to DH, 16 February 1816, JWF Papers, NYPL.

269 "doubtful species": JWF to DH, 16 February 1816, JWF Papers, NYPL.

269 "conspicuous plant": JWF to DH, 16 February 1816, JWF Papers, NYPL.

269 "highest honour": JWF to DH, 7 February 1816, JWF Papers, NYPL.

269 DH elected FRS: Royal Society 2007, n.p.; Banks to JES, 10 January 1818, in Banks 2007, vol. 6, 267.

269 "Mitchill is teased": DH to Pendleton, 23 August 1816, quoted by Robbins 1964, 158.

269 DH cares for Barton: Ewan and Ewan 2007, 836.

270 guide to Boston plants: Bigelow 1814. In 1817, Bigelow would begin publication of a multivolume work on American medicinal plants (Bigelow 1817–20).

270 "Every learned man": FAM to DH, 3 March 1817, Gratz Collection, Historical Society of Pennsylvania.

270 "medicaments for our bodies": TJ to Thomas Cooper, 7 October 1814, *RS-TJ*, vol. 8, 540.

270 "I could not pack": TJ to DH, 13 July 1816, LC-TJ.

270 "aid of manganese": DH to TJ, 1 August 1816, LC-TJ.

271 garden as campus: Brown 1908, 13.

271 "Source of Expense": 7 February 1818, CU-CC TM, vol. 2, 621.

271 "Carriage is in danger": 5 August 1815, CU-CPS TM, vol. 2, 128.

271 state of garden: 5 August 1815, CU-CPS TM, vol. 2, 127–29.

271 DH wants garden back: DH to TJ, 1 August 1816, LC-TJ.

272 Moore as CU trustee: McCaughey 2003, 92, 97.

272 "my Insolvency": 27 September 1816, CU-CPS TM, vol. 2, 235–36.

272 Gentle's visit: Gentle to Moore, 24 October 1816, Columbia College Records.

273 Gentle on Dennison: Gentle to DH, 21 October 1816, Columbia College Records.

273 "preservation of the plants": [n.d.] October 1816, DH to Moore, quoted by Robbins 1964, 98.

273 "I shall again": 20 October 1816, DH to Rawle, quoted by Robbins 1964, 98.

273 Dennison refused: 7 November 1816, CU-CC TM, vol. 2, part 2, 534.

273 "FOR SALE": *National Advocate*, 14 March 1817.

CHAPTER 15: *"Your Fortunate City"*

274 "like magick": entry 23 May 1817, *SPCWP*, vol. 3, 491.

274 "I told him": entry 23 May 1817, *SPCWP*, vol. 3, 492.

274 Peale's visits to New-York Institution: see *SPCWP*, vol. 3, 493–506.

275 "improvement of N. York": *SPCWP*, vol. 3, 524.

275 CWP on museums: *SPCWP*, vol. 3, 511.

276 "a great City": *SPCWP*, vol. 3, 521.

276 "father of Natural History": *SPCWP*, vol. 3, 508.

276 "Rich furniture": *SPCWP*, vol. 3, 509.

276 DH whispered: *SPCWP*, vol. 3, 528.

277 "Adam and eve": quoted by Wulf 2011, 194.

277 strong wind: *SPCWP*, vol. 3, 529.

277 "extraordinary virtue": *SPCWP*, vol. 3, 532.

277 Monroe's visit to New York: see *Papers of James Monroe*, vol. 1, 62–75.

277 entrusted DH: *Papers of James Monroe*, vol. 1, 64.

278 "your fortunate city": CWP to DH, 24 June 1817, CWP Letter Book, 173, Peale Family Papers, APS.

278 "the conservatory alone": Milbert 1828, vol. 1, 31.

278 "In politics all is calm": DC to DH, 3 April 1817, reprinted as frontispiece of Hosack 1829.

278 canal groundbreaking: Cornog 1998, 117.

278 "an eminent Botanist": DC to DH, 18 July 1817, CU-DC.

279 "highly gratifying": DH to JES, 25 July 1817, LS-JES.

279 "Every degree of Respect": Banks to JES, 10 January 1818, Banks 2007, vol. 6, 267.

279 first American edition: Andrew Duncan Jr. 1818, ed. Dyckman. Dyckman was a student of DH by 1810: DH-MB, 22 (medical section).

280 "young man": DH quoted by Britton 1900, 540.

280 "Broadway for garnets": quoted by Robbins 1968, 523.

280 "got my Sheepskin": Torrey to Eaton, 16 April 1818, quoted by Rodgers 1942, 23.

280 "notorious liar": Torrey to Eaton, 21 March 1818, quoted by Ewan 1952, 624.

281 Torrey's collecting trips: see Torrey 1818–1820.

281 Torrey confessed: Robbins 1964, 152.

281 "listen to Hosack": Eaton to Torrey, 6 May 1818, John Torrey Papers, NYBG.

281 very kind letter: JES to Torrey, 5 March 1821, John Torrey Papers, NYBG.

281 botany lectures: Torrey to Eaton, 23 March 1822, quoted by Robbins 1964, 154.

282 "N. York writers": Eaton to Torrey, 24 June 1820, John Torrey Papers, NYBG.

282 Valentine's Day 1818 payment: Robbins Mss.B.H78, APS.

282 "8000 dollars": W. Baits [no first name] to DH, 8 October 1819, Special Ms Collections, CU-CPS.

282 "Evil Consequences": 7 February 1818, CU-CC TM, vol. 2, part 2, 620.

282 "the real Value": 7 February 1818, CU-CC TM, vol. 2, part 2, 622.

282 trustees could lease: 5 April 1819, CU-CC TM, vol. 2, part 2, 665.

283 DH application: 5 June 1819, CU-CC TM, vol. 2, part 2, 688.

283 *"ornamental trees"*: quoted by Brown 1908, 16.

283 asylum foundation stone laid: *Commercial Advertiser*, 2 December 1818; asylum opens: *New-York Evening Post*, 30 May 1821.

283 DH involvement: Robbins 1964, 182.

283 Clinton informed: Vail 1954, 60.

283 bookshop selling: *Columbian*, 24 January 1820. All quotations in this section are from [Verplanck] 1820.

285 "I am glad": DC to Pintard, 27 January 1820, DeWitt Clinton Papers, NYPL.

285 "march of nations": DH 1820a, 8–9.

285 Humboldt elected: N-YHS Minutes, 9 May 1820, vol. 1, 194.

285 Banks and DH correspondence: see, e.g., DH to Banks, 28 October 1818, and 16 January 1819, in Banks 1958, and Banks's inscription to DH in copy of Ross 1819 held by the John Carter Brown Library, Brown University.

286 "admitted by all": DH to Banks, 28 October 1818 in Banks 1958; Hosack 1819, vol. 3, 257.

286 DH awarded medal: 5 January 1819, announced in *Transactions of the Horticultural Society* 1820, vol. 3, Appendix.

286 "weak and crazy": Bard quoted by Mitchill 1821, 38; deaths of Bard and wife: Robbins 1964, 174.

286 "cut down the seven hills": Moore 1818, 49–50; partially quoted by Holloway 2013, 118.

286   sewer system: Moore 1818, 32; DH letter reprinted 54–55.

287   "six or seven thousand": 7 January 1822, CU-CC TM, vol. 3, part 1, 810.

287   Jefferson continued: e.g., TJ to DH, 18 February 1818 and 12 July 1821, LC-TJ. See also TJ to Jonathan Thompson, 25 June 1821, quoted by Hyde 2016, 103.

287   plane trees, etc.: Hosack 1820b, 50.

287   Elgin plants at asylum: Sachsen-Weimar-Eisenach 1828, vol. 1, 195.

287   Gentle: N-Y Hort. Soc. Min., 7 November 1821, 2.

287   Horticultural Society: Mickulas 2002, 40.

287   "Celery" and "chocklate corn": 30 October 1821 and 28 January 1822, N-Y Hort. Soc. Min., 1, 5.

288   Astor's son: 24 August 1824, N-Y Hort. Soc. Min., 48. On membership, see Mickulas 2007, 17–20; and Laird 2014, 193.

288   gratitude to DH: see *New-York Farmer, and Horticultural Repository*, 1828, vol. 1, no. 3, 63.

288   *Camellia hosackia*: name noted in Floy Diary 24 December 1835, and published in Berlèse 1841–1843, vol. 1, plate 90. The species name is no longer in use.

288   On DH's involvement: DH 1824, 15; Robbins 1964, 122.

288   "depredations of deer": *New-York Farmer, and Horticultural Repository*, 1828, vol. 1, no. 2, 32.

288   "fine orchards of Long Island": Douglas Journal, 4.

288   "in ruins": Douglas Journal, 6.

289   Douglas missed Bartram: Bartram died on 22 July 1823, and Douglas arrived on 22 August (Douglas Journal, 8).

289   DC out of office: his term ended 1 January 1823 (Cornog, 143).

289   "saline element": *Essex Register*, 16 October 1823.

289   "state of perfection": Douglas Journal, 23.

290   good wine: SLM to DH, 3 September 1824, reprinted in Appendix to DH 1824b.

290   "Mary Hosack died": quoted by Robbins 1964, 162.

290   "your recent affliction": TJ to DH, 12 May 1824, LC-TJ.

290   "some of it's prominent characters": TJ to DH, 26 October 1823, David Hosack Correspondence, APS.

290   "so justly esteemed": TJ to DH, 23 August 1824, David Hosack Correspondence, APS.

290   "my sense of your eminence": TJ to DH, date unknown, David Hosack Correspondence, APS.

291   DH and Lafayette remarks: quoted by Vail 1954, 373–74. Foner (1998,

47) observes that in all the celebrations of Lafayette during his tour of the United States, "one subject was studiously avoided—the existence of slavery."

291   "your affectionate friend": Lafayette to DH, 11 January 1827, David Hosack Correspondence, APS.

292   successfully nominated: 24 August 1824, N-Y Hort. Soc. Min., 47–48.

292   Sykes's new coffeehouse: *National Advocate*, 26 July 1822.

292   "It is obvious that a garden": DH 1824b, 19.

293   "celebrated Elgin Garden": *New-York Evening Post*, 22 July 1824.

293   "youth of both sexes": DH 1824b, 21.

293   "secrets of nature": DH 1824b, 36.

294   John Francis and James Hosack toasts: *American Farmer*, 10 September 1824.

294   next meeting: 7 September 1824, N-Y Hort. Soc. Min., 53; DH to TJ, 29 September 1824, LC-TJ; DH to Madison, 29 September 1824, *The Papers of James Madison*, Library of Congress.

294   "I love the act": TJ to DH, 9 January 1825, LC-TJ. For Madison's acceptance: Madison to DH, 20 December 1824, *The Papers of James Madison*, Library of Congress.

CHAPTER SIXTEEN: *"Expulsion from the Garden of Eden"*

295   Grace Church: The church was then located at Rector Street and Broadway; it is now farther up Broadway, between Tenth and Eleventh Streets.

295   Irving wrote from Paris: Irving to DH, 30 May 1825, David Hosack Correspondence, APS.

295   Hosack, Coster: Robbins 1964, 166.

295   "His complexion is dark": Royall 1826, 266.

296   "greatest botanists": Royall 1826, 265.

296   "effort of genius": quoted by Robbins 1964, 207.

296   gilt frame: Robbins 1964, 206.

296   furnishings: "Inventory and Appraisal of Personal Property of David Hosack," DH Collection, N-YHS, 1836.

297   "incredible" number of people: Morris 1839, 120.

297   "crowd of literary men": Bryant to Frances Bryant, 23 March 1825, quoted by Muller 2010, 46.

297   "brilliant assemblies": Samuel Gross, quoted by Robbins 1964, 169.

297   "conversation animated": Samuel Gross, quoted by Robbins 1964, 169.

297   "zealously devoted": DH to JES, 14 November 1826, LS-JES.

297 Tocqueville and DH: Tocqueville and Beaumont also visited the Bloomingdale asylum with DH (Pierson 1996 [1938], 86).

297 "very famous" and "learned and pleasing man": The Duke of Sachsen-Weimar-Eisenach refers to Hosack as a "sehr berühmter Arzt" and "dieser gelehrte und gefällige Mann" in his memoirs (Sachsen-Weimar-Eisenach 1828, vol. 1, 193).

297 "remnant of paradise": quoted by Robbins 1964, 171.

298 white muscadine grapes: Robbins 1964, 175.

298 Erie Canal completion: Cornog 1998, 156–57.

298 names of boats: *Daily National Journal*, 11 October 1825.

298 DC remarks: Colden 1825, 320–21.

299 crowd estimate: Colden 1825, 122.

299 more than six thousand: Colden 1825, 122.

299 description of New-York Horticultural Society contingent: Colden 1825, 213.

299 parade and illuminations: Burrows and Wallace 1998, 430–31.

299 "brilliant sparks": pyrotechnics expert Richard Wilcox, quoted by Colden 1825, 270.

299 DH condolences: acknowledged by Thomas Mann Randolph Jr. to DH, 13 August 1826, Misc. Mss., Thomas Jefferson, N-YHS.

300 support for Elgin: DH to JWF, 14 [no month given] 1826, DH Collection, N-YHS.

300 botanical garden at Capitol Hill: The Columbian Institute was disbanded in 1837 and the federal government soon repossessed the land. In 1850, however, a new conservatory was built on the old site to safeguard the thousands of specimens brought back by the Wilkes Expedition, a globe-circling voyage led by Lieutenant Charles Wilkes Jr. from 1838 to 1842. (Titian Peale traveled with the Wilkes Expedition as a naturalist in charge of preserving the animal specimens.) In 1856, this establishment was officially named the United States Botanic Garden. In the 1920s and 1930s, following a redesign of the Mall led by Frederick Law Olmsted and Daniel Burnham, hundreds of the garden's trees were razed and the portable collections were moved south of the new sightline stretching from the Capitol to the Washington Monument and placed on the site that the United States Botanic Garden still occupies today (Solit 1993).

300 "everybody and everything": DH to JWF, 14 [no month given] 1826, DH Collection, N-YHS.

301 Peale still painting: *SPCWP*, vol. 4, 521.

301 Rubens, Rembrandt, Titian: Sellers 1980, 215; *SPCWP*, vol. 4, 543.

301 Mary Stansbury: *SPCWP*, vol. 4, 562.

301 "fish out of water": *SPCWP*, vol. 4, 561.

301 Peale accident and DH: Rubens Peale to Franklin Peale, 9 May 1826, *SPCWP*, vol. 4, 533.

301 "Happiness is certinly": CWP to Rubens Peale, 2 February 1827, *SPCWP*, vol. 4, 574.

301 "diffusion of knowledge": quoted by Ewing 2007, 317.

301 Smithson bequest arrives: Ewing 2007, 324.

301 "great National Museum": quoted by Sellers 1980, 231.

302 tickets to see a mermaid: Sellers 1980, 301.

302 rumor of DC death: Cornog 1998, 180.

302 "a mass of obesity": Francis 1858, 187.

302 sending him medical advice: Cornog 1998, 180.

302 "not afraid to die": quoted by Hosack 1829, 129.

302 audience gave up: 8 November 1828, in Asa Fitch Diary, vol. 1, 109, Yale University.

303 "Permit me Sir": Madison to DH, 28 May 1829, *The Papers of James Madison*, Library of Congress.

303 "trees and plants": Richard Rush to DH, 15 May 1827, reprinted in *New-York Farmer, and Horticultural Repository*, vol. 1, no. 2 (February 1828), 37.

303 next meeting of the New-York Horticultural Society: N-Y Hort. Soc. Min., 29 May 1827, 102.

304 "no public botanic gardens": DH to Richard Rush, 22 January 1828, reprinted in *New-York Farmer, and Horticultural Repository*, vol. 1, no. 2 (February 1828), 39.

304 "pride and ornament": DH to Richard Rush, 22 January 1828, reprinted in *New-York Farmer, and Horticultural Repository*, vol. 1, no. 2 (February 1828), 39.

304 DH and Columbia: 29 April 1828, N-Y Hort. Soc. Min.

304 CWE death: Robbins 1964, 172.

304 place of death: http://library-archives.cumc.columbia.edu/obit/cas par-wistar-eddy.

305 negotiations had failed: 29 July 1828, N-Y Hort. Soc. Min., 127.

305 "university and the Insurance companies": DH to Peter Du Ponceau, 21 June 1834, Gratz Collection, Historical Society of Pennsylvania.

305 DH and Cole encounter 30 May 1829: Parry 1988, 94.

305 species in Cole's *Garden of Eden* and Elgin: Kelly 1994, 27–29.

306 "banquet for hours": quoted by Kelly 1994, 36.

306 "commercial city": Cole to Gilmor, 20 April 1829, quoted by Parry 1988, 92.

306 "fine productions": quoted by Parry 1988, 93.

306 "wild scenery": Cole to Gilmor, 26 April 1829, quoted by Parry 1988, 93.

306 $400: Parry 1988, 93.

306 DH makes offer to Cole: Parry 1988, 93.

307 Cole complained: Cole to Gilmor, 31 May 1829, quoted by Parry 1988, 94.

307 returned in fall 1832: Powell 1990, 62.

307 *Expulsion* in Chambers Street townhouse: "Inventory of Estate" of Magdalena Hosack, N-YHS, 1841.

307 "*Elgin Garden 1829*": see http://sweetgum.nybg.org/science/vh/speci men_details.php?irn=1987314.

307 DH buys Hyde Park, October 1828: Robbins 1964, 173.

308 frozen Hudson: DH to William Bard, 7 March 1829, quoted by Robbins 1964, 176.

308 Magdalena fell in love: Jacob Harvey to his father, 23 July 1830, quoted by Robbins 1964, 180.

308 DH and estate expenses: Pintard to Eliza Pintard Noel Davidson, 10 September 1829, quoted by Robbins 1964, 179.

308 "My father-in-law": Jacob Harvey to his father, 23 July 1830, quoted by Robbins 1964, 180.

308 Trumbull's Niagara panoramas: Ferber 2009, 122–23.

308 paintings by Trumbull and Stuart: "Inventory of Estate" of Magdalena Hosack, N-YHS, 1841.

308 "connoisseur of fine art": Gross 1893, vol. 2, 90.

309 "The noble Hudson": Thacher 1830, Letter I, 149.

309 "not square and formal": Martineau 1838, vol. 1, 54.

309 ever-changing vistas: Parmentier 1828, 184–86. On Parmentier's place in the history of American garden literature, see O'Malley 1992[a], 426.

310 "how ridiculous": Parmentier 1828, 184.

310 acreage of Hyde Park: an advertisement in the *Evening Star* on 11 July 1836 put the acreage at seven hundred fifty; Thacher 1830, Letter II, put it at eight hundred.

310 "the finest seat in America": Downing 1841, 22. On Parmentier, Downing, and the ideals of the Picturesque style of landscape design, see Major 1997, 2014; Cooperman and Hunt 2016; O'Malley 2007; and Pauly 2007, 168–74.

310 "hardly possible": Trollope 1832, vol. 2, 238, quoted by Robbins 1964, 183.

310 estates nearby: Armstead 2012 offers an intimate portrait of the life of a gardener on the Verplanck family estate. On the Hudson estates, see also Lewis 2005.

310 "walking and riding": Hone Diary, vol. 1, 21.

311 "delightful accounts": Irving to DH, 29 May 1832, microfilm 842, David Hosack Correspondence, APS.

311 DH waiting: Martineau 1838, vol. 1, 73; Thacher 1830, Letter I, 148.

311 "truly enchanting": Thacher 1830, Letter I, 148.

311 "sumptuous style" and description of library: Wharton Diary, 24 July 1832, 149ff.

312 finest in the United States: O'Donnell et al. 1992, vol. 1, 29; Robbins 1964, 177.

312 "peep at a book": Downing to Torrey, 28 July 1834, John Torrey Papers, NYBG.

312 Weehawken with Hamilton: Martineau 1838, vol. 1, 56.

312 AH statue and DH: Hone Diary, 24 December 1830, vol. 1, 26.

313 "terrestrial paradise": [anon.] "Letter from a Tourist to the Editor of the American Farmer, dated Albany, July, 1829," *American Farmer*, vol. 11, no. 20, 153.

313 "make a man devout": Martineau 1838, vol. 1, 54.

313 Parmentier's borders: description given in Downing 1841, 373.

313 description of garden and conservatory: Thacher 1830, Letter II, 156.

314 "thousand other beauties": Wharton Diary, 31 July 1832, 158.

314 "remarkable": Martineau 1838, vol. 1, 75; "well filled range of hot-houses": Downing 1836, 101.

314 "wormwood, horehound": Sayers 1837, 327.

315 watermelons: [anon.] "Letter from a Tourist to the Editor of the American Farmer, dated Albany, July, 1829," *American Farmer*, vol. 11, no. 20, 153; citron melons: Wharton Diary, 15 July 1832, 144.

315 took specimens to the city: *New England Farmer, and Horticultural Journal* 8, no. 52 (July 16, 1830), 14. On the development of American horticultural products and practices in this era, see Kevles 2008, 2013; Pauly 2007; and Prentiss 1950.

315 "cooking apparatus" and livestock: Thacher 1830, Letter II, 156.

315 "sagacious about long horns": quoted by AEH 1861, 333.

315 stream and ponds: Thacher 1830, Letter II, 156.

315 "excel in fruits": Harvey to Joseph Harvey, 24 January 1830, quoted by Robbins 1964, 180.

315 "my bees": DH to Thacher, 11 February 1834, microfilm 842, David Hosack Correspondence, APS.

316   DH planned book: Wharton Diary, 9 July 1832, 140.

316   large check: Wharton Diary, 28 July 1832, 156.

316   sat talking: Wharton Diary, 20 July 1832, 147.

316   "withdraw from the *labour*": DH to Vahl, 1 March 1804.

317   JWF request for corrections: see DH to JWF, undated letter sent from Hyde Park, JWF Collection, N-YHS.

317   "If a party": DH quoted by [Francis] 1835, 14.

317   vine and fig-tree: GW to Lafayette, 1 February 1784, quoted by Chernow 2010, 462.

317   published: e.g., *Boston Daily Advertiser*, 18 February 1829.

CHAPTER SEVENTEEN: *"Like a Romance"*

318   lock of his hair: AEH 1861, 335.

318   "superb" fruits and vegetables: Hone Diary, 10 December 1835, vol. 1, 178.

319   below zero for days: *Evening Star*, 23 December 1835.

319   "flashes of lightning": Hone Diary, 16 December 1835, vol. 1, 181.

319   warehouse near the docks: The fire broke out near the corner of Exchange and Pearl Streets (Burrows and Wallace 1998, 596).

319   burning turpentine: *Farmers' Cabinet*, 1 January 1836; Burrows and Wallace 1998, 598.

319   a resident of Flatbush: Hone Diary, 22 December 1835, vol. 1, 188.

319   AH statue shattered: Hone Diary, 22 December 1835, vol. 1, 181.

319   blankets and flowers: *Farmers' Cabinet*, 1 January 1836.

320   "most awful calamity": Hone Diary, 17 December 1835, vol. 1, 180.

320   "singed almost everybody": Washington Irving to Peter Irving, 25 December 1835, quoted by Pierre Munro Irving 1864, vol. 3, 81; quoted by Jones 2008, 323.

320   insurance stock: I am grateful to Eric Hilt for information on DH's holdings; his losses were reported in *Farmer's Cabinet*, 1 January 1836.

320   DH stroke: my account is based on AEH 1861, Philip Hone's diary, and an account published in the *Morristown Jerseyman*, 30 December 1835.

320   DH collapsed: Hone Diary, 18 December 1835, 184.

320   across Chambers Street: Gross 1893, vol. 2, 88; Hone Diary, 19 December 1835, 185.

321   DH improved: Hone Diary, 23 December 1835, 188–89; *Jerseyman*, 30 December 1835.

321   "great man": quoted by *American Journal of Science* 29 (1836), 395.

321   "It is impossible": *Evening Star*, 23 December 1835.

321 "Christmas Day": Hone Diary, 25 December 1835, 189.

321 DH pallbearers: Hone Diary, 25 December 1835, 190.

321 "remains were followed": Francis 1858, 86.

322 "As a physician": Hunt 1836, 162.

322 "garden second to none": Hunt 1836, 161.

322 Astor bought it: Albee et al. 2008, 43. In the same years when Frederick W. Vanderbilt was building at Hyde Park, another grandson of Cornelius Vanderbilt, George Washington Vanderbilt II, was working with Frederick Law Olmsted on the landscape design for Biltmore, his estate in Asheville, North Carolina (Rybczynski 1999, 379–84).

322 DH plants for sale: *Evening Star*, 3 July 1838.

322 "still in Eden": Cole 1836, 12.

322 "If the Garden": quoted by Jones 2008, 385.

323 Shaw sold plants: *Commercial Advertiser*, 22 March 1834.

323 enjoyed a conservatory on the grounds: New-York Hospital 1845, 79.

323 city was laying out cross streets: Brown 1908, 18; entry for 7 October 1839, CU-CC TM, vol. 4, part 1, 2059.

323 fields of Elgin: see transcript of Butler's remarks, 1 November 1939, Buildings and Grounds Collection, CU.

323 college moved 1857: Klein 1939, 294.

323 turn a profit: Klein (1939, 294) notes that Columbia first spent an estimated $500,000 on "taxes, assessments, and opening of the property."

323 By 1870: Brown 1908, 19; Okrent 2003, 12.

324 Carnegie's house: West Fifty-First Street (Nasaw 2006, 293). On the increase in real estate values north of the old Elgin property because of Central Park, see Scobey 2002, chapter 7, and Rosenzweig and Blackmar 1992.

324 medical school: McCaughey 2003, 187. The deal was finalized in 1891.

324 insane-asylum property: Klein 1941, 68.

324 Olmsted pointed out: Olmsted to William Ware, 31 May 1893, Buildings and Grounds Collection, CU, Correspondence: President Low, 1889–1894, Series III: Morningside Heights Campus, Box 16, Folder 1. On the design and construction of this campus, see Bergdoll 1997 and Dolkart 1998.

324 Olmsted reportedly pronounced: see "Editorial," *University Bulletin* 12 (December 1895), 4. The Elgin provenance of the yews is noted in that editorial, in Underwood 1903, 279, and in an article in the *Philadelphia Inquirer*, 26 December 1914, on the impending death of Columbia's Elgin yews.

324 feeling cramped: Kolodin 1966, 18.

324 John D. Rockefeller Jr. residences: Okrent 2003, 46.

325 "reads like a romance": quoted by Okrent 2003, 51.

325 eleven acres: Okrent, 56; *New York Times*, 23 January 1929.

325 construction began: Okrent 2003, 187.

326 rivet ceremony: Okrent 2003, 393–94; *New York Times*, 2 November 1939.

326 "no other piece of land": transcript of Butler's remarks, 1 November 1939, 5, Buildings and Grounds Collection, CU. Over the next decades, roof gardens were laid out on a number of the Rockefeller Center buildings (Deitz 2011, 93–95).

327 $400 million: Dunlap 1993.

327 nearly $2 billion: Bagli 2000.

327 plaque: Okrent 2003, 405. The original plaque recognizing Hosack's work at Elgin was made of granite and placed in the flagstones; at a later date, a new plaque with the same wording was installed at one end of the retaining wall surrounding the Channel Gardens.

   The full text on this plaque reads:

In memory of David Hosack
1769–1835
Botanist, physician, man of science
and citizen of the world
On this site he developed
the famous Elgin Botanic Garden
1801–1811
for the advancement of medical research
and the knowledge of plants.

EPILOGUE

329 Burr stroke: Isenberg 2007, 403–4.

329 Burr reportedly: AEH report of AB reply quoted by Samuel W. Francis 1866, 216.

329 "I know you": ARD quoted by AEH 1861, 327.

330 "know his food": in original as *"qu'il distingue l'aliment du poison"* (Delile 1833, 33).

330 DH mentioned by Tocqueville: Tocqueville and his friend Beaumont had come to the United States to study the prison system, and Hosack, who had spent many years as an attending physician at the

state prison near the village of Greenwich, evidently had shared with them his opinions on the topic. In the study of the American prison system that Tocqueville and Beaumont published in France in 1833, they reproduced a letter on the practice of solitary confinement that Hosack (they spelled it "Hozack") had received in 1830 from a British prison reformer (Beaumont and Tocqueville 1833, vol. 5, item no. 24).

330 "most striking": quoted by Nisbet 2012, 104.

330 "scientific men": quoted by Robbins 1964, 160. On *Hosackia*, see Lindley 1829. Today, the plants called *Hosackia* are classified in the genus *Lotus*, in the *Fabaceae* (or bean) family.

331 "first spark of zeal": quoted by McAllister 1941, 220. By the year of DH's death, Samuel Thomson, a traveling healer from New Hampshire, and his followers had launched a medico-botanical movement of their own, organizing conventions and societies and publishing a journal. Thomsonian medicine was grounded in some of the same principles as DH's practice, but followers of Thomsonianism were often scathingly critical of "regular" medicine (Haller 2000).

331 boys and girls: Hosack 1824b, 21.

331 botany instruction: e.g., *Eastern Argus*, 6 May 1828; *Evening Star*, 26 August 1834; *Vermont Republican and Journal*, 11 June 1831.

331 DH's head gardeners: Sayers 1838 and Gentle 1841.

331 color lithographs in their seed catalogues: Kevles 2011, 2013; Eustis 2014.

332 Downing, Olmsted, Vaux: Rybczynski 2003.

332 Torrey elected president: Mickulas 2002, 28.

333 the Brittons: on the Torrey Botanical Club, their trip to Kew, and the founding of the NYBG, see Mickulas 2007.

333 Morgan's participation: Strouse 1999, 290.

333 officers of the NYBG: Mickulas 2007, 65.

333 landscape of the NYBG: see esp. Mickulas 2007 and Atha et al. 2016.

334 statue of Hosack: Shrady, ed., 1903, 47.

SOURCES AND
BIBLIOGRAPHY

*Note: The titles and dates of the historical newspapers and periodicals
I have consulted regarding particular events and people appear in
the Notes.*

MANUSCRIPT COLLECTIONS CONSULTED

American Philosophical Society
    Barton-Delafield Papers
    Caspar Wistar Papers
    Catharine Wistar Bache Papers
    Bache Family Papers
    David Hosack Correspondence
    David Hosack Letters and Papers
    Peale Family Papers

Archives nationales de France (Pierrefitte-sur-Seine)
    Muséum d'histoire naturelle, Série AJ/15

Bristol (England) Archives
    Sharples Family Papers

*Columbia University, A. C. Long Health Sciences Center, Archives and
    Special Collections*
    Trustees' Minutes, College of Physicians and Surgeons
    Student Notes on Hosack Lectures, 1815–1828

Columbia University, Rare Book and Manuscript Library
    Papers of Aaron Burr (27 microfilm reels)
    Columbia College Records (1750–1861)
    Buildings and Grounds Collection
    DeWitt Clinton Papers
    John Church Hamilton Papers
    Historical Photograph Collections, Series VII: Buildings and
        Grounds
    Trustees' Minutes, Columbia College

Duke University, David M. Rubenstein Rare Book & Manuscript Library
    David Hosack Papers

Harvard University, Botany Libraries
    Jane Loring Gray Autograph Collection

Historical Society of Pennsylvania
    Rush Family Papers, Series I: Benjamin Rush Papers
    Gratz Collection

Library of Congress, Washington, DC
    Thomas Law Papers
    James Thacher Papers
    Alexander Hamilton Papers

Linnean Society of London
    Minutes of the Linnean Society
    Correspondence of James Edward Smith
        http://linnean-online.org/smith_correspondence.html
    Linnean Herbarium
        http://linnean-online.org/view/type/specimen/

Muséum national d'histoire naturelle, Bibliothèque botanique
    Manuscrits de la bibliothèque de Botanique, Cryptogamie
    Collection d'autographes de botanistes constituée par Gustave
        Thuret et Edouard Bornet

New York Academy of Medicine
    David Hosack Letter Book
    David Hosack Memorandum Book

New York Botanical Garden, Mertz Library
    John Torrey Papers
    Minutes of the New-York Horticultural Society

New-York Historical Society
    Bard Family Papers
    DeWitt Clinton Papers
    DeWitt Clinton Diary, July 1, 1802–November 6, 1810
    DeWitt Clinton Diary, November 6, 1810–November 8, 1815
    John Pintard Papers
    John Randel Jr. Field Books
    John Wakefield Francis Collection
    Minutes of the New-York Historical Society
    David Hosack Collection, 1815–1836
    Miscellaneous Manuscripts, Thomas Jefferson

New York Public Library, Manuscripts and Archives Division
    Alexander Anderson Papers, ca. 1775–1870
    John Hartshorne Eddy Diary, 1810
    John W. Francis Papers
    DeWitt Clinton Papers
    New York State Comptroller Memorandum Book, 1799–1826
    Thomas Kelah Wharton Diary and Sketchbook, 1830–1834

New York Society Library
    Circulation Records for Alexander Hosack, 1789–1805

New York State Library (Albany), Manuscripts and Special Collections
    Thomas Cole Papers

Office of the Borough President of Manhattan
    Farm Maps of John Randel Jr.

Princeton University, Firestone Library
    David Hosack Collection
    Fuller Collection of Aaron Burr

Royal Society of London, Archives
    Journal Books of the Royal Society

Université de Montpellier I, Faculté de Médecine
  Fonds Moderne de la Faculté de Médecine, Administration
    Générale (1794–1981)

University of Michigan, William L. Clements Library
  Samuel Latham Mitchill Papers

Yale University
  John Trumbull Collection, Beinecke Library
  John Trumbull Papers, Manuscripts and Archives, Sterling
    Memorial Library
  Asa Fitch Papers, Manuscripts and Archives, Sterling Memorial
    Library

ADDITIONAL PUBLISHED OR ONLINE PAPERS CONSULTED

John Adams Papers, Founders Online, National Archives
  https://founders.archives.gov/about/Adams
The Political Correspondence and Public Papers of Aaron Burr. Edited by
  Mary-Jo Kline. 2 vols. Princeton, NJ: Princeton University Press, 1983.
Alexander Hamilton Papers, Founders Online, National Archives
  https://founders.archives.gov/about/Hamilton
The Law Practice of Alexander Hamilton: Documents and Commentary. Edited
  by Julius Goebel Jr., et al. 5 vols. New York: Columbia University Press,
  1964–1981.
The Papers of Alexander Hamilton. Edited by Harold C. Syrett. 26 vols. New
  York: Columbia University Press, 1961–1987.
Papers of John Jay, Columbia University
  https://dlc.library.columbia.edu/jay
The Correspondence and Public Papers of John Jay. Edited by Henry P. John-
  ston. 4 vols. New York: G. P. Putnam's Sons, 1893.
  http://oll.libertyfund.org/titles/jay-the-correspondence-and-public-
  papers-of-john-jay-4-vols
Thomas Jefferson Papers, Founders Online, National Archives
  https://founders.archives.gov/about/Jefferson
Papers of Thomas Jefferson, Retirement Series. Edited by Jeff Looney. 13 vols.
  Princeton, NJ, and Oxford: Princeton University Press, 2004–2016.
Papers of Thomas Jefferson, Series 1, General Correspondence, Library of
  Congress
  https://www.loc.gov/collections/thomas-jefferson-papers/

The Papers of James Madison, Library of Congress
    https://www.loc.gov/collections/james-madison-papers/
The Papers of James Madison, Digital Edition. Edited by J. C. A. Stagg. University of Virginia.
    http://rotunda.upress.virginia.edu/
The Papers of James Monroe. Edited by Daniel Preston and Marlena C. DeLong, assistant editor. 6 vols. Westport, CT: Greenwood Press, 2003–2017.
The Selected Papers of Charles Willson Peale and His Family. Edited by Lillian B. Miller, Sidney Hart, and Toby A. Appel. 5 vols. New Haven, CT: Yale University Press, 1983.

CITY DIRECTORIES

Webster, Noah, ed. *The New York Directory for 1786.* New York: Trow City Directory Company, 1886 [1786].
Duncan, William, ed. *The New-York Directory, And Register.* New-York: T. and J. Swords, 1792, 1794, 1795.
Longworth, David, ed. *The American Almanack, New-York Register, and City Directory.* New York: David Longworth, 1796.
Longworth, David, ed. *Longworth's American Almanack, New-York Register, and City Directory.* New York: David Longworth, 1797–1805.
Jones, John F., ed. *Jones's New-York Mercantile and General Directory.* New York: Printed for the editor, 1805.
Longworth, Thomas, ed. *Longworth's American Almanack, New-York Register, and City Directory.* New York: Thomas Longworth, 1805–1829.

PLANT DATABASES

C. V. Starr Virtual Herbarium, William and Lynda Steere Herbarium, New York Botanical Garden: sweetgum.nybg.org
International Plant Names Index: www.ipni.org
New York City EcoFlora: https://www.nybg.org/science-project/new-york-city-ecoflora/
The Plant List: www.theplantlist.org
United States Department of Agriculture Plants Database: www.plants.usda.gov

SELECTED WORKS OF DAVID HOSACK

[n.d.] "Notes on Midwifery." Manuscript, New York Academy of Medicine.

1791. *Inaugural Dissertation on Cholera Morbus.* New York: Samuel Campbell.

1794. "Observations on Vision." *Philosophical Transactions of the Royal Society of London* 84: 196–216.

1795. *Syllabus of the Course of Lectures on Botany, Delivered in Columbia College.* New York: John Childs.

1797. "Observations on the Yellow Fever, In a Letter, from Dr. David Hosack of New-York, to his friend in Philadelphia. August 28, 1797," reprinted in *New-York Magazine; Or, Literary Repository* 2, New Series [no issue number] (September 1797): 453–55.

1798a. "A Case of Hydrocele, Cured by Injection." *Medical Repository* 1, no. 3: 419–20.

1798b. "Singular Cases of Disease in Infancy." *Medical Repository* 1, no. 4: 507–10.

1801. *An Introductory Lecture on Medical Education Delivered at the Commencement of the Course of Lectures on Botany and Materia Medica.* New York: T. and J. Swords.

1804. "Lectures on Botany." Manuscript, New York Academy of Medicine.

1806. *A Catalogue of Plants Contained in the Botanic Garden at Elgin.* New York: T. and J. Swords.

1810a. "Description of Elgin Garden, the Property of David Hosack, M.D." *Portfolio and Boston Anthology* (January 1810).

1810b. "Botanical Description of the Canada Thistle or Cnicus Arvensis." *American Medical and Philosophical Register* 1, no. 2 (October 1810): 211–16.

1810–1814. *American Medical and Philosophical Register.* Edited by David Hosack and John W. Francis. New York: C. S. Van Winkle.

1811a. *Hortus Elginensis.* New York: T. and J. Swords.

1811b. *Observations on Croup or Hives, Addressed in a Letter to A. R. Delile.* New York: C. S. Van Winkle.

1811c. *Observations on the Establishment of the College of Physicians and Surgeons in the City of New York.* New York: C. S. Van Winkle.

1811d. *Statement of Facts Relative to the Establishment and Progress of the Elgin Botanic Garden, and the Subsequent Disposal of the Same to the State of New York.* New York: C. S. Van Winkle.

1812a. "A Case of Anthrax, Successfully Treated." *American Medical and Philosophical Register* 2 (April 1812): 388–92.

1812b. "An Account of the Powles Hook Steam Ferry-Boat, in a letter to Dr.

David Hosack, from Robert Fulton, Esq." *American Medical and Philo-sophical Register* 3 (July 1812): 196–203.

1814. "Eulogium on the Late Dr. Rush." *Analectic Magazine* 3 (January 1814): 47–53.

1815. *Remarks on the Treatment of the Typhoid State of Fever.* New York: Van Winkle and Wiley.

1816. *Syllabus of the Courses of Lectures on the Theory and Practice of Physic.* New York: Van Winkle and Wiley.

1819. "Some Account of the Seckle [*sic*] Pear." *Transactions of the Horticul-tural Society of London* 3: 256–58.

1820a. *Inaugural Discourse Delivered to the New-York Historical Society on the Second Tuesday of February, 1820.* New York: C. S. Van Winkle.

1820b. *Observations on the Means of Improving the Medical Police of the City of New-York.* New York: College of Physicians and Surgeons.

1824a. *Essays on Various Subjects of Medical Science.* 3 vols. New York: J. Seymour.

1824b. *Inaugural Discourse Delivered before the New-York Horticultural Society.* New York: J. Seymour.

1824c. "Observations on the Weather and Diseases of the City of New-York from January 1st 1810, to January 1st 1814." In *Essays on Various Subjects of Medical Science,* vol. 2, 334–463. New York: J. Seymour.

1826. *Observations on the Medical Character: Addressed to the Graduates of the College of Physicians and Surgeons of New York, at the Commencement, Held on the 4th of April, 1826.* New York: J. Seymour.

1829. *Memoir of De Witt Clinton.* New York: J. Seymour.

1830. "Address Delivered at the First Anniversary of the New York City Temperance Society, May 11, 1830." Appendix in *First Annual Report of the New York City Temperance Society,* 1–10. New York: Sleight and Robinson.

1838. *Lectures on the Theory and Practice of Physic.* Edited by Henry W. Duca-chet. 3 vols. Philadelphia: H. Hooker.

———

GENERAL BIBLIOGRAPHY

Primary Sources

American Academy of the Fine Arts. *Charter and By-Laws of the American Academy of the Fine Arts, Instituted February 12, 1802.* New York: David Longworth, 1817.

Anderson, Andrew. *An Inaugural Dissertation on the Eupatorium Perfoliatum of Linnaeus*. New York: C. S. Van Winkle, 1813.

Banks, Sir Joseph. *The Banks Letters: A Calendar of the Manuscript Correspondence of Sir Joseph Banks*. Edited by Warren Dawson. London: British Museum, 1958.

————. *Scientific Correspondence of Joseph Banks*. Edited by Neil Chambers. 6 vols. London: Pickering & Chatto, 2007.

Barton, Benjamin Smith. *Collections for an Essay towards a Materia Medica of the United States*. Philadelphia: Way and Groff, 1801.

————. *Elements of Botany: or Outlines of the Natural History of Vegetables*. Philadelphia: William F. M'Laughlin, 1803.

Bartram, John. "Mr. Bartram's Appendix." In *Medicina Britannica*, Thomas Short, 1–7. 3rd ed. Philadelphia: Benjamin Franklin, 1751.

Bartram, William. *The Travels of William Bartram*. Edited by Francis Harper. New Haven, CT: Yale University Press, 1958 [1791].

Bass, Samuel. "Lectures on the Theory and Practice of Physic by Dr. Hosack." Manuscript, College of Physicians and Surgeons, Columbia University, 1817.

Bayley, Richard. *An Account of the Epidemic Fever Which Prevailed in the City of New-York, during Part of the Summer and Fall of 1795*. New York: T. and J. Swords, 1796.

Beaumont, Gustave de, and Alexis de Tocqueville. *Du système pénitentiaire aux États-Unis, et de son application en France*. 6 vols. Paris: H. Fournier jeune, 1833.

Berlèse, Lorenzo. *Iconographie du genre Camellia*. 3 vols. Paris: H. Cousin, 1841–1843.

Bigelow, Jacob. *American Medical Botany*. 3 vols. Boston: Cummings and Hilliard, 1817–1820.

————. *Florula Bostoniensis*. Boston: Cummings and Hilliard, 1814.

Blatchford, Thomas. "Our Alma Mater Fifty Years Ago." Troy, NY: A. W. Scribner, 1861.

Bolton, James. *History of Funguses Growing about Halifax*. 3 vols. Huddersfield, England: Printed for the author, 1788–91.

Brissot de Warville, Jacques-Pierre. *New Travels in the United States of America, Performed in 1788*. Dublin: W. Corbet, 1792.

Burr, Aaron. *Memoirs of Aaron Burr: With Miscellaneous Selections from His Correspondence*. Edited by Matthew Livingston Davis. 2 vols. New York: Harper & Brothers, 1836–1837.

————. *The Private Journal of Aaron Burr, During His Residence of Four*

Years in Europe; with Selections from His Correspondence. Edited by Matthew Livingston Davis. 2 vols. New York: Harper & Brothers, 1838.

————. *The Private Journal of Aaron Burr, Reprinted in full from the original manuscript in the library of Mr. William K. Bixby of St. Louis, Mo.* Edited by William H. Samson. 2 vols. Rochester, NY: [Post Express Printing Co.], 1903.

Clinton, DeWitt. "An Introductory Discourse." In *Transactions of the Literary and Philosophical Society of New-York* 1 (March 1815): 21–77.

————. "Private Canal Journal, 1810." In *The Life and Writings of DeWitt Clinton*. Edited by William W. Campbell, 27–204. New York: Baker and Scribner, 1849.

Colden, Cadwallader D. *Memoir Prepared at the Request of a Committee of the Common Council of the City of New York*. New York: Printed by Order of the Corporation of the City of New York, 1825.

Cole, Thomas. "Essay on American Scenery." *American Monthly*, New Series 1, no. 1 (January 1836): 1–12. Edited by Park Benjamin and Charles F. Hoffman. Boston: Otis, Broaders, and Co.; New York: George Dearborn, 1836.

Coleman, William, ed. *A Collection of the Facts and Documents, Relative to the Death of Major-General Alexander Hamilton*. New York: Hopkins and Seymour, 1804.

College of Physicians & Surgeons. *Exposition of the Transactions Relative to the College of Physicians & Surgeons*. New York: College of Physicians & Surgeons, 1812.

Columbia College in the City of New York. "Editorial." *University Bulletin* 12 (December 1895): 1–7.

Common Council of the City of New York. *Minutes of the Common Council of the City of New York, 1784–1831*. 19 vols. New York: City of New York, 1917.

Coxe, John Redman. *The American Dispensatory*. 2nd ed. Philadelphia: Thomas Dobson, 1810.

Currie, William. *A Treatise on the Synochus Icteroides, or Yellow Fever; As It Lately Appeared in the City of Philadelphia*. Philadelphia: Printed by Thomas Dobson, No. 41, South Second-Street, 1794.

Curtis, William. *Assistant Plates to the Materia Medica*. London: Frys and Couchman, 1786.

————. *Botanical Magazine*. 14 vols. London: Couchman, 1787–1800.

————. *A Catalogue of the British, Medicinal, Culinary, and Agricultural Plants, Cultivated in the London Botanic Garden*. London: B. White, 1783.

————. *Flora Londinensis; or Plates and Descriptions of Such Plants as Grow Wild in the Environs of London.* 6 vols. London: B. White, 1777–1798.

————. *Lectures on Botany.* Edited by Samuel Curtis. 3 vols. London: William Philips/H. D. Symonds, 1803–1805.

————. *Linnaeus's System of Botany.* London: Printed for and sold by the Author and B. White, London, 1777.

————. *Practical Observations on the British Grasses.* London: Couchman, 1790a.

————. *Proposal for a Course of Herbarizing Excursions.* London: [no. pub.], 1792.

————. *Proposals for Opening by Subscription, a Botanic Garden, To be Called the London Botanic Garden.* London: J. Andrews, 1778.

————. *Subscription Catalogue of the Brompton Botanic Garden for the Year 1790.* London: Published by W. Curtis, 1790b.

————. *The Subscription Catalogue of the Brompton Botanic Garden for the Year 1792.* London: Published by W. Curtis, 1792.

Delacoste, [Jean-Baptiste?]. *Catalogue of the Natural Productions and Curiosities, Which Compose the Collections of the Cabinet of Natural History.* New York: Isaac Collins and Son, 1804.

Delile, Alire Raffeneau. *Flore d'Égypte, explication des planches.* Paris: Imprimerie Impériale, 1813.

————. *An Inaugural Dissertation on Pulmonary Consumption.* New York: T. and J. Swords, 1807.

————. *Leçon de botanique.* Montpellier, France: Auguste Ricard, 1833.

Douglas, David. *Journal Kept by David Douglas during his Travels in North America, 1823–1827.* London: William Wesley and Son, 1914.

Downing, Andrew Jackson. "Descriptive Notice of J. W. Knevels Esq.'s Collection of Exotic Plants at Newburgh, N.Y." *American Gardener's Magazine* 2 (1836): 96–102. Boston: Hovey & Co, 1836.

————. *Treatise on the Theory and Practice of Landscape Gardening, Adapted to North America.* New York and London: Wiley and Putnam, 1841.

Duncan, Andrew, Jr. *Edinburgh New Dispensatory.* Worcester, MA: Isaiah Thomas, Jr., 1805.

————. *Edinburgh New Dispensatory.* Edited by Jacob Dyckman. New York: James Eastman, 1818.

Earle, Sir James. *A Treatise on the Hydrocele.* 1st ed. London: J. Johnson, 1791.

————. *A Treatise on the Hydrocele.* 2nd ed. London: J. Johnson, 1796.

Emerson, Ralph Waldo. *Nature.* Boston: James Monroe and Co., 1836.

Field, Henry. *Memoirs Historical and Illustrative of the Botanic Garden at Chelsea.* London: R. Gilbert, 1820.

Floy, Michael, Jr. *The Diary of Michael Floy, Jr., Bowery Village, 1833–1837.* Edited by Richard Albert Edward Brooks. New Haven, CT: Yale University Press, 1941.

[Francis, John Wakefield.] "David Hosack." In *National Portrait Gallery of Distinguished Americans,* edited by James Herring and James B. Longacre, vol. 2, pp. 351–64. New York: Monson Bancroft, 1835.

Francis, John Wakefield. *Old New York; or, Reminiscences of the Past Sixty Years.* New York: C. Roe, 1858.

Francis, Samuel W. *Biographical Sketches of Distinguished Living New York Surgeons.* New York: J. Bradburn, 1866.

Gentle, Andrew. *Every Man His Own Gardener.* New York: Sold, wholesale and retail, by Alexander Smith, Nursery and Seedsman; and by the author, 1841.

Gross, Samuel D. *Autobiography of Samuel D. Gross, M.D.* Edited by Samuel W. Gross. 2 vols. Philadelphia: W. B. Saunders, 1893.

Hamilton, James A. *Reminiscences of James A. Hamilton.* New York: Charles Scribner & Co., 1869.

Hone, Philip. *The Diary of Philip Hone.* Edited by Bayard Tuckerman. 2 vols. New York: Dodd, Mead, 1889.

Hosack, Alexander Eddy. "David Hosack." In *Lives of Eminent American Physicians and Surgeons of the Nineteenth Century,* edited by Samuel D. Gross, 289–337. Philadelphia: Lindsay & Blakiston, 1861.

Hosack, Alexander, Jr. *An Inaugural Essay on the Yellow Fever.* New York: T. and J. Swords, 1797.

Humboldt, Alexander von. *Alexander von Humboldt und die Vereinigten Staaten von Amerika. Briefwechsel.* Edited by Ingo Schwarz. Berlin: Akademie Verlag, 2004.

———. *Briefe aus Amerika, 1799–1804.* Edited by Ulrike Moheit. Berlin: Akademie Verlag, 1993.

Hunt, Freeman. *Letters about the Hudson River.* New York: Freeman Hunt and Co., 1836.

Irving, Pierre Munro, ed. *The Life and Letters of Washington Irving.* 3 vols. New York: G. P. Putnam, 1864.

Irving, Washington. *A History of New York from the Beginning of the World to the End of the Dutch Dynasty . . . by Diedrich Knickerbocker.* 2 vols. New York: Inskeep & Bradford, 1809.

Jefferson, Thomas. *Notes on the State of Virginia.* London: John Stockdale, 1787.

Joly, Nicolas. *Éloge historique d'Alyre Raffeneau Delile.* Toulouse, France: Douladoure Frères, 1859.

Kalm, Peter [Pehr]. *Travels into North America*. Translated by John Reinhold Forster. 2nd ed. 2 vols. London: T. Lowndes, 1772.

Kames, Henry Home, Lord. *The Gentleman Farmer*. 3rd ed. Edinburgh: John Bell, 1788.

Laight, Henry. Diaries, 1795–1803, 1816–1822. Manuscript Collection, New York-Historical Society.

Lewis, Meriwether, and William Clark. *The Lewis and Clark Journals*. Edited by Gary E. Moulton. Lincoln: University of Nebraska Press, 2003.

Lewis, William. *Edinburgh New Dispensatory*. Edinburgh: Charles Elliot, 1786.

———. *Edinburgh New Dispensatory*. Edinburgh: William Creech, 1791.

———. *Edinburgh New Dispensatory*. 3rd American Edition, from 4th Edinburgh Edition. Walpole, NH: D. Carlisle, 1796.

Lindley, John. "Hosackia bicolor." *Edward's Botanical Register* 15, no. 6 (1829): [np; plate 1257]. London: James Ridgway, 1829.

Linnaeus, Carl. *The Elements of Botany . . . Being a Translation of the Philosophia Botanica*. Translated by Hugh Rose. London: T. Cadell, 1775.

———. *Species Plantarum*. Facsimile of 1st ed., 1753. London: Ray Society, 2013.

———. *Systema Naturae*. Originally published 1735. Edited by Johan Friedrich Gmelin. 2 vols. Leipzig: Georg Emanuel Beer, 1788–1793.

Marshal, Andrew. *The Morbid Anatomy of the Brain, in Mania and Hydrophobia*. Edited by Solomon Sawrey. London: Longman, Hurst, Rees, Orme, & Brown, 1815.

Martineau, Harriet. *Retrospect of Western Travel*. 2 vols. London: Saunders and Otley, 1838.

McMahon, Bernard. *The American Gardener's Calendar*. Philadelphia: B. Graves, 1806.

McVickar, John. *A Domestic Narrative of the Life of Samuel Bard*. New York: Paul, 1822.

Michaux, André. *Flora Boreali-Americana*. 2 vols. Paris: Chez Levrault Frères, 1803.

———. *Histoire des Chênes de l'Amérique*. Paris: Chez Levrault Frères, 1801.

Michaux, François André. *Histoire des arbres Forestiers de l'Amérique septentrionale*. 3 vols. Paris: L. Haussmann et d'Hautel, 1810–1813.

———. *The North American Sylva*. Translated by Augustus L. Hillhouse. 2 vols. Paris: C. d'Hautel, 1819.

———. *Travels to the West of the Alleghany* [sic] *Mountains*. London: Richard Phillips, 1805.

————. *Voyage à l'ouest des monts Alléghanys*. Paris: Chez Levrault, Schoell et Compagnie, 1804.

Milbert, Jacques Gérard. *Itinéraire pittoresque du fleuve Hudson et des parties laterales de l'Amérique du Nord*. 2 vols. Paris: Henri Gaugain et Cie, 1828.

————. *Picturesque Itinerary of the Hudson River and the Peripheral Parts of North America*. Translated by Constance D. Sherman. Ridgewood, NJ: The Gregg Press, 1968.

Miller, John. *An Illustration of the Sexual System of Linnaeus*. London: Published for the author, 1779.

————. *An Illustration of the Termini Botanici of Linnaeus*. London: Published for the author, 1789.

[Mitchill, Samuel Latham.] "History of the Worm, Which Spread Terror through New York and the Neighbouring States in June and July 1806." *Medical Repository*, Hexade 2, vol. 4, no. 1 (May–July 1806): 98–100.

Mitchill, Samuel Latham. *A Discourse on the Character and Services of Thomas Jefferson*. New York: G. and C. Carvill, 1826.

————. *A Discourse on the Life and Character of Samuel Bard*. New York: Fanshaw, 1821.

————. *The Picture of New-York*. New York: I. Riley, 1807.

————. *Remarks on the gaseous oxyd* [sic] *of azote or of nitrogene*. New York: T. and J. Swords, 1795.

Moore, Clement Clarke. *Plain Statement, Addressed to the Proprietors of Real Estate, in the City and County of New-York*. New York: J. Eastburn and Co., 1818.

Morris, George Pope. *The Little Frenchman and His Water Lots*. Philadelphia: Lea & Blanchard, 1839.

Mott, Valentine. *Reminiscences of Medical Teaching and Teachers in New York*. New York: Jennings, 1850.

Muhlenberg, Henry. *Catalogus plantarum Americae Septentrionalis*. Lancaster, PA: William Hamilton for the author, 1813.

New-York Hospital. *Charter of the Society of the New-York Hospital, and the Laws Relating Thereto, with the By-Laws and Regulations of the Institution and Those of the Bloomingdale Asylum for the Insane*. New York: John R. M'Gown, 1845.

Nuttall, Thomas. *Genera of North American Plants*. 2 vols. Philadelphia: Printed for the author by D. Heartt, 1818.

Parmentier, André. "Landscapes and Picturesque Gardens." In *New American Gardener*, edited by T. G. Fessenden, 184–86. Boston: J. B. Russell, 1828.

Plantsman's ledger, 1793–1796. Manuscript [author unknown], Mertz Library, New York Botanical Garden.

Prince, William. *Catalogue of American Indigenous Trees, Plants, and Seeds, Cultivated and for Sale at the Linnaean Botanic Garden, Flushing, Long-Island, near New-York.* New York: T. and J. Swords, 1820.

Pursh, Frederick. *Flora Americae Septentrionalis.* 2 vols. London: White, Cochrane, and Co., 1814.

Review of "An Address Delivered Before the New-York Historical Society at its Fortieth Anniversary," by John Romeyn Brodhead. *Knickerbocker* 25, no. 3 (1845): 250–54 [no author].

Review of *History of the American Oaks*, by André Michaux. *Medical Repository* 6, no. 1 (1803): 64–70 [no author].

Ross, John. *A Voyage of Discovery, Made under the Orders of the Admiralty, in His Majesty's Ships Isabella and Alexander. . . .* London: John Murray, 1819.

Rousseau, Jean-Jacques. *Letters on the Elements of Botany Addressed to a Lady.* Translated by Thomas Martyn. 2nd ed. London: Printed for B. White and Son, 1787.

Royal Society. "List of the Fellows of the Royal Society (1660–2007)." London: Royal Society, 2007.

Royall, Anne. *Sketches of History, Life, and Manners in the United States.* New Haven, CT: Printed for the author, 1826.

Rush, Benjamin. "An Account of the Sugar Maple-Tree, of the United States. . . ." *Transactions of the American Philosophical Society* 3 (1793): 64–81.

———. *Medical Inquiries and Observations: Containing an Account of the Bilious and Remitting and Intermitting Yellow Fever, as it Appeared in Philadelphia in the Year 1794*, vol. 4. Philadelphia: Thomas Dobson, 1796.

———. "On the Means of Lessening the Pains and Danger of Child-Bearing." *Medical Repository* 6, no. 1 (1803): 26–31.

Sachsen-Weimar-Eisenach, Bernhard zu. *Reise seiner Hoheit des Herzogs Bernhard zu Sachsen-Weimar-Eisenach durch Nord-Amerika in den Jahren 1825 und 1826.* Edited by Heinrich Luden. 2 vols. Weimar: Wilhelm Hoffman, 1828.

Savigny, John. *A Catalogue of Chirurgical instruments, Made and Sold by J. H. Savigny.* [London,] 1800.

Sawrey, Solomon. "Sketch of the Life of Dr. Marshal." In *The Morbid Anatomy of the Brain, in Mania and Hydrophobia* by Andrew Marshal, pp. i–xxxiv. Edited by Solomon Sawrey. London: Longman, Hurst, Rees, Orme, & Brown, 1815.

Sayers, Edward. *American Flower Garden Companion*. Boston: Joseph Breck and Company; New York: G. C. Thorburn, 1838.

————. "Notes and Observations on Gardens and Nurseries in the Vicinity of Newark, N.J., New York, Hartford, and Boston." *Magazine of Horticulture, Botany, and All Useful Discoveries and Improvements in Rural Affairs* 3, no. 9 (September 1837): 321–30. Boston: Hovey and Co.

Sharples, Ellen. Diary (typescript). Bristol (England) Record Office, 1810–1836.

Short, Thomas. *Medicina Britannica*. London: R. Manby and H. Shute Cox, 1746.

"Sketch of the Weather and Diseases in the Summer and Autumn of 1806." *Medical Repository* 4 ["4th of second hexade"], no. 2 (1807): 214–15 [no author].

Smith, Elihu Hubbard. *The Diary of Elihu Hubbard Smith (1771–1798)*. Edited by James E. Cronin. Philadelphia: American Philosophical Society, 1973.

Speechly, William. *Treatise on the Culture of the Vine*. Dublin: P. Wogan, 1791.

Thacher, James. *American Medical Biography*. 2 vols. Boston: Richardson and Lord, 1828.

————. *American New Dispensatory*. 1st ed. Boston: T. B. Wait and Co., 1810.

————. *American New Dispensatory*. 2nd ed. Boston: T. B. Wait and Co., 1813.

————. "An Excursion on the Hudson, Letter I." *New England Farmer and Horticultural Journal* 9, no. 19 (1830): 148–49.

————. "An Excursion on the Hudson, Letter II." *New England Farmer and Horticultural Journal* 9, no. 20 (1830): 156–57.

"The Linnean Herbarium." In *Hooker's Journal of Botany and Kew Garden Miscellany*, vol. 4, 217–20. London: Reeve and Co., 1852 [no author].

Thoreau, Henry David. *Walden*. Edited by Jeffrey S. Cramer. New Haven, CT: Yale University Press, 2004 [1854].

Thornton, Robert John. "Letter from Doctor Thornton to Dr. Beddoes," dated 7 December 1793. In *Letters from Dr. Withering, of Birmingham, Dr. Ewart, of Bath, Dr. Thornton, of London, and Dr. Biggs, Late of the Isle of Santa-Cruz*, edited by Thomas Beddoes, 22–24. Bristol, UK: Bulgin and Rosser, 1794.

————. *New Illustration of the Sexual System of Linnaeus*. London: T. Bensley, 1807.

————. "Sketch of the Life and Writings of the Late Mr. William Curtis." In *Lectures on Various Subjects as Delivered in the Botanic Garden at Lambeth*, by William Curtis and edited by Samuel Curtis, vol. 3, 1–32. London: H. D. Symonds, 1805.

Torrey, John. "Calendarium Florae for the Vicinity of New York." Manuscript, 1818–1820. Mertz Library, New York Botanical Garden.

————, Caspar Wistar Eddy, and D'Jurco V. Knevels. *A Catalogue of Plants Growing Spontaneously within Thirty Miles of the City of New York*. Albany, NY: Webster and Skinners, 1819.

Trollope, Frances. *Domestic Manners of the Americans*. 2 vols. 4th ed. London: Printed for Whitaker, Treacher, & Co., 1832.

[Verplanck, Gulian.] *Procès Verbal of the Ceremony of Installation of President of the New-York Historical Society as it will be performed February 8, 1820*. New York: [no publisher], 1820.

Webster, Noah, Jr., ed. *A Collection of Papers on the Subject of Bilious Fevers, Prevalent in the United States for a Few Years Past*. New York: Hopkins, Webb, and Co., 1796.

Woodville, William. *Medical Botany*. 3 vols. London: Printed and sold for the author by James Phillips, 1790–93.

Selected Secondary Works

Note: I have included only directly relevant secondary sources here; all the sources I cite can be found in the Notes. A complete bibliography is available at american eden.org.

Ambrose, Stephen E. *Undaunted Courage: Meriwether Lewis, Thomas Jefferson, and the Opening of the American West*. New York: Simon and Schuster, 2005 [1996].

Anderson, Marynita. *Physician Heal Thyself: Medical Practitioners of Eighteenth-Century New York*. New York: Peter Lang, 2004.

Appleby, Joyce. *Inheriting the Revolution: The First Generation of Americans*. Cambridge, MA: Harvard University Press, 2000.

Armstead, Myra B. Young. *Freedom's Gardener: James F. Brown, Horticulture, and the Hudson Valley in Antebellum America*. New York: New York University Press, 2012.

Atha, Daniel, Todd Forrest, Robert F. C. Naczi, Matthew C. Pace, Meryl Rubin, Jessica A. Schuler, and Michael Nee. "The Historic and Extant Vascular Flora of The New York Botanical Garden." *Brittonia* 68, issue 3 (September 2016): 245–77.

Bagli, Charles V. "Era Closes at Rockefeller Center With $1.85 Billion Deal on Sale." *New York Times*, December 22, 2000.

Baird, Eleanora Gordon. "Moses Bartram's Account Book 1778–1788: Notes Made by a Philadelphia Apothecary." *Bartram Broadside* (Spring 2003): 1–8.

Beckert, Sven. *Empire of Cotton*. New York: Vintage, 2015.

Bender, Thomas. *New York Intellect*. Baltimore, MD: Johns Hopkins University Press, 1987.

Bergdoll, Barry, with Hollee Haswell and Janet Parks. *Mastering McKim's Plan: Columbia's First Century on Morningside Heights*. New York: Miriam and Ira D. Wallach Art Gallery, Columbia University, 1997.

Blackmar, Elizabeth. *Manhattan for Rent, 1785–1850*. Ithaca, NY: Cornell University Press, 1989.

Blake, John B. "Yellow Fever in Eighteenth-Century America." *Bulletin of the New York Academy of Medicine* 44, no. 6 (1968): 673–86.

Britton, Nathaniel Lord. "Dr. Torrey as a Botanist." *Bulletin of the Torrey Botanical Club* 27, no. 10 (October 1900): 540–51.

Brodsky, Alyn. *Benjamin Rush: Patriot and Physician*. New York: St. Martin's Press, 2004.

Brooke, John. "King George Has Issued Too Many Pattents for Us: Property and Democracy in Jeffersonian New York." *Journal of the Early Republic* 33, no. 2 (2013): 187–217.

Brown, Addison. *The Elgin Botanic Garden*. Lancaster, PA: New Era Printing Co., 1908.

Brown, Lee Rust. *The Emersonian Museum: Practical Romanticism and the Pursuit of the Whole*. Cambridge, MA: Harvard University Press, 1997.

Buchan, James. *Capital of the Mind: How Edinburgh Changed the World*. London: John Murray, 2003.

Burrows, Edwin G., and Mike Wallace. *Gotham: A History of New York City to 1898*. New York: Oxford University Press, 1998.

Chaplin, Joyce E. "Expansion and Exceptionalism in Early American History." *Journal of American History* 89, no. 4 (2003): 1431–55.

———. *The First Scientific American: Benjamin Franklin and the Pursuit of Genius*. New York: Basic Books, 2006.

———. "Nature and Nation: Natural History in Context." In *Stuffing Birds, Pressing Plants, Shaping Knowledge: Natural History in North America, 1730–1860*, edited by Sue Ann Prince, 75–95. Philadelphia: American Philosophical Society, 2003.

Chernow, Ron. *Alexander Hamilton*. New York: Penguin, 2004.

———. *Washington: A Life*. New York: Penguin, 2010.

Chesney, Sarah. "The Root of the Matter: Searching for William Hamilton's Greenhouse at The Woodlands Estate, Philadelphia, Pennsylvania." In *Historical Archeology of the Delaware Valley, 1600–1850*, edited by Richard Veit and David Orr, 273–96. Knoxville: University of Tennessee Press, 2014.

Cleves, Rachel Hope. *The Reign of Terror in America: Visions of Violence from Anti-Jacobinism to Antislavery*. Cambridge: Cambridge University Press, 2009.

Cooperman, Emily T., and John Dixon Hunt. "The American Translation of the Picturesque." In *Foreign Trends in American Gardens*, edited by Raffaella Fabiani Giannetto, 813–30. Charlottesville: University of Virginia Press, 2016.

Cornog, Evan. *The Birth of Empire: DeWitt Clinton and the American Experience, 1769–1828*. Oxford: Oxford University Press, 1998.

Cox, Robert S. "I Never Yet Parted: Bernard McMahon and the Seeds of the Corps of Discovery." In *The Shortest and Most Convenient Route: Lewis and Clark in Context*, edited by Robert S. Cox, 102–35. *Transactions of the American Philosophical Society*, New Series, 94, no. 5. Philadelphia: American Philosophical Society, 2004.

Cronon, William. *Changes in the Land: Indians, Colonists, and the Ecology of New England*. New York: Hill and Wang, 1983.

Curtis, W. Hugh. *William Curtis*. Winchester, UK: Warren and Son, 1941.

Cutright, Paul Russell. *Lewis and Clark: Pioneering Naturalists*. Urbana: University of Illinois Press, 1969.

Deitz, Paula. *Of Gardens*. Philadelphia: University of Pennsylvania Press, 2011.

Dixon, John M. *The Enlightenment of Cadwallader Colden: Empire, Science, and Intellectual Culture in British New York*. Ithaca, NY: Cornell University Press, 2016.

Dolkart, Andrew S. *Morningside Heights: A History of Its Architecture and Development*. New York: Columbia University Press, 1998.

Drayton, Richard. *Nature's Government: Science, Imperial Britain, and the 'Improvement' of the World*. New Haven, CT: Yale University Press, 2000.

Dunlap, David W. "G.E. Gives Midtown Tower to Columbia University." *New York Times*, June 3, 1993.

Ellis, Joseph J. *Founding Brothers: The Revolutionary Generation*. New York: Knopf, 2000.

Eustis, Elizabeth S. "The Horticultural Enterprise: Markets, Mail, and Media in Nineteenth-Century America." In *Flora Illustrata*, edited by

Susan M. Fraser and Vanessa Bezemer Sellers, 233–63. New Haven, CT: Yale University Press/New York: New York Botanical Garden, 2014.

————, with David Andrews. "Creating a North American Flora." In *Flora Illustrata*, edited by Susan M. Fraser and Vanessa Bezemer Sellers, 155–75. New Haven, CT: Yale University Press/New York: New York Botanical Garden, 2014.

Ewan, Joseph. "Frederick Pursh and His Botanical Associates." *Proceedings of the American Philosophical Society* 96, no. 5 (1952): 599–628.

————, and Nesta Dunn Ewan. *Benjamin Smith Barton: Naturalist and Physician in Jeffersonian America*. St. Louis: Missouri Botanical Garden Press, 2007.

Ewing, Heather. *The Lost World of James Smithson: Science, Revolution, and the Birth of the Smithsonian*. London: Bloomsbury, 2007.

Ferber, Linda S. *The Hudson River School: Nature and the American Vision*. New York: New-York Historical Society/Skira Rizzoli, 2009.

Fleming, Thomas J. *Duel: Alexander Hamilton, Aaron Burr, and the Future of America*. New York: Basic Books, 1999.

Foner, Eric. "Columbia University and Slavery: A Preliminary Report." 2016. https://columbiaandslavery.columbia.edu/content/preliminary-report.

————. *The Story of American Freedom*. New York: W. W. Norton, 1998.

Fox, Claire G. "The Surprising Harvest of Dr. Hosack's Garden." *New York History* 67, no. 2 (April 1986): 199–209.

Fraser, Susan M., and Vanessa Bezemer Sellers, eds. *Flora Illustrata: Great Works from the LuEsther T. Mertz Library of the New York Botanical Garden*. New Haven, CT: Yale University Press/New York: New York Botanical Garden, 2014.

Fraser, Susan M., and Sally Armstrong Leone, eds. *The Trees of North America: Michaux and Redouté's American Masterpiece*. New York and London: New York Botanical Garden and Abbeville Press, 2017.

Freeman, Joanne B. *Affairs of Honor: National Politics in the New Republic*. New Haven, CT: Yale University Press, 2007.

Fry, Joel T. "America's Ancient Garden: The Bartram Botanic Garden, 1728–1850." In *Knowing Nature: Art and Science in Philadelphia, 1740–1840*, edited by Amy R. W. Meyers, 60–95. New Haven, CT: Yale University Press, 2011.

————. "Chronology of William Bartram's Travels, 1773–1777." Appendix C in *William Bartram: The Search for Nature's Design*, edited by Thomas Hallock and Nancy E. Hoffman, 515–18. Athens: University of Georgia Press, 2010.

————. "Inside the Box: John Bartram and the Science and Commerce of the Transatlantic Plant Trade." In *Ways of Making and Knowing: The Material Culture of Empirical Knowledge*, edited by Pamela H. Smith, Amy R. W. Meyers, and Harold Cook, 194–220. The Bard Graduate Center Cultural Histories of the Material World. Ann Arbor: University of Michigan Press, 2014.

————. "An International Catalogue of North American Trees and Shrubs: The Bartram Broadside, 1783." *The Journal of Garden History* 16, no. 1 (January–March 1996): 3–66.

————. "John Bartram and His Garden: Would John Bartram Recognize His Garden Today?" In *America's Curious Botanist: A Tercentennial Reappraisal of John Bartram, 1699–1777*, edited by Nancy E. Hoffmann and John C. Van Home, 155–83. Philadelphia: American Philosophical Society, 2004.

————. "John Bartram & Medicine." Exhibit pamphlet, Bartram's Garden, 2012.

Furstenberg, François. *When the United States Spoke French*. New York: Penguin, 2014.

Gage, Andrew T., and William T. Stearn. *A Bicentenary History of the Linnean Society of London*. London: Academic Press, 1988.

Gaudio, Michael. "Surface and Depth: The Art of Early American Natural History." In *Stuffing Birds, Pressing Plants, Shaping Knowledge: Natural History in North America, 1730–1860*, edited by Sue Ann Prince, 55–74. Philadelphia: American Philosophical Society, 2003.

Gillispie, Charles Coulston. *Science and Polity in France: The Revolutionary and Napoleonic Years*. Princeton, NJ: Princeton University Press, 1980.

Gordon-Reed, Annette, and Peter S. Onuf. *"Most Blessed of the Patriarchs": Thomas Jefferson and the Empire of the Imagination*. New York: Liveright, 2016.

Gronim, Sara Stidstone. "Imagining Inoculation: Smallpox, the Body, and Social Relations of Healing in the Eighteenth Century." *Bulletin of the History of Medicine* 80, no. 2 (2006): 247–68.

————. "What Jane Knew: A Woman Botanist in the Eighteenth Century." *Journal of Women's History* 19, no. 3 (2007): 33–59.

Haller, John S. *The People's Doctors: Samuel Thomson and the American Botanical Movement, 1790–1860*. Carbondale: Southern Illinois University Press, 2000.

Hallock, Thomas, and Nancy E. Hoffman, eds. *William Bartram: The Search for Nature's Design*. Athens: University of Georgia Press, 2010.

Hamel, Paul B., and Mary U. Chiltoskey. *Cherokee Plants and Their Uses: A 400 Year History.* Sylva, NC: Herald Publishing Co, 1975.

Hamilton, Allan McLane. *The Intimate Life of Alexander Hamilton.* New York: Charles Scribner's Sons, 1910.

Harnagel, Edward E. "Doctors Afield: David Hosack and the Duel." *New England Journal of Medicine* 261 (September 3, 1959): 504–5.

Harris, Leslie M. *In the Shadow of Slavery: African Americans in New York City, 1626–1863.* Chicago: University of Chicago Press, 2003.

Hartog, Hendrick. *Public Property and Private Power: The Corporation of the City of New York in American Law, 1730–1803.* Ithaca, NY: Cornell University Press, 1983.

Hatch, Peter J. "Bernard McMahon, Pioneer American Gardener." *Twinleaf* (January 1993).

———. *"A Rich Spot of Earth": Thomas Jefferson's Revolutionary Garden at Monticello.* New Haven, CT: Yale University Press, 2012.

Hedrick, Ulysses Prentiss. *A History of Horticulture in America to 1860.* New York: Oxford University Press, 1950.

Hoge, Robert Wilson. "A Doctor for All Seasons: David Hosack of New York." *American Numismatic Society* 6, no. 1 (Spring 2007).

Holloway, Marguerite. *The Measure of Manhattan: The Tumultuous Career and Surprising Legacy of John Randel Jr.* New York: W. W. Norton, 2013.

Holmes, Richard. *The Age of Wonder: How the Romantic Generation Discovered the Beauty and Terror of Science.* New York: Pantheon, 2008.

Hyde, Elizabeth. "André Michaux and French Botanical Diplomacy in the Cultural Construction of Natural History in the Atlantic World." In *Of Elephants and Roses: Encounters with French Natural History, 1790–1830,* edited by Sue Ann Prince, 87–98. Philadelphia: American Philosophical Society Museum, 2013.

———. "Arboreal Negotiations, or the Cultural Politics of André Michaux's Mid-Atlantic Mission." In *Invaluable Trees: Cultures of Nature, 1660–1830,* edited by Laura Auricchio, Elizabeth Heckendorn Cook, and Giulia Pacini. *Studies on Voltaire and the Eighteenth Century* (August 2012): 185–99. Geneva: Institut et Musée Voltaire, 2012.

———. "Of Monarchical Climates and Republican Soil." In *Foreign Trends in American Gardens,* edited by Raffaella Fabiani Giannetto, 88–109. Charlottesville: University of Virginia Press, 2016.

Isenberg, Nancy. *Fallen Founder: The Life of Aaron Burr.* New York: Penguin, 2007.

Jackson, Benjamin Daydon. "Index to the Linnean Herbarium, with Indication of the Types of Species Marked by Carl von Linné." *Proceedings of the Linnean Society of London* 124 (1912): 1–152.

Jackson, Kenneth T., and David S. Dunbar, eds. *Empire City: New York Through the Centuries*. New York: Columbia University Press, 2002.

Jarvis, Charlie. *Order out of Chaos: Linnaean Plant Names and Their Types*. London: Linnean Society of London, 2007.

Jasanoff, Maya. *Liberty's Exiles: American Loyalists in the Revolutionary World*. New York: Knopf, 2011.

Jeffe, Elizabeth Rohn. "Hamilton's Physician: David Hosack, Renaissance Man of Early New York." *New York Journal of American History* 54, no. 3 (2004): 54–58.

Johnson, Victoria, and Walter W. Powell. "Poisedness and Propagation: Organizational Emergence and the Transformation of Civic Order in Nineteenth-Century New York City." In *Organizations, Civil Society, and the Roots of Development*, edited by Naomi R. Lamoreaux and John Joseph Wallis. Cambridge, MA, and Chicago: National Bureau of Economic Research and University of Chicago Press, 2017.

Jones, Brian Jay. *Washington Irving*. New York: Arcade, 2008.

Juel, H. O., and John W. Harshberger. "New Light on the Collection of North American Plants Made by Peter Kalm." *Proceedings of the Academy of Natural Sciences of Philadelphia* 81 (1929): 297–303.

Kelly, Franklin. *Thomas Cole's Paintings of Eden*. Fort Worth, TX: Amon Carter Museum, 1994.

Kelsay, Isabel Thompson. *Joseph Brant*. Syracuse, NY: Syracuse University Press, 1984.

Kennett, Tom. *The Lord Treasurer of Botany: Sir James Edward Smith and the Linnaean Collections*. London: Linnean Society of London, 2016.

Kevles, Daniel J. "Cultivating Art and Property in American Fruits." *Smithsonian* 42, no. 4 (July/August 2011): 76–82.

———. "Eden and Empire: The Mercantile Making of American Colonial Husbandry." *The Yale Review* 99, no. 1 (January 2011): 104–26.

———. "A Primer of A, B, Seeds: Advertising, Branding, and Intellectual Property in an Emerging Industry." *University of California Davis Law Review* 47, no. 2 (2013): 657–78.

———. "Protections, Privileges, and Patents: Intellectual Property in American Horticulture." *Proceedings of the American Philosophical Society* 152 (June 2008): 207–13.

Klein, Claire. "Columbia and the Elgin Botanic Garden Property." *Columbia University Quarterly* 31, no. 4 (1939): 272–97.

————.“Radio City Towers Rise from the Elgin Botanic Garden: The Rockefeller Center Property.” *Columbia University Quarterly* 33, no. 1 (1941): 58–75.

Klein, Walter A., and George G. Reader. “Epidemic Diseases at the New York Hospital.” *Bulletin of the New York Academy of Medicine* 67, no. 5 (1991): 439–59.

Koerner, Lisbet. *Linnaeus: Nature and Nation*. Cambridge, MA: Harvard University Press, 1999.

Laird, Mark. “Early American Horticultural Traditions: Gardening with Plants from the New World.” In *Flora Illustrata*, edited by Susan M. Fraser and Vanessa Bezemer Sellers, 179–205. New Haven: Yale University Press/New York: New York Botanical Garden, 2014.

————. *A Natural History of English Gardening, 1650–1800*. New Haven, CT: Yale University Press, 2015.

Lamb, Martha J. *History of the City of New York*. Vol. 2. New York: A. S. Barnes and Company, 1880.

Langstaff, J. Brett. *Dr. Bard of Hyde Park*. New York: Dutton, 1940.

Lepore, Jill. *New York Burning: Liberty, Slavery, and Conspiracy in Eighteenth-Century Manhattan*. New York: Knopf, 2005.

Lewis, Andrew J. *A Democracy of Facts: Natural History in the Early Republic*. Philadelphia: University of Pennsylvania Press, 2011.

————. “Gathering for the Republic: Botany in Early Republic America.” In *Colonial Botany: Science, Commerce, and Politics in the Early Modern World*, edited by Londa Schiebinger and Claudia Swan, 66–80. Philadelphia: University of Pennsylvania Press, 2005.

Lewis, Tom. *The Hudson*. New Haven, CT: Yale University Press, 2005.

Lomask, Milton. *Aaron Burr: The Conspiracy and Years of Exile*. New York: Farrar, Straus & Giroux, 1982.

————. *Aaron Burr: The Years from Princeton to Vice President*. New York: Farrar, Straus & Giroux, 1979.

Maehle, Andreas-Holger. *Drugs on Trial: Experimental Pharmacology and Therapeutic Innovation in the Eighteenth Century*. Amsterdam: Rodopi, 1999.

Major, Judith K. *To Live in a New World: A. J. Downing and American Landscape Gardening*. Cambridge, MA: MIT Press, 1997.

————. “Toward an American Landscape Theory.” In *Flora Illustrata*, edited by Susan M. Fraser and Vanessa Bezemer Sellers, 207–31. New Haven, CT: Yale University Press/New York: New York Botanical Garden, 2014.

Mayer, Josephine, and Robert A. East. "The Settlement of Alexander Hamilton's Debts: A Footnote to History." *New York History* 18, no. 4 (1937): 384.

McAllister, Ethel. *Amos Eaton: Scientist and Educator*. Philadelphia: University of Pennsylvania Press, 1941.

McCaughey, Robert A. *Stand, Columbia: A History of Columbia University in the City of New York, 1754–2004*. New York: Columbia University Press, 2003.

McCullough, David. *John Adams*. New York: Simon and Schuster, 2008.

Meacham, Jon. *Thomas Jefferson: The Art of Power*. New York: Random House, 2012.

Merchant, Carolyn. *Reinventing Eden: The Fate of Nature in Western Culture*. New York: Routledge, 2004.

Meyers, Amy R. W., ed. *Knowing Nature: Art and Science in Philadelphia, 1740–1840*. New Haven, CT: Yale University Press, 2011.

———, and Margaret Beck Pritchard, eds. *Empire's Nature: Mark Catesby's New World Vision*. Chapel Hill: University of North Carolina Press, 1998.

Mickulas, Peter. *Britton's Botanical Empire: The New York Botanical Garden and American Botany, 1888–1929*. New York: New York Botanical Garden Press, 2007.

———. "Cultivating the Big Apple: The New-York Horticultural Society, Nineteenth-Century New York Botany, and the New York Botanical Garden." *New York History* 83, no. 1 (Winter 2002): 34–54.

Moore, Wendy. *The Knife Man: Blood, Body Snatching, and the Birth of Modern Surgery*. New York: Broadway Books, 2005.

Mukerji, Chandra. "Dominion, Demonstration, and Domination: Religious Doctrine, Territorial Politics, and French Plant Collection." In *Colonial Botany: Science, Commerce, and Politics in the Early Modern World*, edited by Londa Schiebinger and Claudia Swan, 19–33. Philadelphia: University of Pennsylvania Press, 2005.

———. *Territorial Ambitions and the Gardens of Versailles*. Cambridge, MA: Cambridge University Press, 1997.

Müller-Wille, Staffan. "Walnuts at Hudson Bay, Coral Reefs in Gotland: The Colonialism of Linnaean Botany." In *Colonial Botany: Science, Commerce, and Politics in the Early Modern World*, edited by Londa Schiebinger and Claudia Swan, 34–48. Philadelphia: University of Pennsylvania Press, 2005.

Nasaw, David. *Andrew Carnegie*. New York: Penguin, 2006.

Nisbet, Jack. *David Douglas: A Naturalist at Work*. Seattle, WA: Sasquatch Books, 2012.

O'Brian, Patrick. *Joseph Banks*. Chicago: University of Chicago Press, 1987.

O'Donnell, Patricia, Charles A. Birnbaum, and Cynthia Zaitzevsky. *Cultural Landscape Report for Vanderbilt Mansion Historic Site*. Vol 1. New York: National Park Service, 1992.

O'Malley, Therese. "Appropriation and Adaptation: Early Gardening Literature in America." *Huntington Library Quarterly* 55, no. 3 (Summer 1992[a]): 401–31.

———. "Art and Science in the Design of Botanic Gardens, 1730–1830." In *Garden History: Issues, Approaches, Methods*, edited by John Dixon Hunt, 279–302. Washington, DC: Dumbarton Oaks, 1992[b].

———. "Cultivated Lives, Cultivated Spaces: The Scientific Garden in Philadelphia, 1740–1840." In *Knowing Nature: Art and Science in Philadelphia, 1740–1840*, edited by Amy R. W. Meyers, 36–59. New Haven, CT: Yale University Press, 2011.

———. "From Practice to Theory: The Emerging Profession of Landscape Gardening in Early Nineteenth-Century America." In *Botanical Progress, Horticultural Innovations, and Cultural Changes*, edited by Michel Conan and W. John Kress, 223–37. Washington, DC, and Cambridge, MA: Dumbarton Oaks and Harvard University Press, 2007.

———. *Keywords in American Landscape Design*. With contributions by Elizabeth Kryder-Reid and Anne L. Helmreich. New Haven, CT: Yale University Press, 2010.

———. "'Your Garden Must Be a Museum to You': Early American Botanic Gardens." In *Art and Science in America: Issues of Representation*, edited by Amy R. W. Meyers, 35–59. San Marino, CA: Huntington Library, 1998.

Okrent, Daniel. *Great Fortune: The Epic of Rockefeller Center*. New York: Viking, 2003.

Oshinsky, David. *Bellevue: Three Centuries of Medicine and Mayhem at America's Most Storied Hospital*. New York: Doubleday, 2016.

Parrish, Susan Scott. *American Curiosity: Cultures of Natural History in the Colonial British Atlantic World*. Chapel Hill: University of North Carolina Press, 2006.

Parry, Ellwood C., III. *The Art of Thomas Cole*. Newark: University of Delaware Press, 1988.

Pauly, Philip J. *Fruits and Plains: The Horticultural Transformation of America*. Cambridge: Harvard University Press, 2007.

Peck, Robert McCracken. "Prologue." In *William Bartram: The Search for Nature's Design*, edited by Thomas Hallock and Nancy E. Hoffman, xv–xvi. Athens: University of Georgia Press, 2010.

Peterson, Paul M., Konstantin Romaschenko, Neil Snow, and Gabriel Johnson. "A Molecular Phylogeny of and Classification of *Leptochloa* (Poaceae: Chloridoideae: Chlorideae) *sensu lato* and related genera." *Annals of Botany* 109 (2012): 1317–29.

Pierson, George Wilson. *Tocqueville in America*. Baltimore, MD: Johns Hopkins University Press, 1996 [1938].

Pomerantz, Sidney I. *New York: An American City, 1783–1803*. New York: Columbia University Press, 1938.

Porter, Roy. *The Greatest Benefit to Mankind: A Medical History of Humanity*. New York: W. W. Norton, 1997.

Powell, Earl A. *Thomas Cole*. New York: Abrams, 1990.

Prest, John. *The Garden of Eden: The Botanic Garden and the Re-Creation of Paradise*. New Haven, CT: Yale University Press, 1981.

Quave, Cassandra. "Antibiotics from Nature: Traditional Medicine as a Source of New Solutions for Combating Antimicrobial Resistance." *AMR Control*, July 13, 2016. [Online journal: http://resistancecontrol .info/rd-innovation/antibiotics-from-nature-traditional-medicine-as-a -source-of-new-solutions-for-combating-antimicrobial-resistance/]

Raoult, Didier, Theodore Woodward, and J. Stephen Dumler. "The History of Epidemic Typhus." *Infectious Disease Clinics of North America* 18, no. 1 (2004): 127–40.

Rapport, Mike. *The Unruly City: Paris, London and New York in the Age of Revolution*. New York: Basic Books, 2017.

Robbins, Christine Chapman. *David Hosack: Citizen of New York*. Philadelphia: American Philosophical Society, 1964.

———. "David Hosack's Herbarium and Its Linnaean Specimens." *Proceedings of the American Philosophical Society* 104, no. 3 (June 1960): 293–313.

———. "John Torrey (1796–1873): His Life and Times." *Bulletin of the Torrey Botanical Club* 95, No. 6 (November–December 1968): 515–645.

Robbins, Paula Ivaska. *Jane Colden: America's First Woman Botanist*. Fleischmanns, NY: Purple Mountain Press, 2009.

Robbins, William J., and Mary Christine Howson. "André Michaux's New Jersey Garden and Pierre Paul Saunier, Journeyman Gardener." *Proceedings of the American Philosophical Society* 102, no. 4 (1958): 351–57.

Rodgers, Andrew Denny, III. *John Torrey: A Story of North American Botany*. Princeton, NJ: Princeton University Press, 1942.

Rogers, Elizabeth Barlow. *Green Metropolis: The Extraordinary Landscapes of New York City as Nature, History, and Design*. New York: Knopf, 2016.

Rosenberg, Charles E. "The Therapeutic Revolution." In *The Therapeutic Revolution: Essays in the Social History of American Medicine*, edited by Morris J. Vogel and Charles E. Rosenberg, 3–25. Philadelphia: University of Pennsylvania Press, 1979.

Rosenzweig, Roy, and Elizabeth Blackmar. *The Park and the People: A History of Central Park*. Ithaca, NY: Cornell University Press, 1992.

Rothstein, William G. *American Physicians in the 19th Century: From Sects to Science*. Baltimore, MD: Johns Hopkins University Press, 1985.

Rybczynski, Witold. *A Clearing in the Distance: Frederick Law Olmsted and America in the 19th Century*. New York: Scribner, 2003 [1999].

Sanderson, Eric W., and Marianne Brown. "Mannahatta: An Ecological First Look at the Manhattan Landscape Prior to Henry Hudson." *Northeastern Naturalist* 14 (4): 545–70.

———, with Markley Boyer. *Mannahatta: A Natural History of New York City*. New York: Abrams, 2009.

Savage, Henry, Jr., and Elizabeth J. Savage. *André and François-André Michaux*. Charlottesville: University of Virginia Press, 1986.

Schiebinger, Londa. *Plants and Empire: Colonial Bioprospecting in the Atlantic World*. Cambridge, MA: Harvard University Press, 2004.

Scobey, David. *Empire City: The Making and Meaning of the New York City Landscape*. Philadelphia: Temple University Press, 2002.

Sedgwick, John. *War of Two: Alexander Hamilton, Aaron Burr, and the Duel That Stunned the Nation*. New York: Berkley, 2015.

Sellers, Charles Coleman. "Charles Willson Peale's Portraits of Washington." *Metropolitan Museum of Art Bulletin*, New Series 9, no. 6 (1951): 147–55.

———. *Mr. Peale's Museum: Charles Willson Peale and the First Popular Museum of Natural Science and Art*. New York: W. W. Norton, 1980.

Semonin, Paul. *American Monster*. New York: New York University Press, 2000.

Shorto, Russell. *The Island at the Center of the World: The Epic Story of Dutch Manhattan and the Forgotten Colony That Shaped America*. New York: Random House, 2004.

Shrady, John, ed. *The College of Physicians and Surgeons*. New York: Lewis Publishing Company, 1903.

Skuncke, Marie-Christine. *Carl Peter Thunberg: Botanist and Physician*. Uppsala: Swedish Collegium for Advanced Study, 2014.

Slowick, Nancy. *A Naturalist's Guide to the Southern Palisades*. Palisades, NY: NMS, 2004.

Smith, Beatrice Scheer. "Jane Colden (1724–1766) and Her Botanic Manuscript." *American Journal of Botany* 75, no. 7 (1988): 1090–96.

Smith, Huron H. "Ethnobotany of the Meskwaki Indians." *Bulletin of the Public Museum of the City of Milwaukee* 4 (1928): 175–326.

Solit, Karen D. *History of the United States Botanic Garden*. Washington, DC: U.S. Government Printing Office, 1993.

Sonnedecker, Glenn, ed. *Kremers and Urdang's History of Pharmacy*. 4th ed. Madison, WI: American Institute of the History of Pharmacy, 1986.

Spary, E. C. *Utopia's Garden: French Natural History from Old Regime to Revolution*. Chicago: University of Chicago Press, 2000.

Speck, Frank G. *Catawba Medicines and Curative Practices*. Philadelphia: Publications of the Philadelphia Anthropological Society 1 (1937): 179–97.

Stahr, Walter. *John Jay: Founding Father*. New York: Hambledon and London, 2005.

Starr, Paul. *The Social Transformation of American Medicine*. New York: Basic Books, 2000 [1982].

Stewart, David O. *American Emperor: Aaron Burr's Challenge to Jefferson's America*. New York: Simon and Schuster, 2011.

Stiles, T. J. *The First Tycoon: The Epic Life of Cornelius Vanderbilt*. New York: Vintage, 2010.

Stokes, I. N. Phelps. *The Iconography of Manhattan Island, 1498–1909*. 6 vols. New York: Robert H. Dodd, 1915–1928.

Stradling, David. *The Nature of New York: An Environmental History of the Empire State*. Ithaca, NY: Cornell University Press, 2010.

Strouse, Jean. *Morgan: American Financier*. New York: Random House, 1999.

Sumner, Judith. *American Household Botany: A History of Useful Plants, 1620–1900*. Portland, OR: Timber Press, 2004.

Taylor, Alan. *The Civil War of 1812*. New York: Vintage, 2011.

Taylor, Linda Averill. *Plants Used as Curatives by Certain Southeastern Tribes*. Cambridge, MA: Botanical Museum of Harvard University, 1940.

Thiers, Barbara M., Melissa C. Tulig, and Kimberly A. Watson. "Digitization of The New York Botanical Garden Herbarium." *Brittonia* 68, issue 3 (September 2016): 324–33.

Toll, Ian W. *Six Frigates: The Epic History of the Founding of the U.S. Navy*. New York: W. W. Norton, 2008.

Turner, Nancy Chapman, and Marcus A. M. Bell. "The Ethnobotany of the Southern Kwakiutl Indians of British Columbia." *Economic Botany* 27 (1973): 257–310.

Uglow, Jenny. *The Lunar Men: Five Friends Whose Curiosity Changed the World*. New York: Farrar, Straus and Giroux, 2002.

Vail, Anna Murray. "Botanical Books of Dr. Hosack." *Journal of the New York Botanical Garden* 1, no. 2 (February 1900): 22–26.

Vail, R. W. G. *Knickerbocker Birthday: A Sesqui-centennial History of the New-York Historical Society, 1804–1954*. New York: New-York Historical Society, 1954.

Van Doren, Carl. "The Beginnings of the American Philosophical Society." *Proceedings of the American Philosophical Society* 87, no. 3 (1943): 277–89.

Walls, Laura Dassow. *Emerson's Life in Science: The Culture of Truth*. Ithaca, NY: Cornell University Press, 2003.

———. *Henry David Thoreau: A Life*. Chicago: University of Chicago Press, 2017.

———. *The Passage to Cosmos: Alexander von Humboldt and the Shaping of America*. Chicago: University of Chicago Press, 2009.

Ward, David C. *Charles Willson Peale: Art and Selfhood in the Early Republic*. Berkeley: University of California Press, 2004.

Wilder, Craig Steven. *Ebony and Ivy: Race, Slavery, and the Troubled History of America's Universities*. New York: Bloomsbury, 2013.

Wilentz, Sean. *Chants Democratic: New York City and the Rise of the American Working Class, 1788–1850*. 20th anniversary ed. New York: Oxford University Press, 2004.

———. *Rise of American Democracy*. New York: W. W. Norton, 2005.

Wilf, Steven Robert. "Anatomy and Punishment in Late Eighteenth-Century New York." *Journal of Social History* 22, no. 3 (1989): 507–30.

Wilson, Renate. "Introduction, William Bartram and Eighteenth-Century Medicine: A Collection of Recipes from Post-Revolutionary Philadelphia." In *William Bartram: The Search for Nature's Design*, edited by Thomas Hallock and Nancy E. Hoffman, 440–61. Athens: University of Georgia Press, 2010.

Wood, Gordon S. "The Bleeding Founders." *New York Review of Books*, July 10, 2014.

———. *Empire of Liberty: A History of the Early Republic, 1789–1815*. New York: Oxford University Press, 2009.

Wootton, David. *Bad Medicine: Doctors Doing Harm Since Hippocrates*. Oxford: Oxford University Press, 2006.

Wulf, Andrea. *The Brother Gardeners: Botany, Empire, and the Birth of an Obsession*. New York: Knopf, 2008.

————. *Founding Gardeners: The Revolutionary Generation, Nature, and the Shaping of the American Nation*. New York: Knopf, 2011.

————. *The Invention of Nature: Alexander von Humboldt's New World*. New York: Knopf, 2015.

Yoshpe, Harry B. "Record of Slave Manumissions in New York During the Colonial and Early National Periods." *Journal of Negro History* 26, no. 1 (January 1941): 78–107.

ILLUSTRATION CREDITS

I am grateful to the institutions below for providing permission to use these images from their collections.

IMAGES IN TEXT

ii *David Hosack*, by Charles Heath, engraving, 1816, after oil paintings by Thomas Sully and John Trumbull. Collections of the National Library of Medicine. http://resource.nlm.nih.gov/101433834.

viii *Eupatorium perfoliatum* specimen, n.d. Courtesy of the Linnean Society of London. http://linnean-online.org/9961/.

6 *David Hosack*, by Charles Heath, engraving, 1816, after oil paintings by Thomas Sully and John Trumbull. Collections of the National Library of Medicine. http://resource.nlm.nih.gov/101433834.

17 *View of Columbia College in the City of New York*, by J. Anderson and Cornelius Tiebout, engraving, 1790. The Miriam and Ira D. Wallach Division of Art, Prints and Photographs: Print Collection, New York Public Library. *The New York Public Library Digital Collections.* http://digitalcollections.nypl.org/items/510d47d9-7b08-a3d9-e040 -e00a18064a99.

35 Title page, *Edinburgh New Dispensatory*, by William Lewis. Philadelphia: T. Dobson, 1791. Courtesy of the Wellcome Collection.

40 *William Curtis*, by William Evans, stipple engraving and etching, 1802. © National Portrait Gallery, London.

50 Gentlemen botanizing, by unknown artist, engraving, 1777, from the title page of *Flora Londinensis*, vol. 1, by William Curtis. Biodiversity Heritage Library/Smithsonian Libraries.

54 Hosack's entrance ticket to Dr. Andrew Marshal's anatomy course in London, 1793–1794. Archives and Special Collections, Columbia University Health Sciences Library.

67 *View of New-York*, 1787, lithograph by George Hayward, 1828. The Miriam and Ira D. Wallach Division of Art, Prints and Photographs: Print Collection, New York Public Library. *The New York Public Library Digital Collections.* http://digitalcollections.nypl.org/items/510d47da-282c-a3d9-e040-e00a18064a99.

77 Traveling medicine case, c. 1814, Museum of the City of New York. Bequest of Gherardi Davis in memory of his wife, Alice Davis, 1941. 41.304.4AB.

78 *Samuel Latham Mitchill*, by Rembrandt Peale, oil painting, 1822. Academy of Natural Sciences of Philadelphia, Archive Collection 2011-022.

98 *Samuel Bard*, by James or Ellen Sharples, pastel on paper, n.d. Archives and Special Collections, Columbia University Health Sciences Library.

114 *DeWitt Clinton*, by Cephas G. Childs and Henry Inman, engraving, c. 1828. Prints & Photographs Division, Library of Congress, LC-DIG-pga-05649.

120 *Mastodon*, by Titian Ramsay Peale, wash drawing, 1821. Titian Ramsay Peale Sketches, Mss.B.P31.15d. © American Philosophical Society.

125 Unknown artist, *Philip Hamilton*, c. 1801. Reprinted from Allan McLane Hamilton, *The Intimate Life of Alexander Hamilton* (New York: Charles Scribner's Sons, 1910). New-York Historical Society. Photography © New-York Historical Society.

130 *François André Michaux*, by unknown artist, oil painting, 1819. Reproduced with the kind permission of the Director and the Board of Trustees, Royal Botanic Gardens, Kew.

134 *Loblolly Bay*, by Pierre-Joseph Redouté, from François André Michaux's *North American Sylva*, Vol. 2, Plate 58 (Paris: C. d'Hautel, 1819). Biodiversity Heritage Library/University of Pittsburgh Library System.

140 Charles Gilbert Hine, *Thirteen Trees and Alexander H. House [Hamilton Grange]*, n.d. Platinum print photograph. Charles Gilbert Hine Photograph Collection, image #87991d, New-York Historical Society. Photography © New-York Historical Society.

158 *Alexander von Humboldt*, by Charles Willson Peale, oil painting, 1804. Courtesy of the College of Physicians of Philadelphia.

173 Page from Hosack's memorandum book (garden section). Courtesy of the New York Academy of Medicine Library.

181 *Dr. Hosack's Greenhouses*, by John Trumbull, pencil drawing, 1806. John Trumbull Collection, Beinecke Rare Book and Manuscript Library, Yale University.

COLOR IMAGES

David Hosack, by Thomas Sully, oil painting, 1815. Courtesy of Winterthur Museum. Gift of John Hampton Barnes, Jr., in memory of the warm friendship of his parents, Mr. and Mrs. John Hampton Barnes, with Mr. and Mrs. Henry Francis du Pont, 1977.170.

Hosackia stolonifera (creeping-rooted Hosackia), from *Edward's Botanical Register*, vol. 23 (1837), Plate 1977. Image courtesy of Biodiversity Heritage Library, contributed by Missouri Botanical Garden, Peter H. Raven Library.

View of Bayard's Mount, the Collect Pond, and New York City, by Archibald Robertson, watercolor and chalk, 1798. The Metropolitan Museum of Art, Edward W. C. Arnold Collection of New York Prints, Maps, and Pictures. Bequest of Edward W. C. Arnold, 1954.

Brodway-gatan och Rådhuset i Newyork, by Carl Fredrik Akrell, lithograph, n.d. The Miriam and Ira D. Wallach Division of Art, Prints and Photographs: Print Collection, New York Public Library. The New York Public Library Digital Collections. http://digitalcollections.nypl.org/items/5e66b3e8-b382-d471-e040-e00a180654d7.

Ralph Earl (1751–1801). *Elizabeth Schuyler Hamilton (Mrs. Alexander Hamilton)*. 1787. Museum of the City of New York. Gift of Mrs. Alexander Hamilton and General Pierpont Morgan Hamilton, 1971. 71.31.2.

Alexander Hamilton, by John Trumbull, oil painting, c. 1806. Andrew W. Mellon Collection, National Gallery of Art.

The Grange with Sweetgum, 2018. Photograph courtesy of Daniel Atha, New York Botanical Garden.

Thomas Jefferson, by Rembrandt Peale, oil painting, 1800. © 2018 White House Historical Association.

The Artist in His Museum, by Charles Willson Peale, oil painting, 1822. Courtesy of the Pennsylvania Academy of the Fine Arts, Philadelphia. Gift of Mrs. Sarah Harrison (The Joseph Harrison Jr. Collection).

William Bartram, by Charles Willson Peale, oil painting from life, 1808. Courtesy of the Independence National Historical Park Collection.

Sir Joseph Banks, by Sir Joshua Reynolds, oil painting, 1771–1773. © National Portrait Gallery, London.

Sir James Edward Smith, by John Rising, oil painting, 1793. Courtesy of the Linnean Society of London.

Benjamin Rush, by Thomas Sully, oil painting, c. 1813. Courtesy of the Trout Gallery at Dickinson College, Carlisle, PA.

> AUTHOR NOTE: The Trout Gallery notes that this painting was owned by the Rush family continuously until 2009, when the family donated it. It is unclear whether this is the Sully portrait of Rush commissioned by Hosack or another one done by Sully at the same time. In 1814, Hosack published an engraving of a Sully portrait of Rush and noted that it was "Engraved from the Original Picture in the possession of Dr. D. Hosack." Hosack's engraving of his Rush portrait has the same composition as the Sully oil painting now owned by Dickinson College, suggesting either that Sully painted two paintings of Rush around the same time or that Hosack made a gift of his painting to the Rush family sometime after Rush's death.

John Vanderlyn, *Aaron Burr*, 1802. Oil on canvas, 22¼ in. × 16½ in. Object #1931.58, New-York Historical Society. Photography © New-York Historical Society.

Carl Linnaeus's house in Uppsala. Photo courtesy of Jesper Kårehed, the Linnaean Gardens of Uppsala, Uppsala University.

Linneanum. Photo courtesy of Jesper Kårehed, the Linnaean Gardens of Uppsala, Uppsala University.

Detail from *Map of the Country Thirty Miles Round the City of New York*, by John H. Eddy, engraving, 1811. Courtesy of the David Rumsey Map Collection, www.davidrumsey.com.

Mary Eddy Hosack and David Hosack Jr., by Thomas Sully, oil painting, 1815. Property of Alice B. Lloyd. Photo by Haley W. White. Used with the kind permission of Alice B. Lloyd.

Thomas Cole, American (born in England), 1801–1848. *Expulsion from the Garden of Eden*, 1828. Oil on canvas, 100.96 cm × 138.43 cm (39¾ in. × 54½ in.). Museum of Fine Arts, Boston. Gift of Martha C. Karolik for the M. and M. Karolik Collection of American Paintings, 1815–1865, 47.1188.

INDEX

Page numbers in *italics* indicate illustrations.
Page numbers followed by *n* refer to notes; *n* followed by a number identifies
the page reference for the key phrase endnote.